In Vitro and In Vivo Models of Colorectal Cancer for Clinical Application

In Vitro and In Vivo Models of Colorectal Cancer for Clinical Application

Editors

Marta Baiocchi
Ann Zeuner

MDPI • Basel • Beijing • Wuhan • Barcelona • Belgrade • Manchester • Tokyo • Cluj • Tianjin

Editors

Marta Baiocchi
Oncology and Molecular
Medicine
Istituto Superiore di Sanità,
Roma
Roma
Italy

Ann Zeuner
Oncology and molecular
medicine
Istituto Superiore di Sanità
Roma
Italy

Editorial Office
MDPI
St. Alban-Anlage 66
4052 Basel, Switzerland

This is a reprint of articles from the Special Issue published online in the open access journal *Cancers* (ISSN 2072-6694) (available at: www.mdpi.com/journal/cancers/special_issues/Colorectal_Clinical_Application).

For citation purposes, cite each article independently as indicated on the article page online and as indicated below:

LastName, A.A.; LastName, B.B.; LastName, C.C. Article Title. *Journal Name* **Year**, *Volume Number*, Page Range.

ISBN 978-3-0365-3966-9 (Hbk)
ISBN 978-3-0365-3965-2 (PDF)

Cover image courtesy of Marta Baiocchi

Contents

Preface to "In Vitro and In Vivo Models of Colorectal Cancer for Clinical Application"

Recent years have witnessed an unprecedented expansion of therapeutic options for colorectal cancer (CRC), with several molecularly targeted agents now established in clinical practice. Moreover, many new approaches are currently under evaluation, aimed at targeting different aspects of tumor biology including proliferating and quiescent cancer stem cells, antitumor immunity, tumor–microenvironment interactions, cancer metabolism, and cancer-associated epigenetic modifications. Advancements of high-throughput molecular analyses keep unveiling new details of the CRC molecular landscape contributing to identifying novel targets, to define improved methods for patients'classification and to achieve a better tuning of personalized therapies. However, despite increasing knowledge of the molecular mechanisms underlying CRC, the likelihood of approval (LOA) of new anticancer drugs entering clinical trial is still lower than 10% according to estimates (1, 2). The reason for this disconnection between basic/translational research and the related clinical outcomes resides, at least in part, in the limitations of preclinical models currently available for CRC. In fact, increasingly sophisticated experimental systems recapitulating CRC in vitro and in vivo, such as organoids and patient-derived xenografts (PDXs), are characterized by both intriguing complexities and intrinsic limitations, some of which are currently beginning to be understood. In this scenario, it is equally necessary to fully understand the potential of existing preclinical CRC models and to develop and validate novel tools for anticancer therapies. In this Special Issue, we have collected original research papers and reviews collectively depicting the current state and the perspectives of CRC models for preclinical and translational research. Original research papers published in this issue focus on some of the hottest topics in CRC research such as circulating tumor cells, epigenetic regulation of stemness states, new therapeutic targets, molecular CRC classification and experimental CRC models such as organoids and PDXs. Additionally, four reviews on CRC stem cells, immunotherapy and drug discovery provide an updated viewpoint on key topics linking benchtop to bedside research in CRC.

1. Hay, M., D. W. Thomas, J. L. Craighead, C. Economides and J. Rosenthal (2014). "Clinical development success rates for investigational drugs."Nature Biotechnology 32(1): 40-51.

2. Wong, C. H., K. W. Siah and A. W. Lo (2018). "Estimation of clinical trial success rates and related parameters."Biostatistics 20(2): 273-286.

Marta Baiocchi and Ann Zeuner

Editors

Review

Colorectal Cancer Stem Cells: An Overview of Evolving Methods and Concepts

Maria Laura De Angelis , **Federica Francescangeli** , **Ann Zeuner** and **Marta Baiocchi** *

Department of Oncology and Molecular Medicine, Istituto Superiore di Sanità, Viale Regina Elena 299, 00161 Rome, Italy; marialaura.deangelis@iss.it (M.L.D.A.); federica.francescangeli@iss.it (F.F.); ann.zeuner@iss.it (A.Z.)
* Correspondence: marta.baiocchi@iss.it

Simple Summary: In recent years, colorectal cancer stem cells (cCSCs) have been the object of intense investigation for their promise to disclose new aspects of colorectal cancer cell biology, as well as to devise new treatment strategies for colorectal cancer (CRC). However, accumulating studies on cCSCs by complementary technologies have progressively disclosed their plastic nature, i.e., their capability to acquire different phenotypes and/or functions under different circumstances in response to both intrinsic and extrinsic signals. In this review, we aim to recapitulate how a progressive methodological development has contributed to deepening and remodeling the concept of cCSCs over time, up to the present.

Abstract: Colorectal cancer (CRC) represents one of the most deadly cancers worldwide. Colorectal cancer stem cells (cCSCs) are the driving units of CRC initiation and development. After the concept of cCSC was first formulated in 2007, a huge bulk of research has contributed to expanding its definition, from a cell subpopulation defined by a fixed phenotype in a plastic entity modulated by complex interactions with the tumor microenvironment, in which cell position and niche-driven signals hold a prominent role. The wide development of cellular and molecular technologies recent years has been a main driver of advancements in cCSCs research. Here, we will give an overview of the parallel role of technological progress and of theoretical evolution in shaping the concept of cCSCs.

Keywords: cancer stem cells; colorectal cancer; animal models; in vitro culture; cancer stem cell methods

Citation: De Angelis, M.L.; Francescangeli, F.; Zeuner, A.; Baiocchi, M. Colorectal Cancer Stem Cells: An Overview of Evolving Methods and Concepts. *Cancers* **2021**, *13*, 5910. https://doi.org/10.3390/cancers13235910

Academic Editor: Jürgen Behrens

Received: 15 October 2021
Accepted: 19 November 2021
Published: 24 November 2021

Publisher's Note: MDPI stays neutral with regard to jurisdictional claims in published maps and institutional affiliations.

1. Introduction

According to the stem cell model, most tumors, including colorectal cancer (CRC), contain a small population of cancer stem cells (CSCs) that are deeply implicated in tumor generation and progression, drug resistance, recurrence, and metastasis [1,2]. The characterization of the molecular and functional features of colorectal CSCs (cCSCs) has thus received intense research efforts in recent years due to its promise to reveal new routes of intervention for tumor and metastasis eradication. Methods and concepts for understanding CSCs biology in solid tumors historically derived from studies on normal hematopoiesis and leukemia. In fact, the CSC concept itself originated from landmark research [3,4] that in the early nineties first showed that leukemia is organized as a hierarchical system, mimicking that of the normal hematopoietic system. The neoplastic cell hierarchy in leukemias develops from a small subset of stem cells able to both self-renew and to give rise to a cascade of more differentiated cells. Notably, the golden standard for the identification of leukemic stem cells (LSCs) was established by extending the definition for normal hemopoietic cells as cells functionally capable of initiating neoplasia into recipient mice. A few years later, the experimental methods that led to LSCs identification were translated to solid tumors in general, and in particular to CRC. During the following years, the expanding range of methodological approaches applied to cCSCs biology has paralleled a continuous evolution of the cCSCs concept, providing a significant example of knowledge advancement about a complex biological issue. The main technical approaches that contributed to cCSCs identification and characterization, in particular of human CRC, are summarized in Figure 1, which also underlines the high degree of intersection of their application. Methods for cCSCs identification can be broadly summarized as (1) cell isolation by fluorescence-activated cell sorting, (2) cell culture-based selection systems, (3) transplantation into recipient animals, and (4) lineage tracing techniques. Here, we will revise the major advancements that led to the development of the current cCSC concept, keeping a historical view on the evolution of technologies that allowed cCSC characterization to the present day (Figure 2).

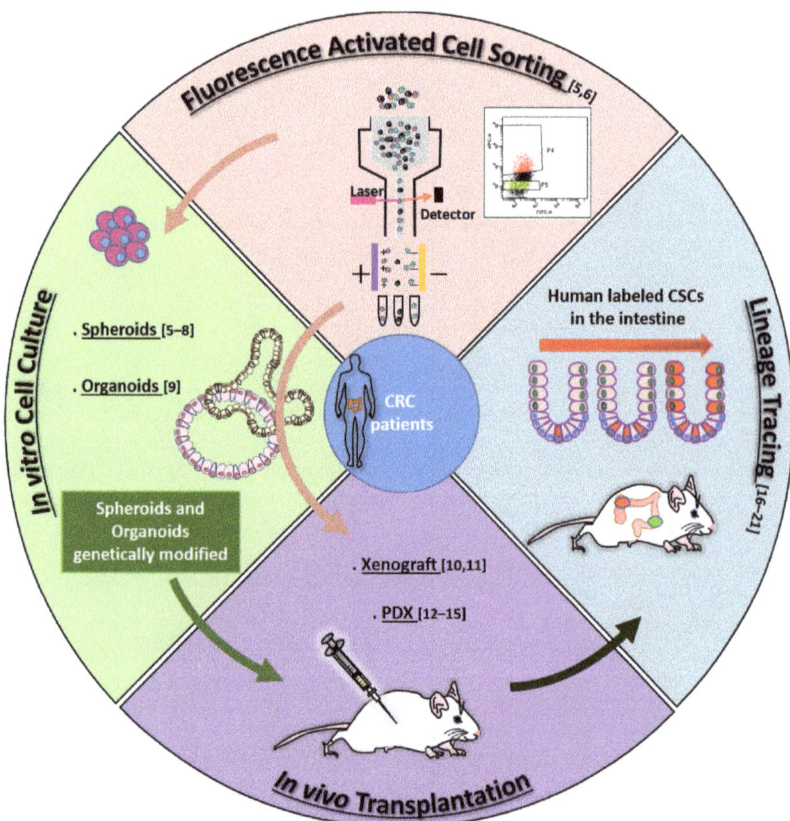

Figure 1. Main methodologies contributing to cCSCs definition: fluorescence-activated cell sorting (FACS) [5,6]; in vitro cell culture [5–9]; in vivo transplantation [10–15]; lineage tracing [16–21]. Arrows across sectors identify the complementary application of different techniques: For example, identification of cCSCs by cell sorting can be validated by in vitro cell culture and/or by in vivo transplantation assays (pink arrows); Xenografting is often performed with cultured and/or genetically labeled cCSCs (light green arrow); human cCSC lineage tracing is mostly performed by transplanting cCSCs previously genetically manipulated in culture (dark green arrow). Numbers in parentheses indicate references.

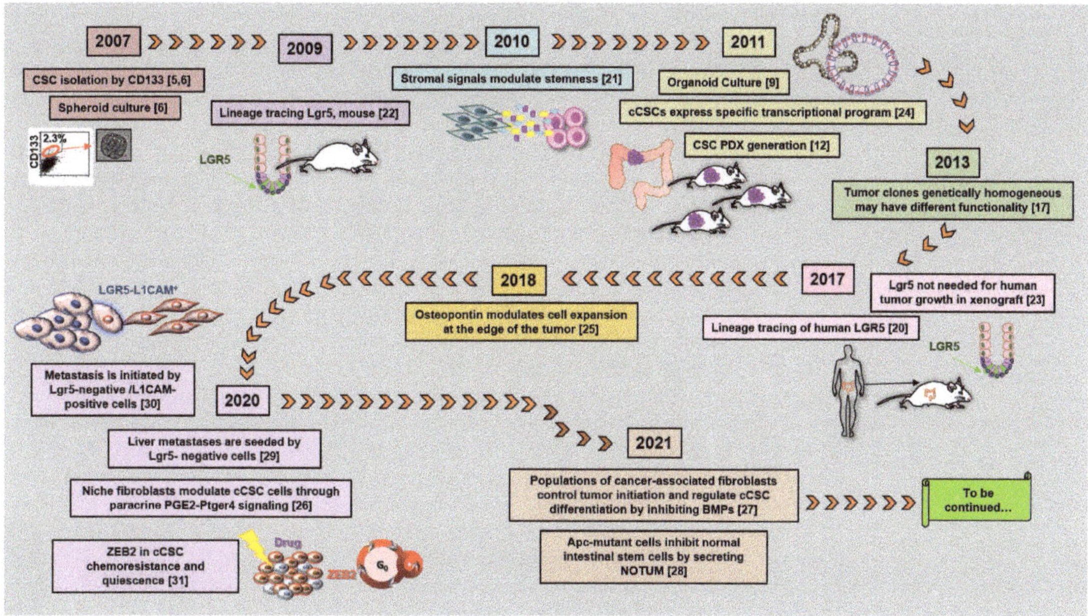

Figure 2. The timetable of landmark studies that have contributed to defining cCSC biological features and functions. Numbers in parentheses indicate references [5,6,9,12,17,20–31].

2. cCSC Definition by Experimental Models

2.1. Identification of cCSCs through the Use of Surface Markers

Fluorescence-activated cell sorting on the base of CD133 expression (also known as Prominin 1) was used in 2007 by two independent groups [5,6] to identify and isolate for the first time putative cCSCs. By using similar approaches, in fact, the two groups demonstrated that the CD133-positive subpopulation isolated by cell sorting from dissociated patient's tumors is enriched in xenograft-initiating cells.

The use of CD133 as a stem cell marker is a representative example of an approach inherited by hematopoiesis studies, in which cell membrane markers have long been used to isolate different cell and progenitor subpopulations. Indeed, CD133 was first identified in 1997 as a marker of normal human hematopoietic stem cells [32] and later indicated as a generic marker of endothelial [33] and neural [34] stem cells. In 2003, CD133 was used to identify glioblastoma stem cells [35], opening the way for its future use as a stem cell marker in multiple solid and hematopoietic tumors.

The efficacy of CD133 as a cCSC marker was shortly thereafter, confirmed by other reports, in particular, Vermeulen et al. demonstrated that CD133-positive cells isolated from patients and grown as spheroids could differentiate in vitro, giving rise to cells belonging to all the main intestinal cell lineages [7]. Lineage tracing experiments in vivo (see below in this review) in turn showed that deregulating β-catenin activity in normal CD133-positive intestinal cells induced neoplastic transformation [36]. However, studies identifying cCSCs through CD133 expression introduced different issues related to cCSC selection from solid cancers based on membrane markers. First, the functional role of cCSC membrane markers is still largely unknown, so that a clear link between their expression and stem cell function can hardly be established. Indeed, analyses of the prognostic value of CD133 expression has given inconsistent results [37,38], and its inhibition had no effect on tumor course [38]. A second issue of concern is that the dissociation of solid tissues may alter the surface density and/or modify the molecular structure of the marker itself, both because dissociation mostly involves the use of proteases and/or because the loss of cell–cell interaction itself can induce reprogramming of membrane–molecule expression. Intrinsic fluctuations of marker expression may also contribute to confusing the outcomes of marker-based CSCs assays. For example, specific mutations in the RAS–RAF axis can alter CD133 expression in CRC cells, independently from their capability to initiate tumors into immunodeficient mice [39]. In addition, cell-state-dependent modifications [40] or glycosylation [41] of CD133 can modify its antibody-binding capability rather than the expression level of the protein itself. Similarly, isolation of cCSCs through CD44, alone or combined with CD166 [42], is complicated by the many alternative splicing-generated CD44 isoforms and by its ubiquitous expression [43]. More recently, the isoform CD44v6 was reported as a strong marker of metastatic cell capability [44]. Interestingly, histological analyses of membrane markers expressed in cells at the edge and in the central zone within human tumors showed that a lower expression of stem cell markers, including CD44 isoforms, CD166, and CD133, by edge cells' correlates with the infiltrating pattern [37,45]. The authors suggested that downregulation of these molecules may be linked to stage-dependent modulation of cancer cell adhesion capability [37]. Flow cytometry isolation of cCSCs has also involved other stemness-related factors, although not expressed on the cell surface. Among these, aldehyde dehydrogenases (ALDH) are a class of detoxifying enzymes responsible for the oxidation of intracellular aldehydes that hold a particular interest for their potential functional role in stemness [46]. However, studies on ALDH as a CSCs marker in colorectal and other tumors gave conflicting results, likely due to the wide number of ALDH isoforms present in different cell types [47].

More generally, a dynamic modulation of membrane markers of cCSCs can easily be expected to relate to the plastic nature of these cells, as discussed more in detail later in this review. As an example, we have recently described in clonal cCSCs a highly regulated, fluctuating expression of Cripto-1, an extracellular GPI-anchored protein, which correlates with the cell's variable clonogenic capability [48].

2.2. In Vitro Cultures of cCSCs

According to the classical definition, stem cells, either normal or neoplastic, must hold the double functional capability to self-renew and to give rise to a differentiated cellular progeny in the long-term. In light of this concept, appropriate methods for in vitro culture of cCSCs allow assessing both the long-term propagation of tumor-initiating capability (indicative of self-renewal) and a conserved ability to generate differentiated cells. Importantly, culture in selective media can also represent an approach for cCSC isolation from freshly dissociated tissues without prior isolation by marker selection.

The isolation and amplification in vitro of intestinal CSCs inherited the same challenges affecting stem cell cultures of other normal and tumor tissues, and primarily the difficulty of preserving the capability of stem cells to self-renew. The achievements in this field were linked to the progressive identification of components for defined media that paralleled an increasing understanding of the molecular and cellular mediators underlying the stemness state. Importantly, an obvious issue that distinguishes cultures of solid tumor stem cells versus leukemia's is that adhesion is essential for the former [49] while blood cells are naturally non-adhering. The requirement for adhesion is addressed by different means in the two most diffused intestinal CSC culture methodologies, i.e., spheroid and organoid cultures.

Spheroids are grown in low-adherence cell culture plasticware, in which, after a few days, dissociated cells from the patient's tissues generate self-adhering floating clusters that can reach up to 1–2 mm size. Depending on the strain, tubular structures develop within single spheres, associated with the presence of differentiated cells [7,8]. cCSC-enriched spheroid cultures can be expanded long-term, retaining both their xenograft-initiating capability and the potential to generate differentiated cells. Spheroid cultures have allowed the identification and first characterization of cCSCs in early studies [6,7].

As mentioned before, spheroid culture in selective media allows direct isolation of cCSCs from dissociated patient tumors. Our group optimized this method, providing a highly efficient workflow for cCSCs isolation from primary tumor fragments, which allowed us to generate a biobank representative of patients' molecular diversity [8]. A distinctive characteristic of spheroid cultures is that they allow great cell expansion; therefore, spheroid biobanks are particularly convenient for studies requiring a high number of cells, such as high-throughput molecular analyses or drug testing [50]. Ours and other groups have identified different potential cCSC-targeted agents through a spheroid-based platform [48,51–53].

Organoid cultures represent an alternative approach for cCSC expansion from patient tissues [54]. Here, the cell requirement for adhesion is met by growing the cells embedded in a basement membrane matrix, usually Matrigel. This system was developed first for normal murine small intestine [55] and then for human tumor intestinal CSCs [9,56]. Organoids preserve a capability to generate tubular, complex structures reproducing the original tumor's architecture more evidently as compared to spheroids. The use of Matrigel, however, renders the system more expensive and time-consuming, also due to the uneasy release of cells from the embedding matrix. Importantly, patient-derived organoids have allowed generating of biobanks that have proved highly representative of gastrointestinal cancer patients' response to drugs and radiation [10,11,57,58].

Both spheroid and organoid cultures containing human cCSCs can be easily genetically modified, thus offering a frame to analyze the role of known or candidate molecular cancer determinants. To mention a particularly interesting group of studies, the organoid culture system coupled with the CRISPR methodology allows dissecting mutational events underlying tumorigenesis. The sequential introduction of mutations into the *APC*, *SMAD4*, *TP53*, and *KRAS* genes, in fact, lead organoids to reproduce the adenoma-carcinoma transition, disclosing a parallel progressive loss of cell requirement for niche factors [59–61]. Recent studies, in turn, have attempted to reconstruct the mutational landscape underlying metastatic tumor capability [62–64]. Importantly, cCSCs cultured both with the spheroid and organoid systems can be used to generate xenografts into immunodeficient mice (see Figure 1). Transplantation of cultured and/or genetically modified cultured intestinal CSCs is an approach of the utmost importance in the understanding of cCSC biology, as described in the following sections.

2.3. cCSC Transplantation Assays

The assessment of tumor-initiating capability, in vivo, into recipient mice (syngeneic when murine stem cells are tested, immunodeficient for human cell assays) is defined as the golden standard for stemness descending from the historical definition established for putative hematopoietic and leukemic stem cells. In this frame, the only ethically feasible approach to assess the stemness of putative human intestinal CSCs is xenotransplantation into immunodeficient mice. Beginning with the discovery of the spontaneous mutant nude mice carrying T-cell deficiency, increasingly immunodeficient murine strains have become available over time, including the SCID (impaired in T and B cells) and the SCID Beige strain that is further impaired in NK activity. Nowadays, the most used immunodeficient strain in human CSC research is NSG (NOD-*scid* IL2rg^{null}), highly deficient in T, B, and NK cell activity [65].

Xenografting of human cCSC either from in vitro cultures or from freshly dissociated tumor samples is widely diffused as a stemness assay. Limiting dilution assay in vivo and linear regression analysis, easily performed by the online software ELDA [66], allows quantitative analysis of cCSC content in a cell population of interest, while serial re-transplantation experiments can be used to assess cCSC capability for long-term propagation. Importantly, serial cCSC transplantation demonstrated that different classes of cCSCs in spheroids are functionally heterogeneous, as they can be distinguished in short versus long-term tumor-initiating cells as well as in metastasis-initiating cells [16].

Xenografting of tissues or freshly dissociated cells from patient tumors is commonly named patient-derived xenografting. By continuous re-transplantation of xenografts from mouse to mouse, this system allows to propagate and expand cCSCs in vivo, thus bypassing any in vitro culture. Patient-derived xenografts (PDX) have been shown to better preserve the original tumor's characteristics, as compared to cultured cCSC grafts. However, this system is expensive, laborious, and time-consuming due to the requirement of high numbers of mice and to the slow development of the grafts [67].

Human cCSC functional assessment by xenografting is not devoid of limits, the first of which is that the estimate of CSC can be influenced by the recipient mouse strain: more immunodeficient strains may detect a higher frequency of stem cells; such an effect has been reported in detail in melanoma [68]. In addition, microenvironment cells, including vascular, immune, and mesenchymal cells that are not human, while the immune system is by definition impaired in immunodeficient mice. Therefore, the contribution of microenvironment cells to tumor development can hardly be evaluated in xenografts. The use of humanized mice, in which a human hematopoietic/lymphoid system is reconstructed by transplanting hematopoietic cells, allows overcoming some of these limitations [69]. Another issue of concern is that in subcutaneous xenografts—the most feasible and therefore most often used system—and the natural organ location of the tumor is also mismatched; indeed, CSCs of the majority of solid cancers, including colon, do not give rise to metastasis upon subcutaneous grafting [70]. Conversely, orthotopic xenografting of cCSCs into the colon results in metastases to the liver and other target organs, but the method requires technical skill [71]. An alternative system, technically easier, is grafting cCSCs into the spleen, whose vascularization leads directly to the liver [71]. This system, however, reproduces only partially the whole process of cellular metastasization, from the primary tumor to the target organ.

Despite these caveats, panels of both cultured cCSC xenografts and PDX have revealed an important therapy-predicting capability [12–15], while cCSC transplantation has contributed to collecting a whole bulk of information on human cCSCs biology. Importantly, CSCs xenografting represents a unique approach for human CSC lineage tracing (See Figure 1).

2.4. Lineage Tracing of cCSC

In general, lineage tracing methods consist in following the development of a given progenitor/stem cell progeny on the basis of a morphological feature or a dye or a molecular/genetic marker that is conserved and transmitted during the developmental process [72,73]. The simplest cell labeling systems consist in staining the cells with supervital colored or fluorescent dyes. However, the development of genetic manipulation techniques, coupled with the increasing availability of high-throughput sequencing methods, are generating an ever-expanding genetic lineage tracing toolbox. Among these, the most direct system is to introduce into the cells of interest a gene for a colored (such as β-galactosidase or alkaline phosphatase) or fluorescent (such as EGFP, RFP, EYFP, tdTomato, and others) or light-emitting marker. An advancement of this approach consists in using conditionally expressed markers: this is obtained by Cre-*Lox* and derived systems, where Cre is a P1 bacteriophage-derived DNA recombinase, and *LoxPs* are its recognition sites. In the Cre-*Lox* tracing systems, the expression of the labeling gene is blocked by a stop cassette flanked by two *LoxP* sites. Concurrently, the cells also carry a *Cre*-recombinase gene, which is activated by a tissue- or cell-stage-specific promoter; therefore, the stop cassette is excised, allowing the expression of the marker gene only in the tissue or the subset of cells that activate the chosen specific promoter. In inducible Cre-*Lox* systems, the expression of *Cre* is further controlled by an inducible element, for example, a tamoxifen-responsive sequence. A pulse of the drug triggers the activation of the labeling gene in the cell population of interest at a time point of choice. This allows, for example, to distinguish stem cells, able to self-renew for prolonged periods of time, from downstream progenitors that may express the same marker but get exhausted over time due to their limited proliferative capability.

Later technologies include multicolor systems such as Confetti and Brainbow, in which conditional, inducible Cre-*Lox* systems induce different fluorescent labeling at random in single cells, allowing to follow the destiny of several individual clones within a tissue. Up to four and ninety different fluorescent wavelengths are generated by the Confetti and Brainbow systems, respectively [73]. Random multifluorescence labeling systems have been widely used to follow specific marker gene-expressing cells, including *Lgr5* (see below), but they are now increasingly being adapted to marker-free labeling. Importantly, dye-free labeling systems have also seen an expanded use in recent years, taking advantage of the high sequencing capability provided by high-throughput technologies: it is thus possible to genetically barcode cells by analyzing the propagation of short sequences/mutations introduced either by lentiviral vectors [16,17] or more recently by CRISPR/Cas9 genome editing [18]. Individual clones are then identified by high-throughput sequencing. Finally, non-labeling approaches based on spontaneous randomly occurring mutations have also been used to follow CRC clonal growth by taking advantage of mitochondrial mutations and/or by single nucleotide/copy number variations (SNV/CNV) [18,19]. Lineage tracing can allow following natural cell development into its own whole organism: in the mouse model, this is achieved by crossing strains carrying loxed labeling gene(s) with strains carrying inducible and/or conditional Cre recombinase. Conversely, tracing of human cells requires the manipulation of cells in vitro, followed by xenotransplantation in an immunodeficient recipient animal. In general, either approach has its own specific limitations, since, as mentioned before, murine cancer models do not fully reproduce the human pathogenesis, while xenotransplanted human cells suffer from microenvironment mismatch. Nevertheless, both murine and human lineage tracing technologies have given landmark information on CSC biology in intestinal adenoma/carcinoma.

Lineage tracing of GFP-labeled CD133 cells in mice allowed confirming their location at the base of the crypt and their capability to give rise to all the intestinal epithelium [36]. In addition, CD133+ cells exhibited massive amplification, generating neoplastic tissue within the intestine upon Cre-dependent, promoter-specific activation of mutant β-catenin, thus demonstrating their stem cell nature [36].

A major contribution of lineage tracing, however, has been in the establishment and progressive definition of the role of Lgr5+ as a stem cell marker in both normal and neoplastic intestinal stem cells. Lgr5 (leucine-rich repeat-containing receptor) is a transmembrane receptor involved in the modulation of the canonical Wnt signaling pathway. As the Wnt pathway is a primary driver of normal and neoplastic intestinal cell development [74], Lgr5 holds the features of a functional marker in intestinal cancer development. However, early studies on Lgr5 role in intestinal tumorigenesis were largely based on lineage tracing, as the scarce availability of efficient commercial antibodies against Lgr5 traditionally hampered the isolation of Lgr5+ cells by FACS.

Labeled Lgr5+ cells were first demonstrated to be capable of giving rise to all the intestinal cell types in the normal mouse intestine by using a tamoxifen-inducible *GFP* integrated into the *Lgr5* locus [75]. Shortly thereafter, the same group also showed that conditional deletion of Apc in the same cells induces them to proliferate and generate adenomas invading the crypt [76]. In addition, in *Apc*-mutant mice, random labeling of single Lgr5+ cells by a tamoxifen-activated multicolor Confetti system allowed to visualize adenomatous clones of different colors growing from the base to the upper edge of the intestinal crypt [22]. Lineage tracing of human cCSC was later achieved by labeling Lgr5+ cells into human organoids upon xenotransplantation: EGFP-labeled Lgr5+ cells proved to be able to initiate xenografts and to differentiate into the main intestinal lineages [20]. Altogether, these studies established Lgr5 as a common marker for both normal and tumor intestinal stem cells.

3. Merging Methodologies and Evolving Concepts: cCSC Plasticity and the Niche

The different approaches discussed above have altogether contributed to defining cCSC as a cell subpopulation driving initiation, development, and growth of CRC. Nevertheless, parallel research has challenged the concept that a fixed phenotype could be attributed to such population(s). Early observations indeed indicated that cells not expressing specific markers might undertake a functional stem cell role, at least under stress circumstances. Among these, Shmelkov et al. followed intestinal murine CD133 expression by a Lac-Z reporter showing that both CD133+ and CD133− cells can initiate metastases into immunodeficient mice [77]. Later, Lgr5+ cells were found dispensable for adenoma formation after irradiation in Apc-mice [78], while a population of keratin-19 (KRT19)-positive, Lgr5-negative cells were shown to be responsible for cancer tissue regrowth following irradiation in mice [79]. More recently, de Sousa et al. analyzed the effects of Lgr5+ cell removal through an inducible diphtheria toxin in a model of murine colon cancer. After killing Lgr5+ cells by toxin activation, tumor growth was not impaired, indicating that other cell type(s) could take over a stem cell function [80]. Similar results were shown in human CSCs, by xenotransplanting organoids carrying an inducible gene for caspase 9 inserted into the *Lgr5* locus. Elimination of Lgr5+ cells by caspase activation blocked tumor growth. Upon removal of the killing stimulus, however, tumor regrowth was driven by a population of Lgr5−/KRT20 positive cells, which regenerated Lgr5+ cells [23].

The observations that stem cell marker-expressing cells may not be the unique drivers of adenoma/carcinoma points to the issue of stem cell plasticity. This concept includes both the capability of intestinal CSCs to acquire different phenotypes and the potential of different tumor cell subpopulations to take on stem cells function under different circumstances [1,2,81]. A scheme comparing the hierarchical model of cCSC versus the emerging model of plastic cCSC is shown in Figure 3.

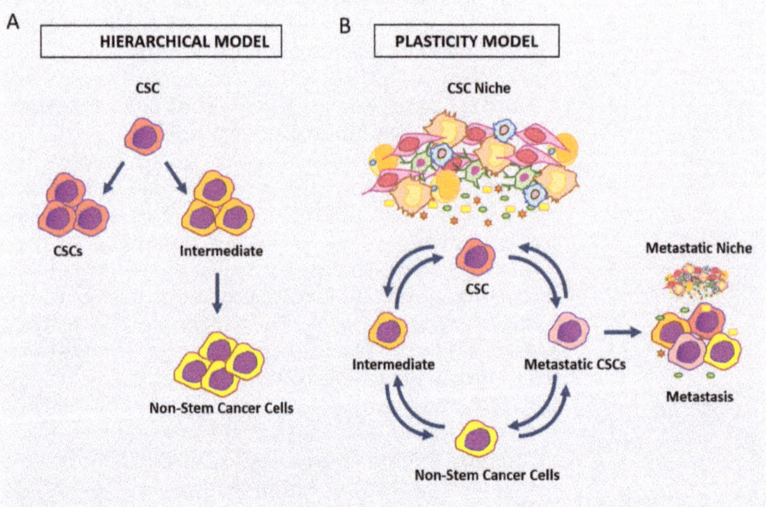

Figure 3. Schematic representation of the hierarchical model of cCSC (**A**) versus the model of plastic cCSC (**B**). According to the plastic model, different classes of cancer cells can dynamically take on stem cell phenotypes/functions.

The plasticity of cCSCs was also supported by other studies tracking stem cells on the base of Wnt activation state rather than through stem cell marker expression. In fact, tracing human xenografted cCSCs by means of a Wnt-dependent GFP reporter showed that cCSCs driving tumor expansion are located at the edge of the tumor. In this system, signals from surrounding stromal cells proved to be instrumental in modulating Wnt activation levels in tumor cells, pointing to a role for stroma in inducing cCSC function [21]. In turn, constitutive NF-kB activation was reported to enhance Wnt activation and stem cell marker expression in mouse crypt cells [82]. More recent phenotype-independent tracking experiments keep extending this concept. Phenotype-independent lineage tracing of human cCSCs confirmed that clonal expansion in colon cancer xenografts mostly arises at the leading edge and that cell position is a main driver of clonal competition during tumor development [83]. Other studies used an improved multicolor marker-independent tracing system (RGB/LeGO) to follow long-term clonal dynamics within xenografts, showing a correlation between clone size and proximity to the edge. These observations further support a persisting role of tumor geometry and cell position in orchestrating clonal competition during tumor growth [84].

The plastic nature of intestinal CSCs is also consistent with the knowledge that genetically homogeneous cells within the tumor can take over different functions, as demonstrated by clonal analysis of expanding human cCSC clones in xenografts by lentiviral marking [17] and by ultra-deep whole-genome sequencing tracking SNV/CNV [18]. In addition, cCSCs are known to express specific transcriptional programs [24,85,86], and a recent report has shown that high levels of ribosomal activity and protein synthesis individuate cCSCs independently from their specific mutational landscape [87]. It is noteworthy that genetic studies on patient's adenoma and CRC tissues collectively indicate that driver mutations are mostly established early in the first stages of tumorigenesis, while limited functional mutational divergence is added during tumor development, thus strengthening the idea that non-genetic events are the main determinants of intestinal CSCs function during tumor development [88–90].

The concept that cCSC plasticity may be related to cell location points to the role of the so-called tumor niche in instructing cCSC behavior. In fact, cells at the tumor's edge reside in close proximity to stromal cells, thus being exposed to stroma-derived signals. Early studies had actually described an instructive role of the niche and of niche-secreted factors, including WNTs, R-spondins, and BMP-inhibitors, in influencing the fate of intestinal CSCs [82,91–94]. In recent years, research on tumor niche has been expanding, disclosing the complexity of the crosstalk that orchestrates the plastic features of cCSCs. In this frame, Lenos et al. recently described a relevant role of stroma-secreted osteopontin as an inducer of cell expansion at the edge of the tumor [25]. Another report identified a specific subpopulation of fibroblasts in the mesenchymal tumor niche, which controls tumor-initiating cells through paracrine PGE2 (Prostaglandin E2)-Ptger4 signaling [26]. In turn, polarized populations of cancer-associated fibroblasts regulate cCSC differentiation and cancer progression by balanced inhibition of BMPs by GREM1 [27].

In this frame, transient epigenetic modifications, including variation in DNA methylation, histone modification, and chromatin accessibility, certainly contribute to sustaining CRC cell stemness [95–97]. Several epigenetic mechanisms capable of affecting CRC cell stemness, including but not limited to Wnt pathway activation/inactivation, have been described [98–103]. In this context, cell position-related epigenetic modulation likely holds a particularly relevant role in cCSC plasticity [95–97]. For example, tumor cells located at the edge or in central areas of the tumor are exposed to varying concentrations of metabolites and oxygen, which can modulate histone modifications and DNA methylation [96]. Stroma-secreted factors acting on cCSC plasticity, among which TGFβ, in turn, are able to induce cell epigenetic modifications [104]. Novel methods for single-cell epigenetic analyses, aided by developing computational systems, are now beginning to shed light on inherited epigenetic states of cellular lineages within CRC [105,106].

Further complexity is added to the picture by the finding that cCSCs are able to deliver inhibitory signals to normal intestinal cells, both directly and by inducing stromal cells to secrete specific factors. In this frame, two studies recently described the capability of Apc-mutated adenoma cells to inhibit stem cell activity of non-mutated cells within the same crypt and adjacent crypts through secretion of soluble Wnt antagonists [107]. Among these factors, a prominent role is played by NOTUM, whose pharmacological inhibition blocks adenoma formation [28]. Other important information has been provided by an innovative tracking system, the Confetti-derived Red2Onco, by which oncogenes such as mutated *KRAS* or *PI3K* are inserted only in RFP+ tumor cells. By allowing tracking of separately normal and mutated intestinal cells, this system has revealed that oncogene-driven signals from mutated cells induce apoptosis and differentiation of surrounding normal intestinal stem cells, both directly and by instructing surrounding stromal cells to secrete inhibitory factors [108].

The plasticity of cCSCs holds a particular interest in view of their capability to initiate metastasis. Several recent studies have described variable stem cell marker expression in metastasis-initiating cCSCs. For example, in the study by de Sousa e Melo already mentioned, while the selective elimination of Lgr5+ cells did not lead to tumor growth stopping, metastasis initiation in the liver was delayed until the Lgr5+ cell re-emerged [80]. By using intravital microscopy on xenotransplanted organoids carrying an inducible Lgr5-EGFP-Confetti, Fumagalli et al. followed metastases seeding and initiation, observing that liver metastases are seeded by Lgr5− cells, although Lgr5 positivity and stem cell marker expression re-emerges in growing metastases [29]. Consistently, Ganesh et al. showed that patients' metastases are initiated by cells overexpressing L1CAM+, that do not concurrently express Lgr5 [30]. Altogether, a picture thus emerges, in which Lgr5 expression is downregulated at some stages of dissemination/seeding, to be then re-expressed during metastasis growth into target organs. Low expression of stem cell markers of metastasis-initiating cCSCs is indeed shared by budding cells at the edge of the tumor [109] and by cCSCs circulating in the blood flow, i.e., putative migrating cells at metastasis target organs [110–113]. The plasticity of metastasis-initiating/circulating cCSCs has been put into relationship with the transition of cCSCs into quiescent/drug-resistant states [31,114], and involves at least at some stages EMT (epithelial to mesenchymal transition, reviewed in [112]). This capability of cCSCs to downregulate epithelial and/or stem cell markers to take on mesenchymal features is driven by specific transcription factors, including SNAI1, SNAI2, ZEB1, ZEB2, and TWIST1 [112,115,116]. The crosstalk between tumor and stromal cells holds a role of utmost importance in CRC and in cCSCs, in particular through TGFβ signaling [62,63,117]. Finally, other mechanisms of stromal/tumor cell modulating metastatic cCSCs are emerging: a recent report details the reciprocal reinforcement between visceral adipose stromal cells and metastatic CD44v6-positive cCSCs, sustained by adipose cell-secreted IL-6 and HGF, and by neurotropin produced by CD44v6+ cells [118].

4. Conclusions and Future Perspectives

Taken together, most recent studies have converged to redefine CRC cell stemness, from the permanent feature of a restricted tumor cell subpopulation to a function that can be undertaken by different cell types under different circumstances. Such a function is modulated by a series of factors awaiting further dissection, but definitely including signals both intrinsic and from the microenvironment [1,2]. Spatial constraints likely also contribute to influencing the clonal development of CRC [119]. It has been proposed that different parameters may dictate clone competition during different developmental phases of tumors, arguing that space constraints may be determinant in leukemia developing into the bone marrow, while stromal signals may have a prevalent role in solid tumors [97]. Even within the same tumor, different parameters may acquire or lose relevance during development: Regarding intestinal cancer, it is easy to hypothesize that cell clonogenicity and stemness may be differently regulated during the initial adenomatous phase, within the restricted crypt environment, and later on, at the expanding invasive edge of carcinoma. Several pieces of evidence indeed indicate that different clonal selection determinants act in adenoma as compared to carcinoma [120,121]. Altogether, the picture of intestinal CSCs represents a fast developing, challenging biological issue.

The wide array of methodological advancements of these years is increasing our knowledge of the cellular events taking place in CRC at a fast speed [122]. New cell culture methods are progressively extending into sophisticated engineering methodologies, among which biomimetic scaffolds and organs on chips [123,124]. Single-cell microfluidics are already contributing to the characterization of circulating cCSCs, and begin to find application in tissue-dissociated cells [97]. Other important developments of in vitro methods aim to reconstruct the contribution of tumor microenvironment components, including co-culture systems of organoids with mesenchymal cells [125] or lymphocytes [126]. Intravital microscopy allows visualizing cells within a whole organ [29,127]. Lineage tracing is increasingly taking advantage by developing high-throughput sequencing techniques [73], allowing unsupervised tagging by lentivirus or by CRISPR, or even by following naturally occurring mutations [128]. Advanced single-cell technologies [129,130], including genome sequencing [131] and scRNA analysis [132,133], are already contributing to colorectal CRC and niche studies [105,134,135] and are expected to gain further strength in the very near future [97].

As a final note, the growing impact of computational methods in research on cancer cell biology, and CSCs in particular, deserves a special mention. The analysis and management of the enormous amount of data generated by multi-omic techniques are in fact made possible only by the continuous development of dedicated algorithms, and most of the recent studies mentioned in this review have heavily taken advantage of increasingly sophisticated computational approaches (see for example [84,88,106,107,132,136]). Recent applications include meta-approaches able to integrate data derived from multiple different analytical tools, such as scDNA- and scRNA-sequencing [97]. Complex systems of lineage tracing reconstruction have been developed, such as the so-called pseudotime projection analyses, able to elaborate cell lineage developmental trajectories based on scRNA expression patterns [128]. Similarly, genetic lineage tracing by scDNA-seq allows the reconstruction of spatial models of clonal development in solid cancers [88,130].

Self-training artificial intelligence (AI) tools, in particular, machine learning (ML) and deep learning (DL), have demonstrated an exceptional power in individuating patterns within wide datasets and are currently not only applied to cancer development research but also evaluated for clinical classification and decision-making [137,138]. The specific capability of DL to discriminate and classify images, in particular, surpassed that of humans in 2005 [138], and it is now widely used to analyze large-size imaging datasets. Most importantly, it is also generating breakthrough innovative imaging technologies, among which label-free cell recognition systems, able to predict fluorescent labels in unlabeled microscopy images [139] or ghost cytometry, that allow sorting of cells on the base of marker-free, image-free cell morphology pattern analysis [140].

In conclusion, it is easy to prophesize that such a wide range of fast-developing methodologies are destined to enlighten an increasing complexity of the cCSCs model in the very near future.

Author Contributions: M.B., Conceptualization and writing; A.Z., review and editing; M.L.D.A., editing, graphics; F.F., graphics. All authors have read and agreed to the published version of the manuscript.

Funding: This research received no external funding.

Conflicts of Interest: The authors declare no conflict of interest.

References

1. Batlle, E.; Clevers, H. Cancer stem cells revisited. *Nat. Med.* **2017**, *23*, 1124–1134. [CrossRef]
2. Zeuner, A.; Todaro, M.; Stassi, G.; De Maria, R. Colorectal Cancer Stem Cells: From the Crypt to the Clinic. *Cell Stem Cell* **2014**, *15*, 692–705. [CrossRef]
3. Bonnet, D.; Dick, J.E. Human acute myeloid leukemia is organized as a hierarchy that originates from a primitive hematopoietic cell. *Nat. Med.* **1997**, *3*, 730–737. [CrossRef]
4. Lapidot, T.; Sirard, C.; Vormoor, J.; Murdoch, B.; Hoang, T.; Caceres-Cortes, J.; Minden, M.; Paterson, B.; Caligiuri, M.A.; Dick, J.E. A Cell Initiating Human Acute Myeloid-Leukemia after Transplantation into Scid Mice. *Nature* **1994**, *367*, 645–648. [CrossRef] [PubMed]
5. O'Brien, C.A.; Pollett, A.; Gallinger, S.; Dick, J.E. A human colon cancer cell capable of initiating tumour growth in immunodeficient mice. *Nature* **2007**, *445*, 106–110. [CrossRef]
6. Ricci-Vitiani, L.; Lombardi, D.G.; Pilozzi, E.; Biffoni, M.; Todaro, M.; Peschle, C.; De Maria, R. Identification and expansion of human colon-cancer-initiating cells. *Nature* **2007**, *445*, 111–115. [CrossRef]

7.   Vermeulen, L.; Todaro, M.; de Sousa Mello, F.; Sprick, M.R.; Kemper, K.; Perez Alea, M.; Richel, D.J.; Stassi, G.; Medema, J.P. Single-cell cloning of colon cancer stem cells reveals a multi-lineage differentiation capacity. *Proc. Natl. Acad. Sci. USA* **2008**, *105*, 13427–13432. [CrossRef] [PubMed]

8.   De Angelis, M.L.; Zeuner, A.; Policicchio, E.; Russo, G.; Bruselles, A.; Signore, M.; Vitale, S.; de Luca, G.; Pilozzi, E.; Boe, A.; et al. Cancer Stem Cell-Based Models of Colorectal Cancer Reveal Molecular Determinants of Therapy Resistance. *Stem Cells Transl. Med.* **2016**, *5*, 511–523. [CrossRef]

9.   Sato, T.; Stange, D.E.; Ferrante, M.; Vries, R.G.J.; Van Es, J.H.; Van Den Brink, S.; Van Houdt, W.J.; Pronk, A.; Van Gorp, J.; Siersema, P.D.; et al. Long-term Expansion of Epithelial Organoids From Human Colon, Adenoma, Adenocarcinoma, and Barrett's Epithelium. *Gastroenterology* **2011**, *141*, 1762–1772. [CrossRef]

10.   Vlachogiannis, G.; Hedayat, S.; Vatsiou, A.; Jamin, Y.; Fernández-Mateos, J.; Khan, K.; Lampis, A.; Eason, K.; Huntingford, I.; Burke, R.; et al. Patient-derived organoids model treatment response of metastatic gastrointestinal cancers. *Science* **2018**, *359*, 920–926. [CrossRef] [PubMed]

11.   Ganesh, K.; Wu, C.; O'Rourke, K.P.; Szeglin, B.C.; Zheng, Y.; Sauvé, C.-E.G.; Adileh, M.; Wasserman, I.; Marco, M.R.; Kim, A.S.; et al. A rectal cancer organoid platform to study individual responses to chemoradiation. *Nat. Med.* **2019**, *25*, 1607–1614. [CrossRef]

12.   Bertotti, A.; Migliardi, G.; Galimi, F.; Sassi, F.; Torti, D.; Isella, C.; Corà, D.; Di Nicolantonio, F.; Buscarino, M.; Petti, C.; et al. A Molecularly Annotated Platform of Patient-Derived Xenografts ("Xenopatients") Identifies HER2 as an Effective Therapeutic Target in Cetuximab-Resistant Colorectal Cancer. *Cancer Discov.* **2011**, *1*, 508–523. [CrossRef]

13.   Lazzari, L.; Corti, G.; Picco, G.; Isella, C.; Montone, M.; Arcella, P.; Durinikova, E.; Zanella, E.R.; Novara, L.; Barbosa, F.; et al. Patient-Derived Xenografts and Matched Cell Lines Identify Pharmacogenomic Vulnerabilities in Colorectal Cancer. *Clin. Cancer Res.* **2019**, *25*, 6243–6259. [CrossRef] [PubMed]

14.   Gao, H.; Korn, J.M.; Ferretti, S.; Monahan, J.E.; Wang, Y.; Singh, M.; Zhang, C.; Schnell, C.; Yang, G.; Zhang, Y.; et al. High-throughput screening using patient-derived tumor xenografts to predict clinical trial drug response. *Nat. Med.* **2015**, *21*, 1318–1325. [CrossRef]

15.   Migliardi, G.; Sassi, F.; Torti, D.; Galimi, F.; Zanella, E.R.; Buscarino, M.; Ribero, D.; Muratore, A.; Massucco, P.; Pisacane, A.; et al. Inhibition of MEK and PI3K/mTOR Suppresses Tumor Growth but Does Not Cause Tumor Regression in Patient-Derived Xenografts of RAS-Mutant Colorectal Carcinomas. *Clin. Cancer Res.* **2012**, *18*, 2515–2525. [CrossRef]

16.   Dieter, S.; Ball, C.; Hoffmann, C.M.; Nowrouzi, A.; Herbst, F.; Zavidij, O.; Abel, U.; Arens, A.; Weichert, W.; Brand, K.; et al. Distinct Types of Tumor-Initiating Cells Form Human Colon Cancer Tumors and Metastases. *Cell Stem Cell* **2011**, *9*, 357–365. [CrossRef] [PubMed]

17.   Kreso, A.; O'Brien, C.A.; van Galen, P.; Gan, O.I.; Notta, F.; Brown, A.M.K.; Ng, K.; Ma, J.; Wienholds, E.; Dunant, C.; et al. Variable Clonal Repopulation Dynamics Influence Chemotherapy Response in Colorectal Cancer. *Science* **2013**, *339*, 543–548. [CrossRef]

18.   Giessler, K.M.; Kleinheinz, K.; Huebschmann, D.; Balasubramanian, G.P.; Dubash, T.D.; Dieter, S.M.; Siegl, C.; Herbst, F.; Weber, S.; Hoffmann, C.M.; et al. Genetic subclone architecture of tumor clone-initiating cells in colorectal cancer. *J. Exp. Med.* **2017**, *214*, 2073–2088. [CrossRef]

19.   Humphries, A.; Cereser, B.; Gay, L.J.; Miller, D.S.J.; Das, B.; Gutteridge, A.; Elia, G.; Nye, E.; Jeffery, R.; Poulsom, R.; et al. Lineage tracing reveals multipotent stem cells maintain human adenomas and the pattern of clonal expansion in tumor evolution. *Proc. Natl. Acad. Sci. USA* **2013**, *110*, E2490–E2499. [CrossRef]

20.   Cortina, C.; Turon, G.; Stork, D.; Hernando-Momblona, X.; Sevillano, M.; Aguilera, M.; Tosi, S.; Merlos-Suárez, A.; Attolini, C.S.-O.; Sancho, E.; et al. A genome editing approach to study cancer stem cells in human tumors. *EMBO Mol. Med.* **2017**, *9*, 869–879. [CrossRef] [PubMed]

21.   Vermeulen, L.; Felipe De Sousa, E.M.; Van Der Heijden, M.; Cameron, K.; De Jong, J.H.; Borovski, T.; Tuynman, J.B.; Todaro, M.; Merz, C.; Rodermond, H.; et al. Wnt activity defines colon cancer stem cells and is regulated by the microenvironment. *Nat. Cell Bol.* **2010**, *12*, 468–476. [CrossRef] [PubMed]

22.   Schepers, A.G.; Snippert, H.J.; Stange, D.E.; van den Born, M.; van Es, J.H.; van de Wetering, M.; Clevers, H. Lineage Tracing Reveals Lgr5+ Stem Cell Activity in Mouse Intestinal Adenomas. *Science* **2012**, *337*, 730–735. [CrossRef]

23.   Shimokawa, M.; Ohta, Y.; Nishikori, S.; Matano, M.; Takano, A.; Fujii, M.; Date, S.; Sugimoto, S.; Kanai, T.; Sato, T. Visualization and targeting of LGR5(+) human colon cancer stem cells. *Nature* **2017**, *545*, 187–192. [CrossRef]

24.   Merlos-Suárez, A.; Barriga, F.; Jung, P.; Iglesias, M.; Céspedes, M.V.; Rossell, D.; Sevillano, M.; Hernando-Momblona, X.; da Silva-Diz, V.; Muñoz, P.; et al. The Intestinal Stem Cell Signature Identifies Colorectal Cancer Stem Cells and Predicts Disease Relapse. *Cell Stem Cell* **2011**, *8*, 511–524. [CrossRef]

25.   Lenos, K.J.; Miedema, D.M.; Lodestijn, S.C.; Nijman, L.E.; van den Bosch, T.; Ros, X.R.; Lourenco, F.C.; Lecca, M.C.; Van Der Heijden, M.; Van Neerven, S.M.; et al. Stem cell functionality is microenvironmentally defined during tumour expansion and therapy response in colon cancer. *Nat. Cell Biol.* **2018**, *20*, 1193–1202. [CrossRef] [PubMed]

26.   Roulis, M.; Kaklamanos, A.; Schernthanner, M.; Bielecki, P.; Zhao, J.; Kaffe, E.; Frommelt, L.-S.; Qu, R.; Knapp, M.S.; Henriques, A.; et al. Paracrine orchestration of intestinal tumorigenesis by a mesenchymal niche. *Nature* **2020**, *580*, 524–529. [CrossRef] [PubMed]

27. Kobayashi, H.; Gieniec, K.A.; Wright, J.A.; Wang, T.; Asai, N.; Mizutani, Y.; Lida, T.; Ando, R.; Suzuki, N.; Lannagan, T.R.; et al. The Balance of Stromal BMP Signaling Mediated by GREM1 and ISLR Drives Colorectal Carcinogenesis. *Gastroenterology* **2021**, *160*, 1224–1239.e30. [CrossRef] [PubMed]

28. Flanagan, D.J.; Pentinmikko, N.; Luopajärvi, K.; Willis, N.J.; Gilroy, K.; Raven, A.P.; Mcgarry, L.; Englund, J.I.; Webb, A.T.; Scharaw, S.; et al. NOTUM from Apc-mutant cells biases clonal competition to initiate cancer. *Nature* **2021**, *594*, 430–435. [CrossRef] [PubMed]

29. Fumagalli, A.; Oost, K.C.; Kester, L.; Morgner, J.; Bornes, L.; Bruens, L.; Spaargaren, L.; Azkanaz, M.; Schelfhorst, T.; Beerling, E.; et al. Plasticity of Lgr5-Negative Cancer Cells Drives Metastasis in Colorectal Cancer. *Cell Stem Cell* **2020**, *26*, 569–578.e7. [CrossRef] [PubMed]

30. Ganesh, K.; Basnet, H.; Kaygusuz, Y.; Laughney, A.M.; He, L.; Sharma, R.; O'Rourke, K.P.; Reuter, V.P.; Huang, Y.-H.; Turkekul, M.; et al. L1CAM defines the regenerative origin of metastasis-initiating cells in colorectal cancer. *Nat. Cancer* **2020**, *1*, 28–45. [CrossRef]

31. Francescangeli, F.; Contavalli, P.; De Angelis, M.L.; Careccia, S.; Signore, M.; Haas, T.L.; Salaris, F.; Baiocchi, M.; Boe, A.; Giuliani, A.; et al. A pre-existing population of ZEB2+ quiescent cells with stemness and mesenchymal features dictate chemoresistance in colorectal cancer. *J. Exp. Clin. Cancer Res.* **2020**, *39*, 2. [CrossRef] [PubMed]

32. Yin, A.H.; Miraglia, S.; Zanjani, E.D.; Almeida-Porada, G.; Ogawa, M.; Leary, A.G.; Olweus, J.; Kearney, J.; Buck, D.W. AC133, a novel marker for human hematopoietic stem and progenitor cells. *Blood* **1997**, *90*, 5002–5012. [CrossRef]

33. Quirici, N.; Soligo, D.; Caneva, L.; Servida, F.; Bossolasco, P.; Deliliers, G.L. Differentiation and expansion of endothelial cells from human bone marrow CD133+cells. *Br. J. Haematol.* **2001**, *115*, 186–194. [CrossRef] [PubMed]

34. Uchida, N.; Buck, D.W.; He, D.; Reitsma, M.J.; Masek, A.; Phan, T.V.; Tsukamoto, A.S.; Gage, F.H.; Weissman, I.L. Direct isolation of human central nervous system stem cells. *Proc. Natl. Acad. Sci. USA* **2000**, *97*, 14720–14725. [CrossRef] [PubMed]

35. Singh, S.K.; Clarke, I.D.; Terasaki, M.; Bonn, V.E.; Hawkins, C.; Squire, J.; Dirks, P.B. Identification of a cancer stem cell in human brain tumors. *Cancer Res.* **2003**, *63*, 5821–5828.

36. Zhu, L.; Gibson, P.H.; Currle, D.S.; Tong, Y.; Richardson, R.J.; Bayazitov, I.T.; Poppleton, H.; Zakharenko, S.; Ellison, D.W.; Gilbertson, R.J. Prominin 1 marks intestinal stem cells that are susceptible to neoplastic transformation. *Nature* **2009**, *457*, 603–607. [CrossRef]

37. Lugli, A.; Iezzi, G.; Hostettler, I.; Muraro, M.G.; Mele, V.; Tornillo, L.; Carafa, V.; Spagnoli, G.; Terracciano, L.; Zlobec, I. Prognostic impact of the expression of putative cancer stem cell markers CD133, CD166, CD44s, EpCAM, and ALDH1 in colorectal cancer. *Br. J. Cancer* **2010**, *103*, 382–390. [CrossRef]

38. Horst, D.; Scheel, S.K.; Liebmann, S.; Neumann, J.; Maatz, S.; Kirchner, T.; Jung, A. The cancer stem cell marker CD133 has high prognostic impact but unknown functional relevance for the metastasis of human colon cancer. *J. Pathol.* **2009**, *219*, 427–434. [CrossRef]

39. Kemper, K.; Versloot, M.; Cameron, K.; Colak, S.; Melo, F.D.S.E.; De Jong, J.H.; Bleackley, J.; Vermeulen, L.; Versteeg, R.; Koster, J.; et al. Mutations in the Ras–Raf Axis Underlie the Prognostic Value of CD133 in Colorectal Cancer. *Clin. Cancer Res.* **2012**, *18*, 3132–3141. [CrossRef]

40. Kemper, K.; Sprick, M.R.; De Bree, M.; Scopelliti, A.; Vermeulen, L.; Hoek, M.; Zeilstra, J.; Pals, S.T.; Mehmet, H.; Stassi, G.; et al. The AC133 Epitope, but not the CD133 Protein, Is Lost upon Cancer Stem Cell Differentiation. *Cancer Res.* **2010**, *70*, 719–729. [CrossRef]

41. Mak, A.B.; Blakely, K.M.; Williams, R.A.; Penttilä, P.-A.; Shukalyuk, A.I.; Osman, K.T.; Kasimer, D.; Ketela, T.; Moffat, J. CD133 Protein N-Glycosylation Processing Contributes to Cell Surface Recognition of the Primitive Cell Marker AC133 Epitope. *J. Biol. Chem.* **2011**, *286*, 41046–41056. [CrossRef]

42. Dalerba, P.; Dylla, S.J.; Park, I.-K.; Liu, R.; Wang, X.; Cho, R.W.; Hoey, T.; Gurney, A.; Huang, E.H.; Simeone, D.M.; et al. Phenotypic characterization of human colorectal cancer stem cells. *Proc. Natl. Acad. Sci. USA* **2007**, *104*, 10158–10163. [CrossRef] [PubMed]

43. Zöller, M. CD44: Can a cancer-initiating cell profit from an abundantly expressed molecule? *Nat. Rev. Cancer* **2011**, *11*, 254–267. [CrossRef] [PubMed]

44. Todaro, M.; Gaggianesi, M.; Catalano, V.; Benfante, A.; Iovino, F.; Biffoni, M.; Apuzzo, T.; Sperduti, I.; Volpe, S.; Cocorullo, G.; et al. CD44v6 Is a Marker of Constitutive and Reprogrammed Cancer Stem Cells Driving Colon Cancer Metastasis. *Cell Stem Cell* **2014**, *14*, 342–356. [CrossRef] [PubMed]

45. Hostettler, L. ABCG5-positivity in tumor buds is an indicator of poor prognosis in node-negative colorectal cancer patients. *World J. Gastroenterol.* **2010**, *16*, 732–739. [CrossRef] [PubMed]

46. Huang, E.; Hynes, M.J.; Zhang, T.; Ginestier, C.; Dontu, G.; Appelman, H.; Fields, J.Z.; Wicha, M.S.; Boman, B.M. Aldehyde Dehydrogenase 1 Is a Marker for Normal and Malignant Human Colonic Stem Cells (SC) and Tracks SC Overpopulation during Colon Tumorigenesis. *Cancer Res.* **2009**, *69*, 3382–3389. [CrossRef]

47. Feng, H.; Liu, Y.; Bian, X.; Zhou, F.; Liu, Y. ALDH1A3 affects colon cancer in vitro proliferation and invasion depending on CXCR4 status. *Br. J. Cancer* **2018**, *118*, 224–232. [CrossRef]

48. Francescangeli, F.; Contavalli, P.; De Angelis, M.L.; Baiocchi, M.; Gambara, G.; Pagliuca, A.; Fiorenzano, A.; Prezioso, C.; Boe, A.; Todaro, M.; et al. Dynamic regulation of the cancer stem cell compartment by Cripto-1 in colorectal cancer. *Cell Death Differ.* **2015**, *22*, 1700–1713. [CrossRef]

49. Kondo, J.; Endo, H.; Okuyama, H.; Ishikawa, O.; Iishi, H.; Tsujii, M.; Ohue, M.; Inoue, M. Retaining cell-cell contact enables preparation and culture of spheroids composed of pure primary cancer cells from colorectal cancer. *Proc. Natl. Acad. Sci. USA* **2011**, *108*, 6235–6240. [CrossRef]

50. Kondo, J.; Ekawa, T.; Endo, H.; Yamazaki, K.; Tanaka, N.; Kukita, Y.; Okuyama, H.; Okami, J.; Imamura, F.; Ohue, M.; et al. High-throughput screening in colorectal cancer tissue-originated spheroids. *Cancer Sci.* **2019**, *110*, 345–355. [CrossRef]

51. Lombardo, Y.; Scopelliti, A.; Cammareri, P.; Todaro, M.; Iovino, F.; Vitiani, L.R.; Gulotta, G.; Dieli, F.; De Maria, R.; Stassi, G. Bone Morphogenetic Protein 4 Induces Differentiation of Colorectal Cancer Stem Cells and Increases Their Response to Chemotherapy in Mice. *Gastroenterology* **2011**, *140*, 297–309. [CrossRef] [PubMed]

52. Francescangeli, F.; Patrizii, M.; Signore, M.; Federici, G.; Di Franco, S.; Pagliuca, A.; Baiocchi, M.; Biffoni, M.; Vitiani, L.R.; Todaro, M.; et al. Proliferation State and Polo-Like Kinase1 Dependence of Tumorigenic Colon Cancer Cells. *Stem Cells* **2012**, *30*, 1819–1830. [CrossRef]

53. Manic, G.; Signore, M.; Sistigu, A.; Russo, G.; Corradi, F.; Siteni, S.; Musella, M.; Vitale, S.; De Angelis, M.L.; Pallocca, M.; et al. CHK1-targeted therapy to deplete DNA replication-stressed, p53-deficient, hyperdiploid colorectal cancer stem cells. *Gut* **2018**, *67*, 903–917. [CrossRef] [PubMed]

54. Lau, H.C.H.; Kranenburg, O.; Xiao, H.; Yu, J. Organoid models of gastrointestinal cancers in basic and translational research. *Nat. Rev. Gastroenterol. Hepatol.* **2020**, *17*, 203–222. [CrossRef]

55. Sato, T.; Vries, R.G.; Snippert, H.J.; Van De Wetering, M.; Barker, N.; Stange, D.E.; Van Es, J.H.; Abo, A.; Kujala, P.; Peters, P.J.; et al. Single Lgr5 stem cells build crypt-villus structures in vitro without a mesenchymal niche. *Nature* **2009**, *459*, 262–265. [CrossRef] [PubMed]

56. Schutgens, F.; Clevers, H. Human Organoids: Tools for Understanding Biology and Treating Diseases. *Annu. Rev. Pathol. Mech. Dis.* **2020**, *15*, 211–234. [CrossRef]

57. Yao, Y.; Xu, X.; Yang, L.; Zhu, J.; Wan, J.; Shen, L.; Xia, F.; Fu, G.; Deng, Y.; Pan, M.; et al. Patient-Derived Organoids Predict Chemoradiation Responses of Locally Advanced Rectal Cancer. *Cell Stem Cell* **2020**, *26*, 17–26.e16. [CrossRef]

58. Narasimhan, V.; Wright, J.A.; Churchill, M.; Wang, T.; Rosati, R.; Lannagan, T.R.; Vrbanac, L.; Richardson, A.B.; Kobayashi, H.; Price, T.; et al. Medium-throughput Drug Screening of Patient-derived Organoids from Colorectal Peritoneal Metastases to Direct Personalized Therapy. *Clin. Cancer Res.* **2020**, *26*, 3662–3670. [CrossRef]

59. Drost, J.; Van Jaarsveld, R.H.; Ponsioen, B.; Zimberlin, C.; Van Boxtel, R.; Buijs, A.; Sachs, N.; Overmeer, R.M.; Offerhaus, G.J.; Begthel, H.; et al. Sequential cancer mutations in cultured human intestinal stem cells. *Nature* **2015**, *521*, 43–47. [CrossRef]

60. Matano, M.; Date, S.; Shimokawa, M.; Takano, A.; Fujii, M.; Ohta, Y.; Watanabe, T.; Kanai, T.; Sato, T. Modeling colorectal cancer using CRISPR-Cas9–mediated engineering of human intestinal organoids. *Nat. Med.* **2015**, *21*, 256–262. [CrossRef]

61. Fujii, M.; Shimokawa, M.; Date, S.; Takano, A.; Matano, M.; Nanki, K.; Ohta, Y.; Toshimitsu, K.; Nakazato, Y.; Kawasaki, K.; et al. A Colorectal Tumor Organoid Library Demonstrates Progressive Loss of Niche Factor Requirements during Tumorigenesis. *Cell Stem Cell* **2016**, *18*, 827–838. [CrossRef]

62. Fumagalli, A.; Drost, J.; Suijkerbuijk, S.J.; Van Boxtel, R.; De Ligt, J.; Offerhaus, G.J.; Begthel, H.; Beerling, E.; Tan, E.H.; Sansom, O.J.; et al. Genetic dissection of colorectal cancer progression by orthotopic transplantation of engineered cancer organoids. *Proc. Natl. Acad. Sci. USA* **2017**, *114*, E2357–E2364. [CrossRef] [PubMed]

63. Sakai, E.; Nakayama, M.; Oshima, H.; Kouyama, Y.; Niida, A.; Fujii, S.; Ochiai, A.; Nakayama, K.I.; Mimori, K.; Suzuki, Y.; et al. Combined Mutation of Apc, Kras, and Tgfbr2 Effectively Drives Metastasis of Intestinal Cancer. *Cancer Res.* **2018**, *78*, 1334–1346. [CrossRef]

64. O'Rourke, K.P.; Loizou, E.; Livshits, G.; Schatoff, E.M.; Baslan, T.; Manchado, E.; Simon, J.; Romesser, P.B.; Leach, B.; Han, T.; et al. Transplantation of engineered organoids enables rapid generation of metastatic mouse models of colorectal cancer. *Nat. Biotechnol.* **2017**, *35*, 577–582. [CrossRef] [PubMed]

65. Shultz, L.D.; Goodwin, N.; Ishikawa, F.; Hosur, V.; Lyons, B.L.; Greiner, D.L. Human Cancer Growth and Therapy in Immunodeficient Mouse Models. *Cold Spring Harb. Protoc.* **2014**, *2014*, 694–708. [CrossRef]

66. Hu, Y.; Smyth, G.K. ELDA: Extreme limiting dilution analysis for comparing depleted and enriched populations in stem cell and other assays. *J. Immunol. Methods* **2009**, *347*, 70–78. [CrossRef]

67. Ibarrola-Villava, M.; Cervantes, A.; Bardelli, A. Preclinical models for precision oncology. *Biochim. Biophys. Acta (BBA)-Bioenerget.* **2018**, *1870*, 239–246. [CrossRef]

68. Quintana, E.; Shackleton, M.; Sabel, M.S.; Fullen, D.R.; Johnson, T.M.; Morrison, S.J. Efficient tumour formation by single human melanoma cells. *Nature* **2008**, *456*, 593–598. [CrossRef]

69. Tian, H.; Lyu, Y.; Yang, Y.-G.; Hu, Z. Humanized Rodent Models for Cancer Research. *Front. Oncol.* **2020**, *10*, 1696. [CrossRef]

70. Hoffman, R.M. Patient-derived orthotopic xenografts: Better mimic of metastasis than subcutaneous xenografts. *Nat. Rev. Cancer* **2015**, *15*, 451–452. [CrossRef] [PubMed]

71. Baiocchi, M.; Biffoni, M.; Ricci-Vitiani, L.; Pilozzi, E.; De Maria, R. New models for cancer research: Human cancer stem cell xenografts. *Curr. Opin. Pharmacol.* **2010**, *10*, 380–384. [CrossRef]

72. Kretzschmar, K.; Watt, F.M. Lineage tracing. *Cell* **2012**, *148*, 33–45. [CrossRef]

73. VanHorn, S.; Morris, S.A. Next-Generation Lineage Tracing and Fate Mapping to Interrogate Development. *Dev. Cell* **2021**, *56*, 7–21. [CrossRef]

74. Schneikert, J.; Behrens, J. The canonical Wnt signalling pathway and its APC partner in colon cancer development. *Gut* **2007**, *56*, 417–425. [CrossRef]
75. Barker, N.; Van Es, J.H.; Kuipers, J.; Kujala, P.; Van Den Born, M.; Cozijnsen, M.; Haegebarth, A.; Korving, J.; Begthel, H.; Peters, P.J.; et al. Identification of stem cells in small intestine and colon by marker gene Lgr5. *Nature* **2007**, *449*, 1003–1007. [CrossRef] [PubMed]
76. Barker, N.; Ridgway, R.A.; Van Es, J.H.; Van De Wetering, M.; Begthel, H.; Born, M.V.D.; Danenberg, E.; Clarke, A.R.; Sansom, O.J.; Clevers, H. Crypt stem cells as the cells-of-origin of intestinal cancer. *Nature* **2009**, *457*, 608–611. [CrossRef]
77. Shmelkov, S.V.; Butler, J.M.; Hooper, A.T.; Hormigo, A.; Kushner, J.; Milde, T.; Clair, R.S.; Baljevic, M.; White, I.; Jin, D.K.; et al. CD133 expression is not restricted to stem cells, and both CD133+ and CD133– metastatic colon cancer cells initiate tumors. *J. Clin. Investig.* **2008**, *118*, 2111–2120. [CrossRef]
78. Metcalfe, C.; Kljavin, N.M.; Ybarra, R.; de Sauvage, F.J. Lgr5+ Stem Cells Are Indispensable for Radiation-Induced Intestinal Regeneration. *Cell Stem Cell* **2014**, *14*, 149–159. [CrossRef] [PubMed]
79. Asfaha, S.; Hayakawa, Y.; Muley, A.; Stokes, S.; Graham, T.A.; Ericksen, R.E.; Westphalen, C.B.; Von Burstin, J.; Mastracci, T.L.; Worthley, D.L.; et al. Krt19(+)/Lgr5(−) Cells Are Radioresistant Cancer-Initiating Stem Cells in the Colon and Intestine. *Cell Stem Cell* **2015**, *16*, 627–638. [CrossRef]
80. De Sousa e Melo, F.; Kurtova, A.V.; Harnoss, J.M.; Kljavin, N.; Hoeck, J.D.; Hung, J.; Anderson, J.E.; Storm, E.E.; Modrusan, Z.; Koeppen, H.; et al. A distinct role for Lgr5(+) stem cells in primary and metastatic colon cancer. *Nature* **2017**, *543*, 676–680. [CrossRef] [PubMed]
81. Gupta, P.B.; Pastushenko, I.; Skibinski, A.; Blanpain, C.; Kuperwasser, C. Phenotypic Plasticity: Driver of Cancer Initiation, Progression, and Therapy Resistance. *Cell Stem Cell* **2019**, *24*, 65–78. [CrossRef] [PubMed]
82. Schwitalla, S.; Fingerle, A.A.; Cammareri, P.; Nebelsiek, T.; Göktuna, S.I.; Ziegler, P.K.; Canli, O.; Heijmans, J.; Huels, D.J.; Moreaux, G.; et al. Intestinal Tumorigenesis Initiated by Dedifferentiation and Acquisition of Stem-Cell-like Properties. *Cell* **2013**, *152*, 25–38. [CrossRef]
83. Lamprecht, S.; Schmidt, E.M.; Blaj, C.; Hermeking, H.; Jung, A.; Kirchner, T.; Horst, D. Multicolor lineage tracing reveals clonal architecture and dynamics in colon cancer. *Nat. Commun.* **2017**, *8*, 1406. [CrossRef] [PubMed]
84. van der Heijden, M.; Miedema, D.M.; Waclaw, B.; Veenstra, V.L.; Lecca, M.C.; Nijman, L.E.; van Dijk, E.; van Neerven, S.M.; Lodestijn, S.C.; Lenos, K.J.; et al. Spatiotemporal regulation of clonogenicity in colorectal cancer xenografts. *Proc. Natl. Acad. Sci. USA* **2019**, *116*, 6140–6145. [CrossRef] [PubMed]
85. Dalerba, P.; Kalisky, T.; Sahoo, D.; Rajendran, P.S.; E Rothenberg, M.; A Leyrat, A.; Sim, S.; Okamoto, J.; Johnston, D.M.; Qian, D.; et al. Single-cell dissection of transcriptional heterogeneity in human colon tumors. *Nat. Biotechnol.* **2011**, *29*, 1120–1127. [CrossRef] [PubMed]
86. Zowada, M.; Tirier, S.; Dieter, S.; Krieger, T.; Oberlack, A.; Chua, R.; Huerta, M.; Ten, F.; Laaber, K.; Park, J.; et al. Functional States in Tumor-Initiating Cell Differentiation in Human Colorectal Cancer. *Cancers* **2021**, *13*, 1097. [CrossRef] [PubMed]
87. Morral, C.; Stanisavljevic, J.; Hernando-Momblona, X.; Mereu, E.; Álvarez-Varela, A.; Cortina, C.; Stork, D.; Slebe, F.; Turon, G.; Whissell, G.; et al. Zonation of Ribosomal DNA Transcription Defines a Stem Cell Hierarchy in Colorectal Cancer. *Cell Stem Cell* **2020**, *26*, 845–861.e12. [CrossRef] [PubMed]
88. Ryser, M.D.; Min, B.-H.; Siegmund, K.D.; Shibata, D. Spatial mutation patterns as markers of early colorectal tumor cell mobility. *Proc. Natl. Acad. Sci. USA* **2018**, *115*, 5774–5779. [CrossRef]
89. Naxerova, K.; Reiter, J.G.; Brachtel, E.; Lennerz, J.K.; van de Wetering, M.; Rowan, A.; Cai, T.; Clevers, H.; Swanton, C.; Nowak, M.A.; et al. Origins of lymphatic and distant metastases in human colorectal cancer. *Science* **2017**, *357*, 55–60. [CrossRef]
90. Sottoriva, A.; Kang, H.; Ma, Z.; Graham, T.A.; Salomon, M.P.; Zhao, J.; Marjoram, P.; Siegmund, K.D.; Press, M.F.; Shibata, D.; et al. A Big Bang model of human colorectal tumor growth. *Nat. Genet.* **2015**, *47*, 209–216. [CrossRef]
91. Brabletz, T.; Jung, A.; Reu, S.; Porzner, M.; Hlubek, F.; Kunz-Schughart, L.A.; Knuechel, R.; Kirchner, T. Variable beta-catenin expression in colorectal cancers indicates tumor progression driven by the tumor environment. *Proc. Natl. Acad. Sci. USA* **2001**, *98*, 10356–10361. [CrossRef]
92. Sato, T.; Van Es, J.H.; Snippert, H.J.; Stange, D.E.; Vries, R.G.; van den Born, M.; Barker, N.; Shroyer, N.F.; Van De Wetering, M.; Clevers, H. Paneth cells constitute the niche for Lgr5 stem cells in intestinal crypts. *Nature* **2011**, *469*, 415–418. [CrossRef] [PubMed]
93. Todaro, M.; Alea, M.P.; di Stefano, A.B.; Cammareri, P.; Vermeulen, L.; Iovino, F.; Tripodo, C.; Russo, A.; Gulotta, G.; Medema, J.P.; et al. Colon Cancer Stem Cells Dictate Tumor Growth and Resist Cell Death by Production of Interleukin-4. *Cell Stem Cell* **2007**, *1*, 389–402. [CrossRef]
94. Van Es, J.H.; Sato, T.; Van De Wetering, M.; Lyubimova, A.; Nee, A.N.; Gregorieff, A.; Sasaki, N.; Zeinstra, L.; Van Den Born, M.; Korving, J.; et al. Dll1(+) secretory progenitor cells revert to stem cells upon crypt damage. *Nat. Cell Biol.* **2012**, *14*, 1099–1104. [CrossRef] [PubMed]
95. Jung, G.; Hernández-Illán, E.; Moreira, L.; Balaguer, F.; Goel, A. Epigenetics of colorectal cancer: Biomarker and therapeutic potential. *Nat. Rev. Gastroenterol. Hepatol.* **2020**, *17*, 111–130. [CrossRef]
96. Wainwright, E.N.; Scaffidi, P. Epigenetics and Cancer Stem Cells: Unleashing, Hijacking, and Restricting Cellular Plasticity. *Trends Cancer* **2017**, *3*, 372–386. [CrossRef]

97. Nam, A.S.; Chaligne, R.; Landau, D.A. Integrating genetic and non-genetic determinants of cancer evolution by single-cell multi-omics. *Nat. Rev. Genet.* **2021**, *22*, 3–18. [CrossRef]

98. Baylin, S.B.; Ohm, J.E. Epigenetic gene silencing in cancer—A mechanism for early oncogenic pathway addiction? *Nat. Rev. Cancer* **2006**, *6*, 107–116. [CrossRef] [PubMed]

99. Jiang, X.; Tan, J.; Li, J.; Kivimäe, S.; Yang, X.; Zhuang, L.; Lee, P.L.; Chan, M.T.; Stanton, L.W.; Liu, E.T.; et al. DACT3 is an epigenetic regulator of Wnt/beta-catenin signaling in colorectal cancer and is a therapeutic target of histone modifications. *Cancer Cell* **2008**, *13*, 529–541. [CrossRef]

100. Grinat, J.; Heuberger, J.; Vidal, R.O.; Goveas, N.; Kosel, F.; Berenguer-Llergo, A.; Kranz, A.; Wulf-Goldenberg, A.; Behrens, D.; Melcher, B.; et al. The epigenetic regulator Mll1 is required for Wnt-driven intestinal tumorigenesis and cancer stemness. *Nat. Commun.* **2020**, *11*, 6422. [CrossRef]

101. Mathur, R.; Alver, B.; Roman, R.M.A.K.S.; Wilson, B.G.; Wang, X.; Agoston, A.T.; Park, B.H.A.P.J.; A Shivdasani, A.K.S.R.R.; Roberts, R.M.B.G.W.X.W.C.W.M. ARID1A loss impairs enhancer-mediated gene regulation and drives colon cancer in mice. *Nat. Genet.* **2017**, *49*, 296–302. [CrossRef]

102. Suzuki, H.; Watkins, D.N.; Jair, K.-W.; E Schuebel, K.; Markowitz, S.D.; Chen, W.D.; Pretlow, T.P.; Yang, B.; Akiyama, Y.; Van Engeland, M.; et al. Epigenetic inactivation of SFRP genes allows constitutive WNT signaling in colorectal cancer. *Nat. Genet.* **2004**, *36*, 417–422. [CrossRef] [PubMed]

103. Puig, I.; Tenbaum, S.P.; Chicote, I.; Arqués, O.; Martínez-Quintanilla, J.; Cuesta-Borrás, E.; Ramírez, L.; Gonzalo, P.; Soto, A.; Aguilar, S.; et al. TET2 controls chemoresistant slow-cycling cancer cell survival and tumor recurrence. *J. Clin. Investig.* **2018**, *128*, 3887–3905. [CrossRef]

104. Tam, W.L.; Weinberg, R.A. The epigenetics of epithelial-mesenchymal plasticity in cancer. *Nat. Med.* **2013**, *19*, 1438–1449. [CrossRef] [PubMed]

105. Bian, S.; Hou, Y.; Zhou, X.; Li, X.; Yong, J.; Wang, Y.; Wang, W.; Yan, J.; Hu, B.; Guo, H.; et al. Single-cell multiomics sequencing and analyses of human colorectal cancer. *Science* **2018**, *362*, 1060–1063. [CrossRef]

106. Roerink, S.F.; Sasaki, N.; Lee-Six, H.; Young, M.D.; Alexandrov, L.B.; Behjati, S.; Mitchell, T.J.; Grossmann, S.; Lightfoot, H.; Egan, D.A.; et al. Intra-tumour diversification in colorectal cancer at the single-cell level. *Nature* **2018**, *556*, 457–462. [CrossRef]

107. van Neerven, S.M.; de Groot, N.E.; Nijman, L.E.; Scicluna, B.P.; van Driel, M.S.; Lecca, M.C.; Warmerdam, D.O.; Kakkar, V.; Moreno, L.F.; Braga, F.A.; et al. Apc-mutant cells act as supercompetitors in intestinal tumour initiation. *Nature* **2021**, *594*, 436–441. [CrossRef] [PubMed]

108. Yum, M.K.; Han, S.; Fink, J.; Wu, S.H.; Dabrowska, C.; Trendafilova, T.; Mustata, R.; Chatzeli, L.; Azzarelli, R.; Pshenichnaya, I.; et al. Tracing oncogene-driven remodelling of the intestinal stem cell niche. *Nature* **2021**, *594*, 442–447. [CrossRef]

109. Lugli, A.; Zlobec, I.; Berger, M.D.; Kirsch, R.; Nagtegaal, I.D. Tumour budding in solid cancers. *Nat. Rev. Clin. Oncol.* **2020**, *18*, 101–115. [CrossRef]

110. Gorges, T.M.; Tinhofer, I.; Drosch, M.; Röse, L.; Zollner, T.M.; Krahn, T.; Von Ahsen, O. Circulating tumour cells escape from EpCAM-based detection due to epithelial-to-mesenchymal transition. *BMC Cancer* **2012**, *12*, 178. [CrossRef]

111. Grillet, F.; Bayet, E.; Villeronce, O.; Zappia, L.; Lagerqvist, E.L.; Lunke, S.; Charafe-Jauffret, E.; Pham, K.; Molck, C.; Rolland, N.; et al. Circulating tumour cells from patients with colorectal cancer have cancer stem cell hallmarks in ex vivo culture. *Gut* **2017**, *66*, 1802–1810. [CrossRef] [PubMed]

112. Lambert, A.W.; Weinberg, R.A. Linking EMT programmes to normal and neoplastic epithelial stem cells. *Nat. Rev. Cancer* **2021**, *21*, 325–338. [CrossRef] [PubMed]

113. Gazzaniga, P.; Raimondi, C.; Nicolazzo, C.; Carletti, R.; Di Gioia, C.; Gradilone, A.; Cortesi, E. The rationale for liquid biopsy in colorectal cancer: A focus on circulating tumor cells. *Expert Rev. Mol. Diagn.* **2015**, *15*, 925–932. [CrossRef] [PubMed]

114. De Angelis, M.L.; Francescangeli, F.; La Torre, F.; Zeuner, A. Stem Cell Plasticity and Dormancy in the Development of Cancer Therapy Resistance. *Front. Oncol.* **2019**, *9*, 626. [CrossRef]

115. Mizukoshi, K.; Okazawa, Y.; Haeno, H.; Koyama, Y.; Sulidan, K.; Komiyama, H.; Saeki, H.; Ohtsuji, N.; Ito, Y.; Kojima, Y.; et al. Metastatic seeding of human colon cancer cell clusters expressing the hybrid epithelial/mesenchymal state. *Int. J. Cancer* **2020**, *146*, 2547–2562. [CrossRef]

116. Blaj, C.; Schmidt, E.M.; Lamprecht, S.; Hermeking, H.; Jung, A.; Kirchner, T.; Horst, D. Oncogenic Effects of High MAPK Activity in Colorectal Cancer Mark Progenitor Cells and Persist Irrespective of RAS Mutations. *Cancer Res.* **2017**, *77*, 1763–1774. [CrossRef]

117. Calon, A.; Espinet, E.; Palomo-Ponce, S.; Tauriello, D.V.; Iglesias, M.; Céspedes, M.V.; Sevillano, M.; Nadal, C.; Jung, P.; Zhang, X.H.; et al. Dependency of colorectal cancer on a TGF-beta-driven program in stromal cells for metastasis initiation. *Cancer Cell* **2012**, *22*, 571–584. [CrossRef]

118. Di Franco, S.; Bianca, P.; Sardina, D.S.; Turdo, A.; Gaggianesi, M.; Veschi, V.; Nicotra, A.; Mangiapane, L.R.; Iacono, M.L.; Pillitteri, I.; et al. Adipose stem cell niche reprograms the colorectal cancer stem cell metastatic machinery. *Nat. Commun.* **2021**, *12*, 5006. [CrossRef]

119. West, J.; Schenck, R.O.; Gatenbee, C.; Robertson-Tessi, M.; Anderson, A.R.A. Normal tissue architecture determines the evolutionary course of cancer. *Nat. Commun.* **2021**, *12*, 2060. [CrossRef]

120. Saito, T.; Niida, A.; Uchi, R.; Hirata, H.; Komatsu, H.; Sakimura, S.; Hayashi, S.; Nambara, S.; Kuroda, Y.; Ito, S.; et al. A temporal shift of the evolutionary principle shaping intratumor heterogeneity in colorectal cancer. *Nat. Commun.* **2018**, *9*, 2884. [CrossRef]

121. Reiter, J.G.; Hung, W.-T.; Lee, I.-H.; Nagpal, S.; Giunta, P.; Degner, S.; Liu, G.; Wassenaar, E.C.E.; Jeck, W.R.; Taylor, M.S.; et al. Lymph node metastases develop through a wider evolutionary bottleneck than distant metastases. *Nat. Genet.* **2020**, *52*, 692–700. [CrossRef] [PubMed]

122. McKinley, K.L.; Castillo-Azofeifa, D.; Klein, O.D. Tools and Concepts for Interrogating and Defining Cellular Identity. *Cell Stem Cell* **2020**, *26*, 632–656. [CrossRef] [PubMed]

123. Yin, X.; Mead, B.E.; Safaee, H.; Langer, R.; Karp, J.M.; Levy, O. Engineering Stem Cell Organoids. *Cell Stem Cell* **2016**, *18*, 25–38. [CrossRef] [PubMed]

124. Sarvestani, S.K.; DeHaan, R.K.; Miller, P.G.; Bose, S.; Shen, X.; Shuler, M.L.; Huang, E.H. A Tissue Engineering Approach to Metastatic Colon Cancer. *iScience* **2020**, *23*, 101719. [CrossRef] [PubMed]

125. Neal, J.T.; Li, X.; Zhu, J.; Giangarra, V.; Grzeskowiak, C.L.; Ju, J.; Liu, I.H.; Chiou, S.-H.; Salahudeen, A.A.; Smith, A.R.; et al. Organoid Modeling of the Tumor Immune Microenvironment. *Cell* **2018**, *175*, 1972–1988.e16. [CrossRef] [PubMed]

126. Dijkstra, K.K.; Cattaneo, C.M.; Weeber, F.; Chalabi, M.; Van De Haar, J.; Fanchi, L.F.; Slagter, M.; Van Der Velden, D.L.; Kaing, S.; Kelderman, S.; et al. Generation of Tumor-Reactive T Cells by Co-culture of Peripheral Blood Lymphocytes and Tumor Organoids. *Cell* **2018**, *174*, 1586–1598.e12. [CrossRef] [PubMed]

127. Huang, Q.; Garrett, A.; Bose, S.; Blocker, S.; Rios, A.C.; Clevers, H.; Shen, X. The frontier of live tissue imaging across space and time. *Cell Stem Cell* **2021**, *28*, 603–622. [CrossRef] [PubMed]

128. Kester, L. and A. van Oudenaarden, Single-Cell Transcriptomics Meets Lineage Tracing. *Cell Stem Cell* **2018**, *23*, 166–179. [CrossRef] [PubMed]

129. Lawson, D.A.; Kessenbrock, K.; Davis, R.; Pervolarakis, N.; Werb, Z. Tumour heterogeneity and metastasis at single-cell resolution. *Nat. Cell Biol.* **2018**, *20*, 1349–1360. [CrossRef]

130. Wagner, D.E.; Klein, A.M. Lineage tracing meets single-cell omics: Opportunities and challenges. *Nat. Rev. Genet.* **2020**, *21*, 410–427. [CrossRef]

131. Baker, A.M.; Cross, W.; Curtius, K.; Al Bakir, I.; Choi, C.-H.R.; Davis, H.L.; Temko, D.; Biswas, S.; Martinez, P.; Williams, M.; et al. Evolutionary history of human colitis-associated colorectal cancer. *Gut* **2018**, *68*, 985–995. [CrossRef]

132. Li, H.; Courtois, E.T.; Sengupta, D.; Tan, Y.; Chen, K.H.; Goh, J.J.L.; Kong, S.L.; Chua, C.; Hon, L.K.; Tan, W.S.; et al. Reference component analysis of single-cell transcriptomes elucidates cellular heterogeneity in human colorectal tumors. *Nat. Genet.* **2017**, *49*, 708–718. [CrossRef]

133. Barriga, F.; Montagni, E.; Mana, M.; Mendez-Lago, M.; Hernando-Momblona, X.; Sevillano, M.; Guillaumet-Adkins, A.; Rodriguez-Esteban, G.; Buczacki, S.; Gut, M.; et al. Mex3a Marks a Slowly Dividing Subpopulation of Lgr5+ Intestinal Stem Cells. *Cell Stem Cell* **2017**, *20*, 801–816.e7. [CrossRef] [PubMed]

134. Tikhonova, A.N.; Lasry, A.; Austin, R.; Aifantis, I. Cell-by-Cell Deconstruction of Stem Cell Niches. *Cell Stem Cell* **2020**, *27*, 19–34. [CrossRef] [PubMed]

135. Baslan, T.; Hicks, J. Unravelling biology and shifting paradigms in cancer with single-cell sequencing. *Nat. Rev. Cancer* **2017**, *17*, 557–569. [CrossRef]

136. Lenos, K.J.; Lodestijn, S.C.; Lyons, S.; Bijlsma, M.F.; Miedema, D.M.; Vermeulen, L. A marker-independent lineage-tracing system to quantify clonal dynamics and stem cell functionality in cancer tissue. *Nat. Protoc.* **2019**, *14*, 2648–2671. [CrossRef]

137. He, J.; Baxter, S.L.; Xu, J.; Xu, J.; Zhou, X.; Zhang, K. The practical implementation of artificial intelligence technologies in medicine. *Nat. Med.* **2019**, *25*, 30–36. [CrossRef] [PubMed]

138. Mukherjee, S.; Yadav, G.; Kumar, R. Recent trends in stem cell-based therapies and applications of artificial intelligence in regenerative medicine. *World J. Stem Cells* **2021**, *13*, 521–541. [CrossRef]

139. Christiansen, E.M.; Yang, S.J.; Ando, D.M.; Javaherian, A.; Skibinski, G.; Lipnick, S.; Mount, E.; O'Neil, A.; Shah, K.; Lee, A.K.; et al. In Silico Labeling: Predicting Fluorescent Labels in Unlabeled Images. *Cell* **2018**, *173*, 792–803.e19. [CrossRef] [PubMed]

140. Ota, S.; Horisaki, R.; Kawamura, Y.; Ugawa, M.; Sato, I.; Hashimoto, K.; Kamesawa, R.; Setoyama, K.; Yamaguchi, S.; Fujiu, K.; et al. Ghost cytometry. *Science* **2018**, *360*, 1246–1251. [CrossRef]

Article

Functional States in Tumor-Initiating Cell Differentiation in Human Colorectal Cancer

Martina K. Zowada [1,2,3,†], Stephan M. Tirier [4,5,6,†], Sebastian M. Dieter [1,2,7,†], Teresa G. Krieger [4,5,8], Ava Oberlack [1], Robert Lorenz Chua [4,5,8], Mario Huerta [1,2], Foo Wei Ten [4,5,8], Karin Laaber [1,2,3], Jeongbin Park [4,8], Katharina Jechow [4,5,8], Torsten Müller [9,10], Mathias Kalxdorf [9,10], Mark Kriegsmann [11], Katharina Kriegsmann [12], Friederike Herbst [1,2], Jeroen Krijgsveld [9,10], Martin Schneider [13], Roland Eils [4,5,8,14], Hanno Glimm [1,2,15,16,*,‡], Christian Conrad [4,5,8,*,‡] and Claudia R. Ball [1,2,15,16,*,‡]

1   Translational Functional Cancer Genomics, National Center for Tumor Diseases (NCT) Heidelberg and German Cancer Research Center (DKFZ) Heidelberg, 69120 Heidelberg, Germany; martina.zowada@nct-heidelberg.de (M.K.Z.); sebastian.dieter@nct-heidelberg.de (S.M.D.); ava.oberlack@med.uni-muenchen.de (A.O.); mario.huerta@nct-heidelberg.de (M.H.); karin.laaber@nct-heidelberg.de (K.L.); friederike.herbst@nct-heidelberg.de (F.H.)

2   Department of Translational Medical Oncology, NCT Dresden and DKFZ, 01307 Dresden, Germany

3   Faculty of Biosciences, Heidelberg University, 69120 Heidelberg, Germany

4   Division of Theoretical Bioinformatics, DKFZ Heidelberg, 69120 Heidelberg, Germany; s.tirier@dkfz-heidelberg.de (S.M.T.); teresa.krieger@charite.de (T.G.K.); robert-lorenz.chua@charite.de (R.L.C.); foo-wei.ten@charite.de (F.W.T.); j.park@dkfz-heidelberg.de (J.P.); katharina.jechow@bihealth.de (K.J.); roland.eils@charite.de (R.E.)

5   Center for Quantitative Analysis of Molecular and Cellular Biosystems (BioQuant), Heidelberg University, 69120 Heidelberg, Germany

6   Division of Chromatin Networks, DKFZ Heidelberg, 69120 Heidelberg, Germany

7   German Cancer Consortium (DKTK), 69120 Heidelberg, Germany

8   Digital Health Center, Berlin Institute of Health (BIH) and Charité-Universitätsmedizin Berlin, 10117 Berlin, Germany

9   Division of Proteomics of Stem Cells and Cancer, DKFZ Heidelberg, 69120 Heidelberg, Germany; torsten.mueller@dkfz-heidelberg.de (T.M.); mathiaskalxdorf@gmail.com (M.K.); j.krijgsveld@dkfz-heidelberg.de (J.K.)

10   Medical Faculty, Heidelberg University, 69120 Heidelberg, Germany

11   Institute of Pathology, Heidelberg University Hospital, 69120 Heidelberg, Germany; Mark.Kriegsmann@med.uni-heidelberg.de

12   Department of Hematology, Oncology and Rheumatology, Heidelberg University Hospital, 69120 Heidelberg, Germany; Katharina.Kriegsmann@med.uni-heidelberg.de

13   Department of General, Visceral and Transplantation Surgery, Heidelberg University Hospital, 69120 Heidelberg, Germany; Martin.Schneider@med.uni-heidelberg.de

14   Department for Bioinformatics and Functional Genomics, Institute for Pharmacy and Molecular Biotechnology (IPMB), Heidelberg University, 69120 Heidelberg, Germany

15   Center for Personalized Oncology, University Hospital Carl Gustav Carus Dresden at Technische Universität (TU) Dresden, 01307 Dresden, Germany

16   DKTK, 01307 Dresden, Germany

*   Correspondence: hanno.glimm@nct-dresden.de (H.G.); christian.conrad@bihealth.de (C.C.); claudia.ball@nct-dresden.de (C.R.B.); Tel.: +49-351-458-5540 (H.G.); +49-30-450-543-097 (C.C.); +49-351-458-5536 (C.R.B.)

†   Shared first authorship.

‡   Shared last authorship.

Citation: Zowada, M.K.; Tirier, S.M.; Dieter, S.M.; Krieger, T.G.; Oberlack, A.; Chua, R.L.; Huerta, M.; Ten, F.W.; Laaber, K.; Park, J.; et al. Functional States in Tumor-Initiating Cell Differentiation in Human Colorectal Cancer. *Cancers* **2021**, *13*, 1097. https://doi.org/10.3390/cancers13051097

Academic Editors: Marta Baiocchi and Ann Zeuner

Received: 12 February 2021
Accepted: 28 February 2021
Published: 4 March 2021

Publisher's Note: MDPI stays neutral with regard to jurisdictional claims in published maps and institutional affiliations.

Simple Summary: Different types of cells with tumor-initiating cell (TIC) activity contribute to colorectal cancer (CRC) progression and resistance to anti-cancer treatment. In this study, we aimed to understand whether different cell types exist within a patient-derived tumor culture, distinguishable by different patterns of their gene expression. By mRNA sequencing of patient-derived CRC cultures at the single-cell level, we defined expression programs that closely resemble differentiated cell populations of the normal intestine. Here, cell type-associated subpopulations showed differences in functional properties such as cell growth and energy metabolism. Subsequent functional analyses in vitro and in vivo demonstrated that metabolic states are linked to TIC activity in primary CRC cultures. We also show that TIC activity is dependent on oxidative phosphorylation, which may therefore represent a target for novel therapies.

Abstract: Intra-tumor heterogeneity of tumor-initiating cell (TIC) activity drives colorectal cancer (CRC) progression and therapy resistance. Here, we used single-cell RNA-sequencing of patient-derived CRC models to decipher distinct cell subpopulations based on their transcriptional profiles. Cell type-specific expression modules of stem-like, transit amplifying-like, and differentiated CRC cells resemble differentiation states of normal intestinal epithelial cells. Strikingly, identified subpopulations differ in proliferative activity and metabolic state. In summary, we here show at single-cell resolution that transcriptional heterogeneity identifies functional states during TIC differentiation. Furthermore, identified expression signatures are linked to patient prognosis. Targeting transcriptional states associated to cancer cell differentiation might unravel novel vulnerabilities in human CRC.

Keywords: colorectal cancer; tumor-initiating cells; tumor heterogeneity; patient-derived cancer models; single-cell RNA-sequencing; tumor metabolism; transcriptional programs; tumor cell differentiation

1. Introduction

In many tumor entities, tumor formation and progression are driven by a cellular subfraction with tumor-initiating cell (TIC) activity [1–3]. In colorectal cancer (CRC), the TIC compartment is organized as a functional cellular hierarchy with extensively self-renewing long-term TICs driving serial tumor propagation in vivo. Long-term TICs generate highly proliferative, short-lived tumor transient-amplifying cells with limited or no self-renewal capacity giving rise to the bulk of post-mitotic tumor cells [4]. Remarkably, this functional heterogeneity within individual CRCs is not primarily driven by genetic events, suggesting that epigenetic or extrinsic factors contribute to functional cellular heterogeneity [5].

Lineage-tracing experiments demonstrate that in CRC the population of highly self-renewing TICs expresses *LGR5* and generates progeny differentiating towards mucosecreting- and absorptive-like phenotypes [6]. Thus, CRCs harbor a subfraction of stem-like TICs and maintain a hierarchical organization reminiscent of the normal intestinal epithelium [7]. Moreover, a gene signature specific for intestinal stem cells has been suggested to predict disease relapse [8], indicating a potential clinical relevance of stem-like TICs for CRC patients. However, prospective validation in an independent cohort is still not available.

Recent evidence suggests that the epigenome of an individual CRC is already formed by the cell-of-origin. Methylation analyses demonstrate maintenance of the cell-of-origin differentiation state during tumor progression, and identified three CRC subclasses of intestinal crypt differentiation of the cell-of-origin. Importantly, patients with a stem-like methylation signature showed significantly reduced overall survival [9].

While the hierarchical organization of normal and malignant stem cell systems has previously been thought to be fixed and unidirectional, evidence for plasticity in these systems is accumulating [10–12]. Lineage-tracing experiments in CRC highlight that more differentiated cells can repopulate a free stem-like niche and acquire TIC activity upon ablation of the active stem-like population [6,13,14]. Similarly, pronounced plasticity drives pancreatic cancer by clonal succession of transient TIC activity [15].

Current understanding of TIC heterogeneity in CRC is mainly derived from serial syngeneic or xenogeneic transplantation models, where TICs have been retrospectively identified by interpreting the kinetics of genetically marked or pre-enriched bulk cells [4,8,16–18]. While this allowed deep insights into functional heterogeneity within tumors, such retrospective experimental strategies from bulk samples hamper direct assignment of transcriptional states in individual cells.

To characterize molecular underpinnings of functional CRC intra-tumor heterogeneity at the single-cell level, we here asked whether distinct functional programs within individual cells from patient-derived CRC models can be assigned to specific cellular subpopulations.

2. Results

2.1. Transcriptional Heterogeneity of Patient-Derived CRC Spheroid Cultures

To assess whether heterogeneous transcriptional programs can be detected in CRC tumor spheroids at the single-cell level, we performed single-cell RNA-sequencing (scRNA-seq) using a nanowell platform [19]. As patient-derived spheroid cultures contain purely tumor cells, thereby allowing to study tumor cell heterogeneity in high resolution, and recapitulate the histology of the original tumor after xenotransplantation into immunodeficient NOD.Cg-$Prkdc^{scid}Il2rg^{tm1Wjl}$/SzJ (NSG) mice [4], we sequenced 12 three-dimensional tumor spheroid cultures (P1–P12) derived from primary tumors (n = 6 patients) or metastases (n = 6 patients) of 12 different CRC patients. These patient tumors and derived spheroids cover known subtypes (microsatellite stable or microsatellite instable tumors) and driver mutations (loss of *APC* and/or *TP53*, activating mutations in *KRAS*; Table 1). On average, 389 cells (range: 141–736) were sequenced per patient, resulting in 4663 single-cell profiles with an average of more than 4000 detected genes per cell (Table 2).

Table 1. Patient overview. Patient-derived colorectal cancer spheroids (P1–P12), organoids (O1–O3), xenografts (X1, X2), and primary colorectal cancer samples (T1–T3) used for single-cell RNA-sequencing. X indicates mutation, - indicates wild type. f, female; m, male; met, metastasis; MS, microsatellite; MSI, microsatellite instable; MSS, microsatellite stable; N/A, not available.

Patient	Sex	Origin	Site	Stage (UICC)	MS status	*TP53*	*APC*	*KRAS*
P1	m	liver met	Rectum	IV	MSS	X	X	X
P2	m	lung met	Caecum	IV	MSS	X	X	X
P3	f	liver met	Rectum	IV	MSS	X	X	X
P4	f	liver met	Ascending colon	IV	MSI	X	X	X
P5	f	primary	Transverse colon	IV	MSS	-	-	-
P6	f	primary	Caecum	IV	MSS	X	-	-
P7	m	liver met	Sigmoid	IV	MSS	-	X	X
P8	m	liver met	Caecum	IV	MSS	X	-	-
P9	m	primary	Rectum	IV	MSS	X	-	X
P10	m	primary	Sigmoid	IIIB	MSS	N/A	N/A	N/A
P11	m	primary	Rectum and caecum	IIIB	MSS	-	-	-
P12	m	primary	Rectum and transverse colon	II	MSI	X	X	X
O1	f	liver met	Sigmoid	IV	MSS	X	X	X
O2	f	liver met	Caecum	IV	MSS	X	X	X
O3	f	liver met	Ascending colon	IV	MSI	-	X	-
X1	m	primary	Rectum	I	N/A	N/A	N/A	N/A
X2	m	primary	Ascending colon	II	MSI	N/A	N/A	-
T1	m	primary	Sigmoid	III	N/A	N/A	N/A	N/A
T2	m	primary	Ascending colon	IV	N/A	N/A	N/A	N/A
T3	m	primary	Ascending colon	IVa	MSS	N/A	N/A	X

Table 2. Single-cell RNA-sequencing analysis. Top: Colorectal cancer (CRC) spheroids (P1–P12). Cultures with an *LGR5* score (=*LGR5* read counts/cell number) >1 are considered *LGR5*+. Bottom: patient-derived organoids (PDOs; O1–O3), patient-derived xenografts (PDXs; X1, X2), and tumors (T1–T3). Cell numbers for X1 and X2 indicate human cells. Cell numbers in brackets indicate epithelial cells used for analysis of T1–T3. *Hs*, *Homo sapiens*; *Mm*, *Mus musculus*; QC, quality control.

CRC Spheroids				
Patient	Mean Reads Per Cell	Cell Number after QC	Mean Detected Genes Per Cell	*LGR5* Score
P1	348,016	325	3535	12.85
P2	261,595	309	4072	0.23
P3	460,471	551	4537	6.43
P4	1,061,813	263	4186	87.72
P5	334,099	502	3943	4.61
P6	1,276,856	141	5116	0.03
P7	359,362	434	4335	10.18
P8	190,170	197	4174	3.38
P9	527,407	464	4354	0.00
P10	391,680	736	3418	3.35
P11	505,439	308	4036	1.43
P12	454,258	433	3977	0.00

CRC PDOs, PDXs, Tumors				
Sample	Mean Reads Per Cell	Cell Number after QC	Mean Detected Genes Per Cell (*Hs*)	Mean Detected Genes Per Cell (*Mm*)
O1	120,218	5550	5542	-
O2	169,086	3003	5425	-
O3	73,415	8785	4176	-
X1	238,836	1475	1841	2281
X2	237,891	1070	4598	2415
T1	1,333,884	362 (136)	3646	-
T2	847,472	538 (77)	4090	-
T3	623,942	724 (40)	2474	-

Unsupervised clustering of single-cell profiles [20] revealed grouping of cells according to the patient-of-origin (Figure 1a). Hierarchical clustering based on the top 10 differentially expressed genes per patient showed that cells primarily cluster, with one exception, by the tumor site they originate from, but not by microsatellite status (Figure 1a,b).

Within individual patient-derived spheroids, top differentially expressed genes (Wilcoxon rank sum test: adjusted *p*-value < 0.05; log fold-change > 0.25) between patients contained WNT signaling components and downstream targets (e.g., *FRZB*, *DKK1*, *TCF4*, *SOX2*) and normal tissue-associated differentiation markers (e.g., *MUC12*, *MUC17*, *SPINK1*, *SPINK4*, *DEFA5*, *DEFA6*; Figure 1b). Thus, beyond patient tumor-specific alterations, differentiation state-associated expression programs can be attributed to transcriptional profiles derived from single CRC cells.

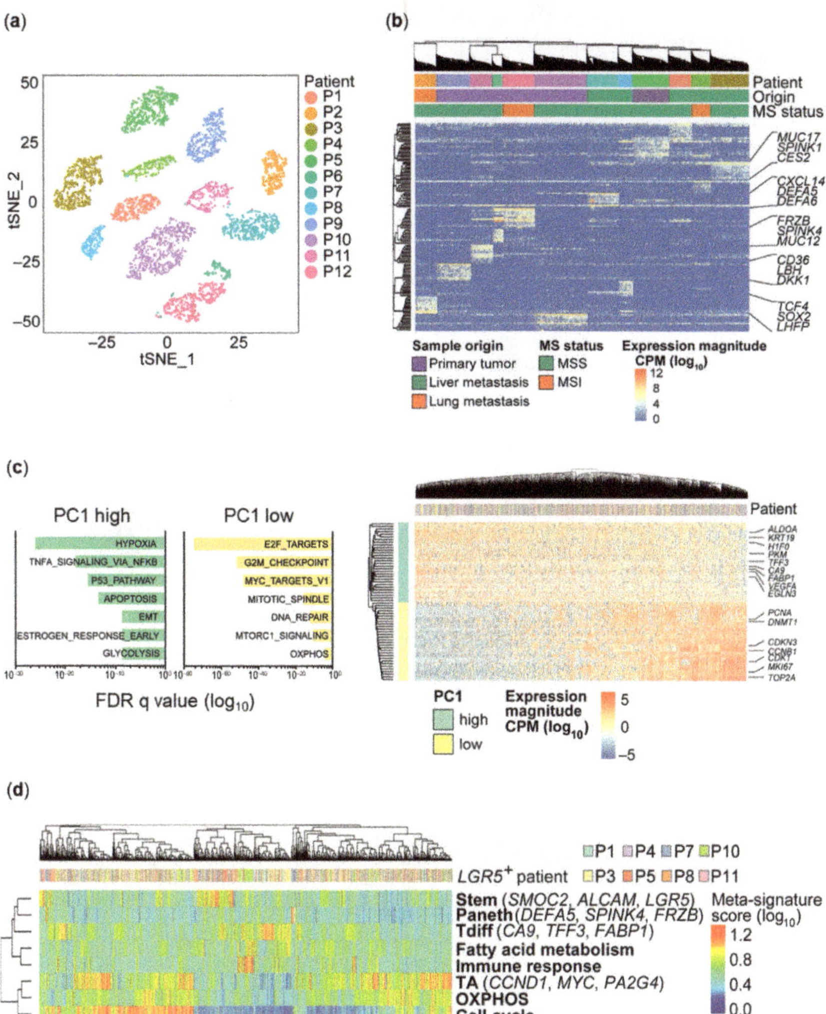

Figure 1. Identification of transcriptional subpopulations by single-cell RNA-sequencing (scRNA-seq). (**a**) Two-dimensional t-distributed stochastic neighbor embedding (tSNE) visualization of scRNA-seq expression profiles. (**b**) Hierarchical clustering and heatmap visualization of single-cell gene expression using the top 10 differentially expressed genes per patient (*n* = 12). CPM, counts-per-million; MS, microsatellite; MSI, microsatellite instable; MSS, microsatellite stable. (**c**) Principal component analysis of scRNA-seq data corrected for inter-patient variability. Left: Gene set enrichment analysis (GSEA) for the first principal component (PC1; hallmark gene sets). Gene sets are ranked by false discovery rate (FDR) q values. Right: Heatmap showing gene expression magnitude of the top 30 genes with highest and lowest PC scores for PC1. OXPHOS, oxidative phosphorylation. (**d**) Heatmap reflecting hierarchical clustering of core meta-signature scores of eight *LGR5*+ CRC spheroid cultures determined by non-negative matrix factorization. Brackets indicate marker genes for specific signatures. TA, transit-amplifying; Tdiff, terminally differentiated.

2.2. Distinct Cell Types and Cell States in Individual CRC Spheroids

To identify heterogeneous gene expression programs shared across patients in single cells from individual tumor spheroid cultures, we corrected for inter-patient variability by calculating relative expression levels for each patient individually [21,22]. Principal component analysis (PCA) of the combined dataset revealed an anti-correlated transcriptional pattern independent of patient origin with genes either involved in cell growth, proliferation, and oxidative phosphorylation (OXPHOS), or hypoxia and glycolysis. Notably, the hypoxia/glycolysis signature contains intestinal differentiation markers (e.g., *TFF3*, *FABP1*, *KRT19*; Figure 1c), indicating an association of distinct metabolic states with tumor cell differentiation and proliferation, as recently described for the normal intestinal crypt [23].

As the activation of continuous gene expression programs may not be captured by discrete clustering, we adapted a previously described computational approach based on non-negative matrix factorization (NNMF) [24,25] to more precisely identify transcriptional programs heterogeneously expressed across patients (Figure 1d, Figure S1a–e). In order to focus on tumors that display preserved hierarchical organization, we focused on the eight cultures with detectable *LGR5* transcript levels (*LGR5* score = total *LGR5* transcript counts/cell number > 1; Table 2), as *LGR5* represents an established marker for intestinal stem cells and CRC TICs, whereas the phenotype and the role of potential *LGR5*-negative stem cells and TICs are much less defined [8,14,26,27]. Thus, four cultures with very low or non-detectable *LGR5* transcript abundance were excluded for this analysis. We identified 13 heterogeneous gene expression programs that could be classified into two partially overlapping categories: one (A) linked to 'cell types' or lineages analogous to the normal intestinal epithelium, and the other (B) associated with 'cell states' (Figure S1f).

Category A identified distinct cells harboring marker expression similarities to normal intestinal stem cells (e.g., *LGR5*, *AXIN2*), Paneth cells (e.g., *DEFA5*, *DEFA6*), or transit-amplifying (TA) cells (e.g., *PA2G4*, *CCND1*) in the healthy human intestine, suggesting that distinct cell types can be identified based on individual gene expression programs. As the analyzed cells derive from the colon and only resemble the cell types of the normal intestine, we refer to these cell type-associated subpopulations as stem-like, Paneth-like, TA-like, and terminally differentiated (Tdiff)-like. Category B comprised expression programs enriched for genes involved in cell cycle regulation (e.g., *CDK1*, *MKI67*), immune/stress response (e.g., *CEACAM6*, *CXCL2*), or metabolic functions (e.g., OXPHOS (e.g., *PRDX3*, *ATP5O*), fatty acid metabolism (e.g., *CES2*, *RETSAT*), and hypoxia/glycolysis (e.g., *HILPDA*, *VEGFA*)). Similar to PCA results (Figure 1c), one expression program (Tdiff) was enriched for both, genes associated with hypoxia/glycolysis and differentiation markers (e.g., *TFF3*, *KRT20*; Figure S1f).

Next, each individual cell was scored for inferred expression programs using the averaged expression of the top genes per factor identified by NNMF. To reduce redundancy, signatures showing similar enrichment and clustering patterns were combined, resulting in eight meta-signatures (Figure S1a–f, Table S1). Clustering of meta-signature scores allowed identification of discrete and overlapping transcriptional programs (Figure 1d). Similar to PCA, cell cycle, OXPHOS, and TA signatures showed a pronounced overlap, indicating a highly proliferative cell fraction—potentially corresponding to the TA-like compartment—driven by MYC and characterized by high OXPHOS. In contrast, stem-like, Paneth-like, and Tdiff-like cells did not show significant overlap with the cell cycle signature (Figure 1d), suggesting reduced or absent proliferative activity. This indicates that scRNA-seq and matrix factorization analysis are capable of distinguishing functionally distinct cell populations based on transcriptional profiles.

To analyze the cell type composition in all eight $LGR5^+$ cultures individually, we used the NNMF-inferred signature scores (stem, TA, Paneth, Tdiff) to assign cells to one of the four cell types which allowed us to assess the extent of active cell type-specific transcriptional programs. Despite different cell type compositions, we observed presence of stem-like, TA-like, and Tdiff-like cells in all, and rare, but detectable Paneth-like cells in six out of the eight $LGR5^+$ cultures (Figure S1g). This indicates that individual CRC tumors display similar cellular diversity resembling normal intestinal cell types even with different clinico-pathological features (Table 1).

We next assessed whether the signatures identified in our patient-derived in vitro models can also be identified in patient tumors. We therefore applied our signatures

(Table S1) on publicly available expression data of colon adenocarcinoma (COAD) patients (The Cancer Genome Atlas (TCGA) cohort; $n = 328$ patients) [28]. Correlations among cell type and cell state signatures in the spheroid scRNA-seq data (Figure 1d) were detectable in patient whole transcriptome data. Clustering of the TCGA cohort based on signature expression resulted in six clusters of patients (cl1–cl6) with different combinations of low or high expression of individual signatures. Significantly different progression-free survival ($p = 0.043$) and numerically decreased overall survival ($p = 0.059$) were observed between groups of clusters, indicating a relevance of signature expression for patient prognosis (Figure 2a–c, Table S2).

We further compared the association of cl1–cl6 with consensus molecular subtypes (CMS1–CMS4) [29,30]. CMS1 tumors were mostly represented in cl3 (49%), CMS2 tumors displayed mostly cl2 (37%) and cl4 (24%), CMS3 tumors were predominantly found in cl1 (36%). CMS4 tumors were spread across cl4 (14%), cl5 (43%), and cl6 (20%). CMS4 has been shown to have poor progression-free survival [30]. Accordingly, cl4, cl5, and cl6 (33%, 60%, and 65% CMS4 contribution, respectively) showed the worst progression-free survival. cl6 comprised the majority of patients with the shortest overall survival of CMS4, whereas cl4 displayed worse progression-free survival than cl1, cl2, and cl3 but similar overall survival (Figure 2c,d).

In line with previously published data reporting an intestinal stem cell-specific gene signature linked to LGR5 and EPHB2 expression related to CRC relapse [8], high expression of our stem signature defined by 200 genes (Table S1) in the TCGA cohort displayed decreased progression-free survival ($p = 0.068$) compared to patients with low expression (Figure 2e).

Taken together, our six clusters exhibit a better prognostic value for progression-free survival ($p = 0.043$) than previously reported subtypes linked to cancer-associated fibroblasts [31] ($p = 0.15$), CMSs [29,30] ($p = 0.18$), or our stem signature alone ($p = 0.068$). Indeed, when our clusters were added to multivariable clinico-molecular survival models, we still observed a significant discriminative contribution by our cluster combinations in predicting recurrence, but no significant contribution was appreciated when adding CMSs or cancer-associated fibroblasts to our model. On the other hand, incorporating stroma cells like cancer-associated fibroblasts can substantially improve the overall survival prediction (Table S2). These results underscore the relevance of combinations of cell type and cell state signature expression for COAD outcome, and demonstrate a prognostic value of cell type and cell state signatures inferred from spheroid single-cell transcriptomes.

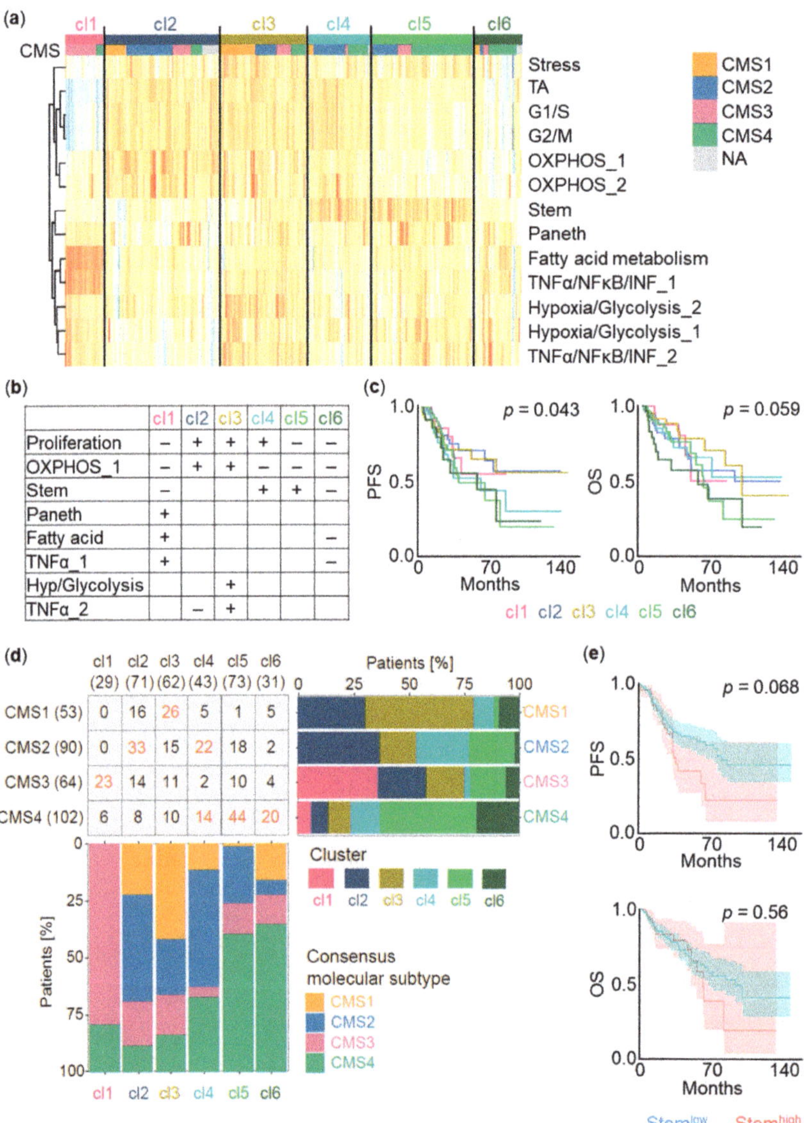

Figure 2. Analysis of single-cell RNA-sequencing (scRNA-seq) signature expression in The Cancer Genome Atlas (TCGA) colon adenocarcinoma (COAD) cohort. (**a**) Heatmap reflecting signature expression in the identified clusters of patients (cl1–cl6). Rows correspond to scRNA-seq signatures (*n* = 13). Columns correspond to TCGA COAD samples (*n* = 328). Clusters are further classified according to consensus molecular subtypes (CMS1–CMS4). NA, not assigned; OXPHOS, oxidative phosphorylation; TA, transit-amplifying. (**b**) Cell type and cell state signature expression levels defining cl1–cl6. Hyp, hypoxia. (**c**) Kaplan–Meier survival curves displaying progression-free survival (PFS) and overall survival (OS) for cl1–cl6. *p*-Values of the comparison cl6 + cl5 + cl4 versus cl3 + cl2 + cl1 for PFS and cl6 + cl5 versus cl4 + cl3 + cl2 + cl1 for OS. (**d**) Representation of CMS1–CMS4 within cl1–cl6 and of cl1–cl6 within CMS1–CMS4. Numbers indicate amount of patients classified under individual categories. Numbers marked in red highlight dominant combinations. Patients not assigned to a CMS (*n* = 19 patients) were excluded. (**e**) PFS and OS of TCGA COAD patients with high versus low expression of the stem signature. Shaded areas indicate 95% confidence intervals.

2.3. Cell Cycle and Proliferative Activity of Human CRC Cells

scRNA-seq suggests the existence of cell types with different proliferative activity within individual spheroid cultures and stem-like, TA-like, and Tdiff-like subpopulations. We therefore asked whether subfractions of cells with different cell cycle and proliferative activity exist within CRC tumors in vivo and whether they are functionally relevant.

To assess proliferative heterogeneity of tumor cells, we utilized a genetic label-retaining strategy based on expression of tetracycline-regulated (Tet-off) histone 2B (H2B) green fluorescent protein (GFP) [32] (Figure S2a). Upon doxycycline addition, nuclear H2B-GFP expression is suppressed and subsequently diluted with each cell division, allowing identification of subpopulations according to proliferative history.

To evaluate whether proliferatively inactive cells within established tumors possess TIC capacity, we transduced tumor spheroid cultures derived from seven different patients with an H2B-GFP-encoding lentiviral vector prior to xenotransplantation into NSG mice ($n = 14$; 1–4 mice per culture). After successful tumor formation, H2B-GFP expression was suppressed by doxycycline administration for two weeks. Analysis of H2B-GFP expression in established tumors by flow cytometry revealed presence of fast (GFP$^-$), slow (GFP^{low}), and rare dividing (GFP^{high}) cell fractions, demonstrating proliferative heterogeneity of CRC cells in vivo. To assess whether heterogeneously proliferating cell fractions are associated with TIC activity, cells from 12 out of 14 primary xenografts were sorted into fast, slow, and rare dividing subfractions and serially transplanted into secondary mice ($n = 33$). Importantly, all subfractions contained cells with TIC activity irrespective of transplanted cells' proliferative history prior to re-transplantation (fast: 5/12; slow: 4/9; rare: 5/12 mice with tumors), showing that TIC activity is not strictly linked to proliferative active cell fractions but also present in proliferatively inactive populations within tumors (Figure S2a–c). In summary, these data show that proliferatively inactive TICs exist within established tumors in vivo. We therefore conclude that within individual tumors, TIC activity can be present in cells with heterogeneous proliferative activity and is therefore not restricted to a specific proliferative state of individual cells.

2.4. Divergent Cell Type-Associated Energy Metabolic Preferences

Prominent heterogeneously expressed transcriptional programs in individual spheroid cells were related to energy metabolism. Whereas a glycolysis/hypoxia signature could be assigned to Tdiff-like cells (Figure S1f), OXPHOS strongly overlapped with MYC-target and cell cycle signatures, both identifying cells belonging to the putative TA-like cell compartment (Figure 1c,d). Thus, we hypothesized that metabolic preferences distinguish functionally distinct cell subpopulations and focused on these for further validation.

Consistently, we observed clearly overlapping TA-like, OXPHOS, and cell cycle signatures (Figure 1d), but no obvious association between stem-like and OXPHOS or cell cycle signatures. Of note, in the normal intestinal epithelium, intestinal stem cells actively cycle and constantly produce progeny, but their relative abundance compared to non-cycling Tdiff cells is very low [33]. Thus, we reasoned that differential metabolic trends in stem-like and Paneth-like cells could be masked by much higher or lower expression of individual metabolic signatures in highly cycling cells or the rare dividing Tdiff-like subpopulation. To overcome this, we performed pairwise comparisons of cell state signatures across CRC subpopulations that resemble normal intestinal cell types as identified by differential NNMF signature expression.

Cell cycle scores were strongly increased in TA-like cells compared to stem-like, Paneth-like, and Tdiff-like cells ($p < 0.000001$; respectively). The greatest differences in metabolic states existed between Tdiff-like and TA-like subpopulations, demonstrating that the majority of TA-like cells had high OXPHOS scores, whereas Tdiff-like cells showed high scores for hypoxia and glycolysis, but low scores for OXPHOS. Albeit less pronounced, similar and highly significant differences were detectable for stem-like and Paneth-like cells. In comparison to Paneth-like cells, stem-like cells showed increased OXPHOS scores and decreased glycolysis/hypoxia scores ($p < 0.000001$, respectively; Figure 3a).

In addition to the overall high OXPHOS scores, the stem-like signature was associated with enhanced expression of *OXR1* and *PON2*. Being essential for protection against oxidative stress, these genes may counteract higher reactive oxygen species (ROS) levels resulting from enhanced OXPHOS rates [34,35]. Another gene included in the stem-like signature is *MAP2K6*—an essential p38 signaling component [36] known to be associated with high OXPHOS levels in intestinal stem cells [23] (Figure S1f, Table S1).

Collectively, these results demonstrate an association between tumor cell differentiation and metabolic identities in this three-dimensional in vitro CRC model.

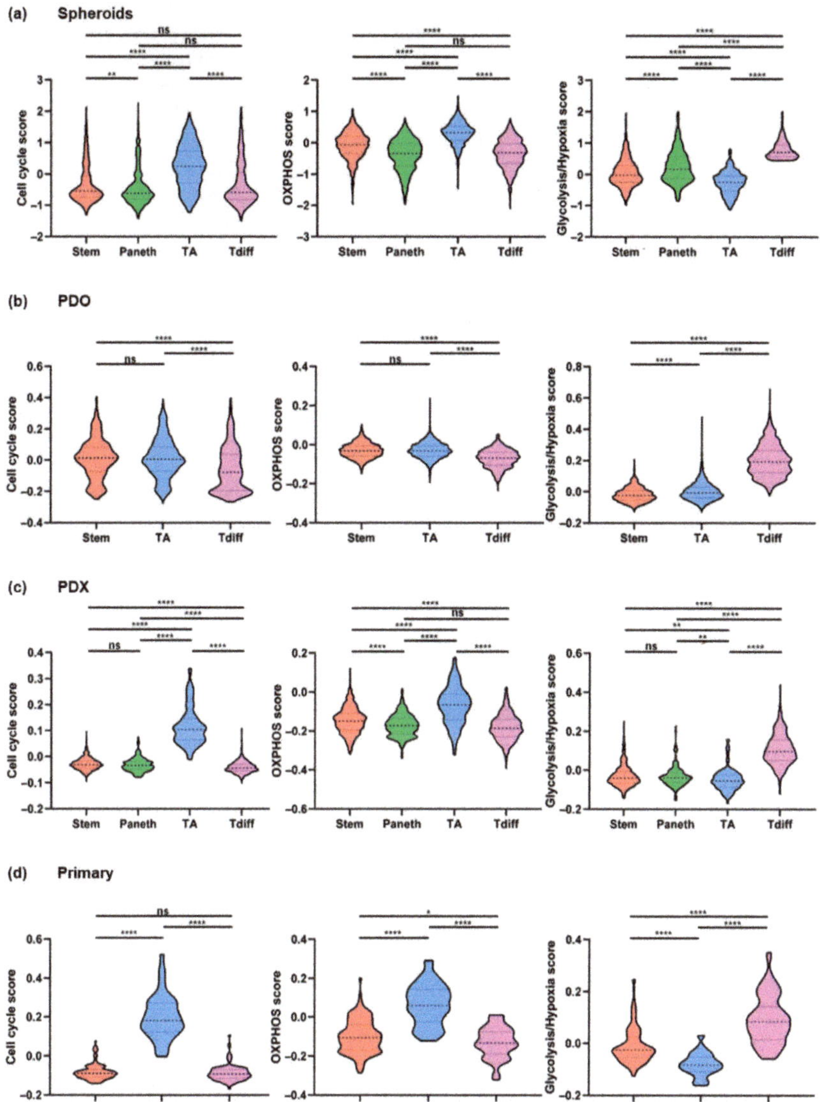

Figure 3. Cell state scores for cell type-specific cell subsets inferred by non-negative matrix factorization. (**a–d**) Cell cycle, oxidative phosphorylation (OXPHOS), and glycolysis/hypoxia scores (signatures G2/M, OXPHOS_2, and Hypoxia/Glycolysis_2, respectively) for cells of (**a**) spheroids (*n* = 8 *LGR5*⁺ cultures), (**b**) patient-derived organoid (PDO; O1), (**c**) patient-derived xenograft (PDX; X1), and (**d**) merged tumor epithelial cells (T1–T3) classified under active cell type-specific meta-signatures: stem-like (spheroids: *n* = 467; PDO: *n* = 944; PDX: *n* = 667; primary: *n* = 124), Paneth-like (spheroids: *n* = 357; PDX: *n* = 189), transit-amplifying (TA)-like (spheroids: *n* = 554; PDO: *n* = 3,967; PDX: *n* = 100; primary: *n* = 49), terminally differentiated (Tdiff)-like cells (spheroids: *n* = 486; PDO: *n* = 639; PDX: *n* = 424; primary: *n* = 80). *p*-Values were calculated based on the Mann–Whitney Test (two-tailed). * *p* < 0.05; ** *p* < 0.01; **** *p* < 0.0001; ns, not significant. Dashed lines indicate medians. Upper and lower dotted lines indicate 75% and 25% percentiles, respectively.

2.5. Cell States in Patient-Derived CRC Organoids, Xenografts, and Primary Tumors

To assess whether cell types and transcriptional programs identified in tumor spheroids are present in other *LGR5*⁺ CRC models and patient tumors, we analyzed three patient-derived organoids (PDOs; O1–O3), two patient-derived xenografts (PDXs; X1, X2), and cells from three primary tumor samples (T1–T3) by droplet-based scRNA-seq [37,38] (Table 1). 3003–8785 cells passed quality control per PDO. A mean of 4176–5542 genes per cell were detected. For the PDXs, 1475 (X1) and 1070 cells (X2) passed quality control, with a mean of 1841 and 4598 detected human genes per cell, respectively (Table 2).

To distinguish functionally distinct subpopulations, *LGR5* levels were determined and sufficient levels detected in O1, O3, and X1. Since the absolute cell numbers after quality control in the primary tumors were low (T1: 362; T2: 538; T3: 724 cells), the three primary samples were merged and analysis focused on epithelial cells only (total: 253 cells). As observed in spheroids, clustering of cells from *LGR5*⁺ PDO, PDX, and primary tumor cells revealed subpopulations of stem-like, TA-like, or Tdiff-like cells. Additionally, a prominent fraction of Paneth-like (deep crypt secretory-like, *REG4*⁺) cells [39] was detected in the PDX (Figure S3a–c).

Importantly, applying the signatures identified by NNMF of spheroid scRNA-seq data (Table S1) revealed similar trends for heterogeneous metabolic states associated with distinct cell types, that is, OXPHOS in stem-like and TA-like, and glycolysis/hypoxia in Tdiff-like cells (Figure 3b–d). This shows that transcriptional states and cellular composition identified in spheroids are representative of further patient-derived CRC models as well as patient tumors.

2.6. Spatial Distribution of OXPHOS and Distinct Cell Types in CRC Spheroids

To analyze spatial organization of metabolic states, we stained spheroids with mitochondrial live-dyes for visualization of mitochondrial membrane potential (MMP) and OXPHOS activity. Histological examination of spheroids ($n = 3$ cultures) revealed crypt-like structures formed by partially polarized cells around lumina, morphologically showing some degree of differentiation. Cells within individual spheroids demonstrated highly heterogeneous MMP, with MMP^{high} cells consistently localized at outer 'budding' regions of spheroids and around crypt-like structures (Figure S4a).

Multiplexed RNA fluorescence in situ hybridization (FISH) for intestinal cell type markers *LGR5* (stem-like), *DEFA5* (Paneth-like), and *FABP1* (Tdiff-like) resulted in discrete staining of individual cells by either a single or none of the markers, indicating existence of distinct intestinal cell types in all three patient cultures. Cellular subtypes also showed tendencies for spatial localization. *DEFA5*⁺ cells were primarily detectable in inner regions of spheroids. *LGR5*⁺ cells preferably localized towards outer regions. Frequently, *DEFA5*⁺ cells were identified in proximity to *LGR5*⁺ cells (Figure 4a,b). In the intestinal crypt, *LGR5*⁺ cells reside at the crypt base [27,40], and—in line with our observations—imaging of intestinal organoids has shown localization of *LGR5*⁺ cells close to Paneth cells [41].

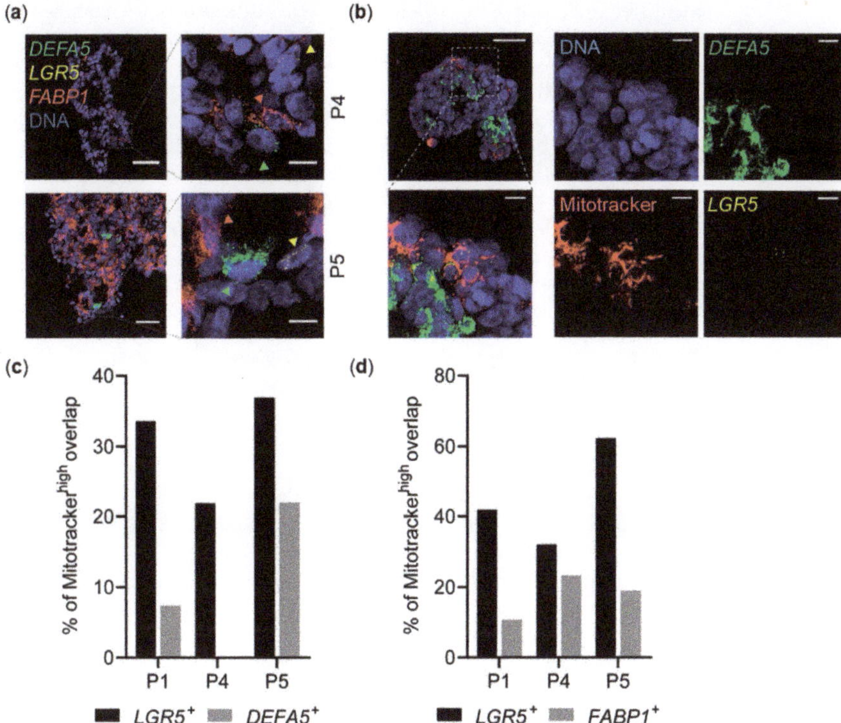

Figure 4. Spatial distribution of metabolic activity and distinct cell types in individual colorectal cancer (CRC) spheroids. (**a**) Histological sections of CRC spheroids co-stained for representative lineage-specific marker genes by RNA fluorescence in situ hybridization (FISH). Left: Overview images. Scale bar, 50 μm. Right: Magnified images representing dashed box regions in overview images (4× digital zoom). Scale bar, 10 μm. *DEFA5*: Paneth-like, *FABP1*: terminally differentiated-like, *LGR5*: stem-like cells. Colored arrowheads mark associated subtypes in magnified images. Images represent z projections from 10 μm slices and DNA is counterstained by 6′-diamidino-2-phenylindole (DAPI). (**b**) Histological section of a spheroid (P1) stained for cell type-specific marker genes (RNA-FISH) and mitochondria (Mitotracker). Top left: Merged overview image. Scale bar, 50 μm. Bottom left: Magnification of dashed region in top left image (4× digital zoom). Scale bar, 10 μm. Center and right: Single channels. Scale bar, 10 μm. Images represent z projections from 10 μm slices and DNA is counterstained by DAPI. (**c**,**d**) Fraction of Mitotracker 'ON' cells as determined by automated image analysis pipeline. (**c**) *LGR5+* or *DEFA5+* cells (total number of cells analyzed: P1: *n* = 7379; P4: *n* = 2670; P5: *n* = 2213). (**d**) *LGR5+* or *FABP1+* cells (total number of cells analyzed: P1: *n* = 3403; P4: *n* = 1580; P5: *n* = 1601).

To correlate MMP with specific cell types, we combined mitochondrial staining and multiplexed RNA-FISH, showing *DEFA5+* and *FABP1+* cells to be largely excluded from MMP^high regions, whereas *LGR5+* cells were primarily located in MMP^high regions. Matching our scRNA-seq results, quantitative image analysis in thousands of single cells revealed that the fraction of *LGR5+* cells located in MMP^high regions is indeed much higher compared to *DEFA5+* and *FABP1+* cells in all examined cultures (*n* = 3; Figure 4c,d, Figure S4b,c).

Hence, in situ RNA fluorescence microscopy further confirmed cell type-specific metabolic preferences of putative stem-like, Paneth-like, and Tdiff-like cell subtypes in CRC. In addition, metabolic activities of cellular subtypes are associated with specific spatial localization within spheroids.

2.7. Heterogeneous Energy Metabolism in Patient Tumors

Identification of cell type-specific metabolic preferences in patient-derived CRC cultures raised the question whether heterogeneously expressed metabolic signatures can be identified directly in CRC patient tumors. To address this, we analyzed primary tumors ($n = 25$ patients) and liver metastases ($n = 25$ patients) by immunohistochemistry for expression of LDH-A and CA9 (Figure S5a,b) as marker genes of hypoxia/glycolysis and Tdiff signatures (Figure S1f, Table S1).

Within the majority of examined specimens, immunohistochemical analysis revealed that only subfractions of all cells express LDH-A and CA9, indicating existence of metabolic heterogeneity within individual patient tumors. Despite high expression of the proliferation marker MKI67, previously reported to preferentially mark TA-like cells [42], regions of CA9 expression were largely overlapping with MKI67$^-$ areas in most patient tumors, suggesting that tumor cells with expression of the hypoxia/glycolysis signature were indeed less proliferative, and actively cycling TA-like cells might prefer OXPHOS to generate energy (Figure S5c).

2.8. Heterogeneous Energy Metabolism in Patient-Derived Models

To assess whether cellular subfractions with distinct OXPHOS activities can be identified in viable cells, we used an MMP dye for flow cytometry allowing distinction of cells with different mitochondrial activity based on fluorescence intensity. Indeed, heterogeneous fluorescence intensities allowed separation of populations with different MMP (MMP^{low}, MMP^{high}; Figure 5a).

To understand whether heterogeneous metabolic activity is relevant in patient CRC tumors, we determined OXPHOS activity of two freshly purified patient tumors and a patient tumor expanded as PDX in vivo by flow cytometry-based MMP analysis. In all samples, two cell populations with distinct MMP were identified (Figure 5a), indicating that heterogeneous mitochondrial activity also exists in PDXs and patient tumors.

This finding was further supported by proteomic analysis of MMP^{low} and MMP^{high} populations of *LGR5*$^+$ (i.e., *LGR5* score > 1) spheroid cultures (P1, P4, P7, P11) which revealed differentially abundant proteins between the two populations. Interestingly, three proteins contributing to the stem-like signature (PROX1, GRN, DEFA6; Table S1) were significantly higher abundant in MMP^{high} compared to MMP^{low} (Figure S5d).

2.9. Increased Spheroid and Tumor Formation Capacity in OXPHOS^{High} Cells

scRNA-seq data suggested that subfractions of MMP^{low} and MMP^{high} spheroid cells preferentially harbor Tdiff-like and Paneth-like (MMP^{low}) or stem-like and TA-like tumor cells (MMP^{high}). As spheroid and tumor forming capacity is supposed to be restricted to stem-like tumor cells [43], we calculated spheroid-forming cell (SFC) frequencies in vitro and TIC frequencies in vivo by limiting dilutions of sorted MMP^{low} and MMP^{high} cell fractions.

SFC frequencies were strongly increased in MMP^{high} cell fractions compared to MMP^{low} fractions or bulk spheroid cells in four out of five cultures (Figure 5b). Spheroid cells (P1) sorted according to JC-1 aggregation—a different MMP indicator—also demonstrated increased SFC frequency in the MMP^{high} subpopulation (MMP^{high}: 1/26; MMP^{low}: 1/46; Figure S5e).

As increased mitochondrial OXPHOS is linked to enhanced ROS levels [23], we further assessed the association of SFC frequency and OXPHOS by staining spheroid cells (P4) with a live-dye fluorescent upon ROS oxidation. In vitro limiting dilutions revealed substantial enrichment of SFCs in the sorted ROS^{high} compared to the ROS^{low} subpopulation (ROS^{high}: 1/9; ROS^{low}: 1/117; Figure S5f).

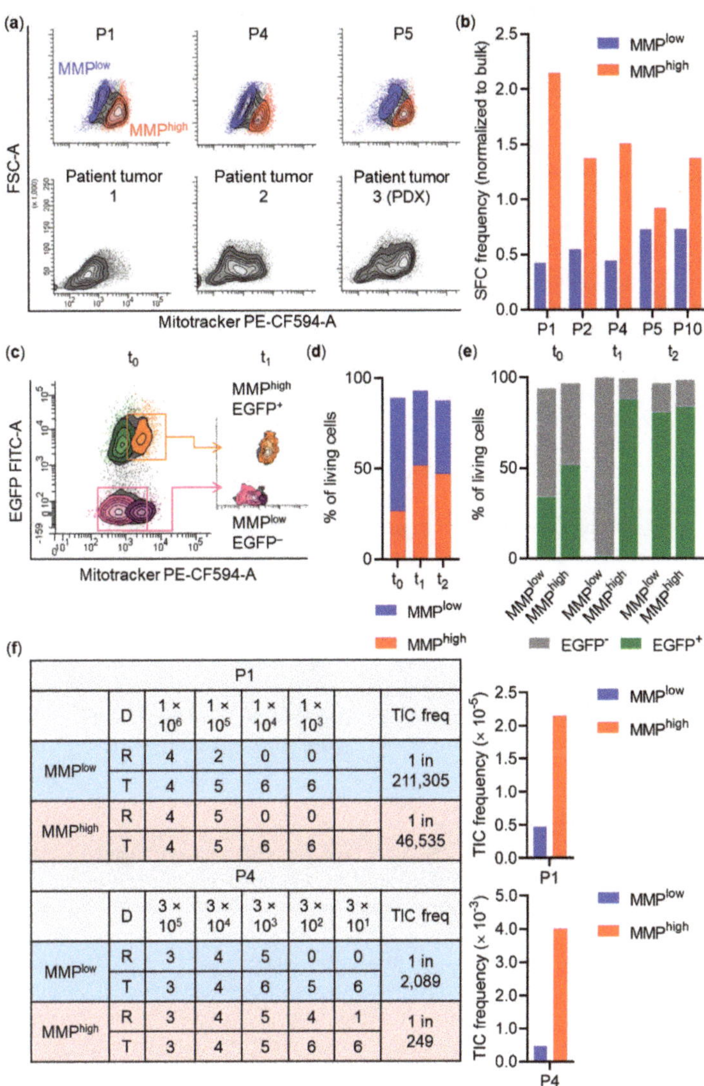

Figure 5. Association of tumor-initiating cell (TIC) activity and mitochondrial membrane potential (MMP). (**a**) Top: Heterogeneous MMP staining (Mitotracker) of spheroid cells assessed by flow cytometry (representative plots shown). Colored cell populations indicate sorted fractions. Bottom: Heterogeneous MMP staining pattern in tumor cells. Axis scale numbers are representative for all plots. PDX, patient-derived xenograft. (**b**) Spheroid-forming cell (SFC) frequencies of MMP sorted spheroid cells (P1, P4: $n = 3$; P5: $n = 2$; P2, P10: $n = 1$) determined by in vitro limiting dilutions. SFC frequencies were calculated based on sphere formation 5–7 days after seeding and normalized to bulk (P1, P4, P5: $n = 2$; P2, P10: $n = 1$). (**c**) Experimental layout of co-cultivation experiments. x- and y-axis are displayed biexponentially. Results for a representative spheroid culture (P1) are shown. Axis scale numbers are representative for all plots. EGFP, enhanced green fluorescent protein. (**d,e**) Composition of (**d**) MMP^{low} and MMP^{high} cells or (**e**) EGFP$^-$ and EGFP$^+$ cells in the MMP^{low} and MMP^{high} subpopulations over time. t_0: before, t_1: directly after, t_2: 21 days after sort. (**f**) Results of limiting dilutions in vivo. Left: Overview of dose (D) and response (R). Right: TIC frequencies of MMP sorted spheroid cells. TIC frequencies were calculated based on tumor formation seven weeks (P1: $n = 42$ mice) or five weeks (P4: $n = 48$ mice) after transplantation. Freq, frequency; T, tested.

We then asked whether MMP^{high} cells exhibit a growth advantage in competition with MMP^{low} cells. Spheroid cultures (P1, P4, P5) were transduced with a lentiviral vector encoding for enhanced green fluorescent protein (EGFP) under the control of the human phosphoglycerate kinase (PGK) promoter in order to allow follow up of sorted populations by assigning presence or absence of EGFP expression to the metabolic state at the time of sort. To achieve this, ~40–50% EGFP$^+$ cultures were stained for MMP (t_0) and cells were sorted as co-cultures of MMP^{high}EGFP$^+$ and MMP^{low}EGFP$^-$ spheroid cells (1:1 ratio; t_1). After three weeks (t_2), despite similar relative contributions of MMP^{low} and MMP^{high} fractions, co-cultures were nearly completely EGFP$^+$, indicating a growth advantage of MMP^{high} compared to MMP^{low} cells (Figure 5c–e).

To quantify TIC frequency in MMP^{high} and MMP^{low} subpopulations in vivo, spheroid cultures ($n = 2$) were sorted according to MMP. Descending cell numbers of each population were subcutaneously injected into NSG mice. For P1, 42 mice with four different dilutions (10^3–10^6 cells), for P4, 48 mice with five different dilutions (3×10^1–3×10^5 cells) were transplanted. For all mice where endpoint criteria have not been reached before, tumor formation was assessed simultaneously at defined endpoints (P1: seven weeks; P4: five weeks after transplantation). Importantly, in both tested cultures, calculated TIC frequencies were substantially increased in MMP^{high} compared to MMP^{low} cells (P1: 1/46,535 vs. 1/211,305; P4: 1/249 vs. 1/2089 for MMP^{high} vs. MMP^{low}, respectively; Figure 5f), demonstrating strong enrichment of stem-like tumor cells in the MMP^{high} population.

Cell type-specific metabolic preferences might represent a targetable metabolic vulnerability in CRC. To test this hypothesis, we assessed the impact of carbonyl cyanide *m*-chlorophenyl hydrazine (CCCP), a drug perturbing adenosine triphosphate synthesis by transporting protons across the mitochondrial inner membrane [44], on SFC frequency ($n = 3$ spheroid cultures). Upon 4 h pretreatment with 25 µM CCCP, a lower SFC frequency of CCCP compared to dimethyl sulfoxide (DMSO) treated cells was observed for all cultures tested (P1: 1/19 vs. 1/14; P4: 1/18 vs. 1/9; P5: 1/68 vs. 1/14 for CCCP vs. DMSO treated, respectively; Figure S5g), indicating sensitivity of stem-like cells towards OXPHOS inhibition.

3. Discussion

We here analyzed functional CRC intra-tumor heterogeneity at single-cell level and demonstrate that distinct functional programs within individual CRC cells can be assigned to specific cellular subpopulations.

In healthy tissues including normal intestine, functional cellular heterogeneity is established by differentiation processes of stem and progenitor cell populations, which control the tissues' functionality in a demand-dependent manner [27]. Similarly, in CRC and other solid tumors as well as in hematological malignancies, functional heterogeneity of tumor and non-tumor cells in the surrounding microenvironment exists and acts as driver of tumor progression [31,45]. Although the majority of tumor cells in CRC cycles actively, we identified proliferatively inactive cells in patient-derived cultures and within established xenograft tumors in vivo—in line with recent data on the healthy intestine [46]. Nevertheless, these cells eventually can re-enter the cell cycle and exhibit TIC activity, suggesting that cells escape from a quiescent state, possibly driven by cellular plasticity as described before for CRC [47]. Accordingly, slow or non-cycling cells were suggested to exhibit increased chemoresistance and drive relapse following initial successful therapy [17,48].

This has striking parallels to the normal intestine, where ablation of stem cells under pathological conditions (e.g., irradiation) can be compensated by a reserve pool of stem cells that are rare during homeostasis but can regenerate all different cell populations including stem, progenitor, and differentiated cell types upon activation, thereby maintaining a functional intestine after tissue injury [49].

The complex composition of different subpopulations within normal and malignant intestinal epithelium and their dynamic interactions are poorly understood. Their characterization has been hampered by the dependency of experimental approaches on purifying cell populations, which cannot fully distinguish between or comprehensively capture distinct cell types and intermediates and might fail to detect rare and poorly characterized cell populations. Recent studies shed light on this complexity by utilizing single-cell approaches to detect and characterize rare cell types in the normal intestine and CRC [42,46,49–53]. We here demonstrate that scRNA-seq further allows the identification of cell type-specific expression modules in CRC and enables identification of functional states during TIC differentiation based on transcriptional heterogeneity.

In line with observations in other entities, transcriptional programs across multiple CRC patients were dominated by inter-patient heterogeneity, most likely due to individual genetic and epigenetic alterations [21,25,54]. Interestingly, most patient-derived spheroid cells clustered according to primary tumor or metastasis site, suggesting either a stable effect of tumor environment on transcriptional programs or selection of tumor cells with specific expression profiles.

Gene sets most heterogeneously expressed within individual spheroids and PDOs included genes specifically expressed in distinct cell types of the normal intestinal epithelium (e.g., a gene set including *LGR5* for stem-like, a gene set including *KRT20* for Tdiff-like cells) [55]. This further supports the notion that, in CRC, there exist functionally distinct cell types that phenotypically reflect those of the normal intestinal epithelium [6,8]. Still, in contrast to the normal intestinal epithelium where distinct cellular subpopulations can be discriminated in high resolution by scRNA-seq technologies [55], gene expression within identified subfractions of CRC was less distinct. However, individual subfractions shared transcriptional traits, potentially reflecting continuous cell type transitions after malignant transformation comparable to reports on hematopoietic stem cell differentiation [56]. In glioblastoma, bulk RNA-sequencing of individual tumors was used to analyze transcriptional heterogeneity and identified different tumor subtypes, while scRNA-seq revealed different proportions of cell types within individual tumors underlying transcriptional heterogeneity rather than distinct homogeneous tumor subtypes [54]. This is in line with our data showing cellular diversity of cell types and cell states within individual patient tumors. Of note, four out of 12 spheroid cultures did not meet inclusion criteria for NNMF analysis due to low *LGR5* scores. Accordingly, previous findings show that, while *LGR5*⁺ tumor cells can be detected in tumors from all CRC subtypes independent of their cellular composition [53], up to a third of individual CRCs tumors may lack detectable *LGR5* levels [14]. Furthermore, *LGR5* plasticity has recently been shown to drive CRC metastasis [57]. In the presented study, we only focused on patient-derived cultures with high expression of *LGR5*. Future analyses of the hierarchical organization of *LGR5⁻* cultures and existing cellular subpopulations in comparison to the cellular subpopulations and cellular states described in this study could further widen the understanding of cellular heterogeneity in CRC.

Our approach to decipher transcriptional programs heterogeneously expressed in functionally distinct CRC cell subfractions identified heterogeneous gene expression programs related to cell cycle, immune response, and metabolic states like OXPHOS and glycolysis. Given the considerable functional and proliferative differences between distinct cell populations, cell-to-cell variability in energy turnover and demand appears likely. A recent study has linked decreased biosynthetic capacities to differentiation [58]. As OXPHOS can be more efficient in energy production [59], highly proliferative TA-like tumor cells might prefer OXPHOS over glycolysis to generate energy. Even though such cell type-specific metabolic identities are known from the normal intestinal epithelium [23], distinct metabolic preferences within normal and malignant stem cell systems are not uniform across different tissue types and tumor entities, and are not necessarily correlated with proliferation activity in general. For example, TICs in hepatocellular carcinoma [60], breast cancer [61], osteosarcoma [62], and nasopharyngeal carcinoma [63] rely on glycolysis for tumor formation, while TICs in pancreatic ductal adenocarcinoma [64], glioma [65], and acute myeloid leukemia [66] prefer OXPHOS. Importantly, tumor cells can also alternate between glycolysis and OXPHOS, thereby adapting to metabolic challenges [67].

Here, we were able to assign the metabolic demand of OXPHOS to functionally relevant stem-like and TA-like cells and observed substantial enrichment of self-renewing and proliferating SFCs and TICs in OXPHOS^{high} cell subfractions. As a consequence, inhibition of OXPHOS impaired spheroid formation in vitro identifying OXPHOS as a novel druggable target in CRC. Since high OXPHOS levels were detected in stem-like and TA-like cell compartments, targeting OXPHOS as treatment strategy might eliminate the most self-renewing and proliferating cell types simultaneously.

Interestingly, stem-like tumor cells demonstrated overexpression of *OXR1* and *PON2*, both involved in protection against ROS accumulating as co-product of OXPHOS [34]. Further studies are needed to address whether expression of *OXR1* and *PON2* may be involved in a mechanism by which this long-lived and thus vulnerable population of stem-like tumor cells protects itself against ROS-mediated damage.

In our proteomic analysis, proteins significantly higher abundant in the MMP^{high} subpopulation included PROX1, one of the top markers of the stem signature and usually expressed in the enteroendocrine lineage [51]. Interestingly, *PROX1* has been reported to be positively correlated with *LGR5* expression in CRC [43] and linked to stem cell maintenance and metastasis [68,69]. Another protein significantly more abundant in MMP^{high} was DEFA6, a protein expressed in normal Paneth and Paneth-like tumor cells [70]. Its moderate expression in the stem-like cell population might reflect a continuous rather than a stepwise process underlying transition from stem-like to Paneth-like cell subsets (and potentially vice versa) in CRC. While Paneth cells constitute the niche for LGR5$^+$ cells in the small intestinal epithelium, this function is performed by REG4-expressing deep crypt secretory cells in the colon [39,71,72]. *REG4* was also included in the NNMF Paneth-like signature, suggesting that both cell types might contribute to this signature.

Of note, the expression signatures identified by scRNA-seq of patient-derived CRC spheroids have shown a prognostic relevance for CRC patients comparable to previously reported subtypes linked to cancer-associated fibroblasts [31] or CMSs [29], indicating that cell types and cell states might indeed be biologically distinct and of potential clinical relevance for CRC patients.

In summary, we here show that distinct functional cell states during TIC differentiation can be identified by single-cell transcriptomes. Targeting differentiation of cancer cells and associated transcriptional states might represent a novel therapeutic strategy for human CRC.

4. Materials and Methods

4.1. Primary CRC Spheroids and Organoids

Human CRC samples (male and female patients) were obtained from Heidelberg University Hospital in accordance with the Declaration of Helsinki. Informed consent on tissue collection was received from each patient, as approved by the University Ethics Review Board on 19 May 2009 (323/2004) and 7 June 2013 (S-649/2012). Tumor sample processing and purification procedures were described previously [4,73,74].

For generation of three-dimensional spheroid cultures, cells freshly isolated from patient material or PDXs were cultivated in ultra-low attachment flasks (Corning, Corning, NY, USA) in serum-free culture medium (Advanced DMEM/F-12 supplemented with 0.6% glucose, 2 mM L-glutamine (all ThermoFisher, Waltham, MA, USA), 5 mM HEPES, 4 µg/mL heparin (all Sigma-Aldrich, St. Louis, MO, USA), 4 mg/mL bovine serum albumin (PAN-Biotech, Aidenbach, Germany)). Growth factors (20 ng/mL epidermal growth factor, 10 ng/mL fibroblast growth factor basic (all R&D Systems, Minneapolis, MN, USA)) were added twice per week.

To dissociate tumor spheroids, cells were pelleted, resuspended in 0.25% trypsin-EDTA (ThermoFisher, Waltham, MA, USA), and incubated for 10–30 min at 37 °C. The reaction was stopped by adding 20% fetal bovine serum (PAN-Biotech, Aidenbach, Germany) in phosphate-buffered saline (PBS; ThermoFisher, Waltham, MA, USA). Cells were pelleted, resuspended in medium, and filtered through a 40 µm cell strainer (Corning, Corning, NY,

USA). To avoid secondary cell culture artifacts, like hypoxic cores in large spheroids [75], cultures were dissociated at defined, pretested time points 6–14 days before individual experiments.

For generation of three-dimensional organoid cultures, purified cells were seeded in 10 µL drops of Cultrex reduced growth factor basement membrane extract (R&D Systems, Minneapolis, MN, USA) into not-treated 6-well plates (Corning, Corning, NY, USA). Organoids were cultured as previously described with minor modifications [76,77] and in the absence of WNT, R-spondin and Noggin, thereby selecting for tumor cells with activation of WNT/β-catenin signaling and inhibition of BMP signals [78,79]. In brief, cells were cultured in serum-free culture medium (Advanced DMEM/F-12 supplemented with B-27 supplement, 2 mM L-glutamine, 100 µg/mL streptomycin, 100 U/mL penicillin (all ThermoFisher, Waltham, MA, USA), 10 mM HEPES, 10 mM nicotinamide, 1.25 mM N-acetyl-L-cysteine, 1 µM SB 202190, 500 nM A 83-01, 10 nM gastrin, 10 nM prostaglandin E_2 (all Sigma-Aldrich, St. Louis, MO, USA), 100 µg/mL primocin (InvivoGen, San Diego, CA, USA)). 20 ng/mL of epidermal growth factor was added three times per week and medium was exchanged weekly. After seeding, 10 µM Y-27632 (StemCell Technologies, Vancouver, BC, Canada) was added. To dissociate tumor organoids, cells were taken up in 0.25% trypsin-EDTA diluted 1:1 in PBS and incubated for 10–20 min at 37 °C. To enhance dissociation, organoids were mechanically disrupted by pipetting. The reaction was stopped by adding 20% fetal bovine serum in PBS. Cells were washed twice with PBS before reseeding.

Spheroid and organoid cultures were authenticated using Multiplex Cell Authentication by Multiplexion (Heidelberg, Germany) as described [80]. The SNP profiles matched known profiles or were unique. The purity of spheroid and organoid cultures was validated using the multiplex cell contamination test by Multiplexion (Heidelberg, Germany) as described recently [81]. No mycoplasma, SMRV or interspecies contamination was detected. To assure pure epithelial cell content and exclude contaminations with murine or hematopoietic cells, established cultures were tested for EPCAM, H2kd, and CD45 expression by flow cytometry.

4.2. Laboratory Animals

Male and female immunodeficient NSG mice purchased from The Jackson Laboratory (Bar Harbor, ME, USA) were further expanded in the Centralized Laboratory Animal Facilities of the DKFZ, Heidelberg. Animals were group-housed in standard individually ventilated cages with wood chip embedding (LTE E-001, ABEDD, Vienna, Austria), nesting material, autoclaved tap water and ad libitum diet (autoclaved mouse/rat housing diet 3437, Provimi Kliba, Kaiseraugst, Switzerland). Room temperature and relative humidity were adjusted to 22.0 ± 2.0 °C and 55.0 ± 10.0%, respectively, in accordance with Appendix A of the European Convention for the Protection of Vertebrate Animals used for Experimental and Other Scientific Purposes from 19 March 1986. According to FELASA recommendations, all animals were housed under strict specific pathogen-free conditions. The light/dark cycle was adjusted to 14 h lights on and 10 h lights off with the beginning of the light and dark period set at 6 am and 8 pm, respectively. The age of transplanted mice was at least seven weeks. All animal experimentation performed in this study was conducted according to national guidelines and was reviewed and confirmed by an institutional review board/ethics committee headed by the responsible animal welfare officer. The Regional Authority of Karlsruhe, Germany finally approved the animal experiments as the responsible national authority (approval numbers G228/12 (29 January 2013), G49/14 (26 June 2014), G233/15 (17 November 2015)).

4.3. scRNA-seq of Spheroids

To generate single-cell suspensions, cells were trypsinized as described. Trypsinization was enhanced by applying shear forces with a pipette every 5 min. After stopping the reaction, cells were washed twice with PBS and filtered through a 15–20 µm cell strainer

(PluriSelect, Leipzig, Germany). To count and test for cell viability using an automated cell counter (Countess, ThermoFisher, Waltham, MA, USA), single-cell suspensions were stained with Hoechst and propidium iodide (ReadyProbes Cell Viability Imaging Kit, ThermoFisher, Waltham, MA, USA) for 10 min at room temperature. Only samples with at least 85% viability were used for further processing. For isolation of single cells, reverse transcription, and cDNA amplification, the Rapid Development Kit (Wafergen, Fremont, CA, USA; compare: SMARTer iCELL8 3' DE Reagent Kit, TakaraBio, Kusatsu, Japan) for in-chip reverse transcription-PCR amplification with the iCELL8 system (TakaraBio, Kusatsu, Japan) [19] was used. The cell suspension was diluted to 25 cells/μL. Cells were dispensed from a 384-well source plate into a nanowell chip (SmartChip v1/v2 kit, TakaraBio, Kusatsu, Japan; P7: v2; others: v1) containing uniquely barcoded oligo-dT primers for each well, resulting in up to 30% of wells containing single cells following Poisson distribution. Wells were imaged using an automated fluorescence microscope (BX43, Olympus, Shinjuku, Japan) and image processing was performed using CellSelect (TakaraBio, Kusatsu, Japan). Additional manual curation for multiplets and dead cells was performed. 50 nL RT/Amp solution was dispensed into nanowells (master mix: 56 μL 5 M betaine (Sigma-Aldrich, St. Louis, MO, USA), 24 μL 25 mM dNTP mix (TakaraBio, Kusatsu, Japan), 3.2 μL 1 M magnesium chloride (ThermoFisher, Waltham, MA, USA), 8.8 μL 100 mM dithiothreitol, 61.9 μL 5× SMARTScribe first-strand buffer, 33.3 μL 2× SeqAmp PCR buffer, 4.0 μL 100 μM RT E5 oligo, 8.8 μL 10 μM Amp primer (all TakaraBio, Kusatsu, Japan), 1.6 μL 100% Triton X-100 (ThermoFisher, Waltham, MA, USA), 28.8 μL SMARTScribe reverse transcriptase, 9.6 μL SeqAmp DNA polymerase (all TakaraBio, Kusatsu, Japan)). In-chip RT/Amp amplification was performed for 18 amplification cycles (Bio-Rad, Hercules, CA, USA; modified for iCELL8 chips). Libraries were pooled, concentrated (DNA Clean&Concentrator-5, Zymo Research, Irvine, CA, USA), purified (0.6× Ampure XP beads, Beckman Coulter, Brea, CA, USA), and assessed for DNA quality (Bioanalyzer and High Sensitivity DNA Kit, Agilent, Santa Clara, CA, USA). Next generation sequencing libraries were constructed following manufacturer's instructions using the Nextera XT DNA Library Prep Kit (Illumina, San Diego, CA, USA) and sequenced using NextSeq500 (Illumina, San Diego, CA, USA; high-output mode, paired-end; v1 chip: 21 × 70 bp; v2 chip: 24 × 67 bp).

4.4. scRNA-seq of Tumors, PDXs, PDOs

To generate single-cell suspensions, cells were trypsinized as described. After stopping the reaction, cells were washed with PBS and filtered through a 40 μm cell strainer. Cells were washed, resuspended in PBS supplemented with 0.05% bovine serum albumin, and filtered through a 20 μm cell strainer. Single-cell suspensions were loaded following the Chromium Single Cell 3' Library Kit v2 (10× Genomics, Pleasanton, CA, USA) protocol to generate cell and gel bead emulsions. Reverse transcription, cDNA amplification, and sequencing library generation were performed according to manufacturer's protocol. Each library was sequenced in one lane of the NextSeq500 (Illumina, San Diego, CA, USA; high-output mode, paired-end, 26 × 49 bp).

4.5. Preprocessing and Analysis of iCELL8 Data

scRNA-seq data were preprocessed using an automated in-house workflow (Roddy; https://github.com/TheRoddyWMS/Roddy). FastQC was used to evaluate read quality. Assignment of iCELL8 library barcodes to corresponding nanowells was performed with the Je demultiplexing suite [82]. Sequences were trimmed for primer sequences, poly-A/T tails, and low-quality ends using Cutadapt with the '–nextseq-trim' option. Mapping to the reference genome hs37d5 was performed (STAR aligner). Quantification of mapped BAM files was performed using featureCounts (reference annotation gencode v19). Only scRNA-seq libraries matching the following criteria were used: (i) >100,000 reads, (ii) >1000 detected genes, (iii) <15% mitochondrial reads. Strong PCA outliers as well as libraries with top 5% of reads for every patient independently were removed. As previously published [25], expression levels based on raw read counts were quantified as

$$E_{i,j} = \log_2\left(\frac{CPM_{i,j}}{10} + 1\right), \tag{1}$$

with $CPM_{i,j}$ as the counts-per-million for gene i in sample j. Aggregate expression of each gene across all cells was calculated as

$$E_a = E_{i,j} = \log_2(mean[E_{i,1...n}] + 1) \tag{2}$$

with genes with $Ea < 3.5$ being excluded to focus on highly or intermediately expressed genes.

Combined filtered and normalized data of all patients were used for evaluation of inter-patient gene expression differences. The R package Seurat [38] was used for identification of highly variable genes, PCA, clustering, two-dimensional visualization, and differential expression analysis (Wilcoxon rank sum test: adjusted p-value < 0.05; log fold-change > 0.25).

Before combining the data of all patients, relative expression levels were calculated individually for each patient using a mean-centering approach

$$Er_{i,j} = E_{i,j} - mean[E_{i,1...n}] \tag{3}$$

to eliminate global inter-patient gene expression shifts.

PCA was applied and—for visualization—the top 30 genes with low and high scores in the first principal component were clustered using average group linkage (UPGMA) by the 'aheatmap' function from R's 'NMF' package. Gene set enrichment analysis [83] was performed on the top 300 genes with highest and lowest PC scores.

Transcriptional signatures shared across patients were identified using NNMF [24] of mean-centered data of all patients defined as $LGR5^+$ ($n = 8$ patients; Table 2). Analysis was performed in MATLAB (MathWorks, Natick, MA, USA; 'nnmf') with a factor number of $k = 25$ and negative events set to 0. To exclude patient-specific signatures, pairwise overlaps in frequency distributions of cell scores for individual factors were determined and factors with overlaps <50% in at least five patients were excluded. Biological relevance of factors and their associated genes was analyzed manually and by gene set enrichment analysis [83]. Factors potentially driven by technical artifacts were excluded. Signature scores were defined as averaged expression of the top 200 genes per factor. To reduce redundancy for visualization, signatures showing similar enrichment and clustering patterns were combined to meta-signatures (Figure S1a–e, Table S1).

Meta-signature scores (calculated based on the combined gene lists of the comprised signatures) were clustered using complete linkage of Euclidean distances. NNMF analysis was repeated with various numbers of factors resulting in identification of similar core signatures.

To test whether cell type-specific transcriptional programs (stem-like, TA-like, Paneth-like, Tdiff-like) are active in individual cells or—in other words—to differentiate between cells that belong to the four cell type-specific subpopulations, we adapted the above described cell scoring approach based on the expression of inferred NNMF meta-signatures [25] and used control random gene sets as background model to control for technical confounders as library complexity. Cell type-specific transcriptional programs were defined as active if their expression in individual cells was >1 standard deviation above the mean across all cells. Inferred cell state-specific signatures were scored for cells of a particular cell type to assess the degree to which cell states are active in specific cell types.

4.6. Preprocessing and Analysis of 10x Data

For 10× 3' libraries generated from cells derived from PDOs, PDXs, and primary tumor samples, raw sequencing data were processed using CellRanger (10× Genomics, Pleasanton, CA, USA; version 2.1.1). Transcripts were aligned with the 10× reference human genome hg19 1.2.0 and the mouse genome mm10 1.2.0. Quality control and downstream analysis were performed with Seurat (https://github.com/satijalab/seurat; version 3.0.0). Only cells matching the following criteria were used for downstream analysis: PDOs: (i) >2000 detected genes, (ii) <10% mitochondrial reads; PDXs: (i) >500 detected genes, (ii) <10% mitochondrial reads for *Homo sapiens*, and (i) >1000 and <4500 detected genes, (ii) <10% mitochondrial reads for *Mus musculus*; primary tumor samples: (i) >200 and <6000 detected genes, (ii) <15% mitochondrial reads. Only human cells from the PDXs and epithelial cells (*EPCAM*$^+$, *VIL1*$^+$, *CEACAM5*$^+$, *VIM*$^-$, *SPARC*$^-$) from the primary tumor samples were analyzed.

Subsequent downstream analysis was performed with standard Seurat workflow, including log-normalization and scaling as well as PCA and clustering using the top 2000 variable genes. Datasets were visualized using two-dimensional t-distributed stochastic neighbor embedding maps [84]. The three primary tumor samples were aligned using canonical correlation analysis implemented in Seurat [85]. In brief, this method identifies pairwise correspondences between single cells across different datasets belonging to specific biological states, termed 'anchors'. These anchors are the basis of harmonizing datasets. Differentially expressed genes between identified clusters were identified using Wilcoxon rank sum test. Identified clusters were scored for cell state signatures using the 'AddModuleScore' function (Seurat), using gene signatures from NNMF analysis (Table S1).

4.7. Patient Clustering and Survival Analysis

Cell type and cell state signatures obtained from spheroid scRNA-seq data (Table S1) were evaluated in a patient survival analysis. Bulk transcriptomic data for COAD patients with available survival data were collected from TCGA (level 3 RNA-seq, $n = 328$ patients) [28] and log-transformed. For each TCGA patient, the mean expression of gene signatures was calculated and used to cluster bulk transcriptomes by complete linkage of Euclidean distances. Patients were grouped according to different combinations of cell type and cell state signatures. In a new clustering process, the sample space was progressively subdivided using the main signatures defining each cluster of patients: First, OXPHOS_1, G1/S, G2/M, and stem signatures separate cl2 and cl3 (high) from the rest (low; Euclidean distances). Then, hypoxia/glycolysis_1 and TNFα_2 signatures distinguish cl2 (low) from cl3 (high; Euclidean distances). Next, fatty acid and TNFα_1 signatures separate cl1 (high) from cl4, cl5, and cl6 (Euclidean distances). Subsequently, stem and TA signatures separate cl6 (stem^{low}) from cl4 and cl5 (stem^{high}; correlation). Finally, G1/S, G2/M, and OXPHOS_1 signatures also distinguish between cl4 (medium) and cl5 (low; Euclidean distances). Complete linkage of Euclidean distances was used to cluster stem^{high} and stem^{low} patients. Kaplan–Meier survival curves were generated using 'survival' and 'survminer' libraries in R. We performed Cox proportional hazards modeling and multivariable models with and without cell type and cell state clusters were compared by performing analysis of variance (ANOVA). 'CMScaller' [29] was used to stratify the TCGA COAD cohort. To generate the contingency table, patients that could not be assigned to a CMS ($n = 19$ patients) were excluded.

4.8. Genetic Labelling of Spheroids

For tracking of cells within tumors, lentiviral vector particles encoding for tetracycline-regulated (Tet-off) H2B-GFP were produced in HEK293T cells, concentrated by ultracentrifugation, and titrated on HeLa cells as described [4,5]. Patient-derived spheroid cultures ($n = 7$) were transduced with a multiplicity of infection of 1–20 aiming at transduction efficiencies of ~ 40% to avoid multiple vector integrations. Within 24 h after transduction, 4×10^5–1.7×10^6 transduced cells were transplanted under the kidney capsule of NSG mice ($n = 14$, 1–4 mice per spheroid culture) anesthetized by 1.75% isoflurane (Abbott, Chicago, IL, USA) in the breathing air. Mice were checked daily for tumor growth, and starting two weeks prior to tumor harvesting, doxycycline (Genaxxon, Ulm, Germany) was added to the drinking water of tumor-bearing mice to shut down H2B-GFP expression.

Mice were sacrificed, xenograft tumors were digested as described [5,73], cells were stained with 200 nM TOTO-3 (ThermoFisher, Waltham, MA, USA) in Hank's Balanced Salt solution (Sigma-Aldrich, St. Louis, MO, USA) supplemented with 2% fetal bovine serum for dead cell exclusion, and tumor cells were sorted according to GFP expression intensity (AriaII and FACS Diva, Becton Dickinson, Franklin Lakes, NJ, USA). GFP signal was detected in the FITC channel (488 nm laser; 505 LP, 525/50 filter). TOTO-3 signal was detected in the APC channel (633 nm laser; 670/30 filter). Samples were gated for cells (FSC-A vs. SSC-A), singlets (FSC-A vs. FSC-W, SSC-A vs. SSC-W), and living cells (FSC-A vs. APC-A). Populations with high, medium, and low/absent GFP expression were sorted (SSC-A vs. FITC-A), reanalyzed to test for sort efficiency, and serially transplanted into secondary recipient mice (1×10^2–4.5×10^4 cells; n = 33 mice). Mice were monitored daily for tumor formation and sacrificed when tumors reached the maximum tolerable size.

4.9. RNA-FISH

For combinatory stainings of mitochondrial activity and mRNA, undissociated spheroids were stained for 3 h with 100 nM Mitotracker Red CMXRos solution (ThermoFisher, Waltham, MA, USA). For histological preparation, cells were fixed in 4% formaldehyde (ThermoFisher, Waltham, MA, USA) for 20 min at 4 °C, washed twice with PBS, and incubated in 30% sucrose overnight at 4 °C. Samples were embedded (Richard-Allan Scientific Neg-50 Frozen Section Medium, ThermoFisher, Waltham, MA, USA) and frozen in the gaseous phase of liquid nitrogen. Histological sections (10 μm slices) were prepared on a cryostat (Leica, Wetzlar, Germany) and mounted on Superfrost Plus slides (ThermoFisher, Waltham, MA, USA). For RNA-FISH, the RNAscope Multiplex Fluorescent v2 (Bio-Techne, Minneapolis, MN, USA) was used according to manufacturer's instructions with probes targeting mRNAs of *LGR5*, *DEFA5*, and *FABP1*. Alexa488, Atto550, or Atto647 were used as fluorescent dyes. Cryosections were stained with 6′-diamidino-2-phenylindole (DAPI) and mounted in SlowFade Gold Antifade solution (ThermoFisher, Waltham, MA, USA). Images were acquired by confocal laser scanning microscopy (SP8, Leica, Wetzlar, Germany) in 15 z stacks (z range: 20 μm).

For quantitative analysis of RNA-FISH/Mitotracker imaging data, we developed a single-cell image analysis pipeline to relate metabolic activity (Mitotracker) to intestinal subtypes (RNA-FISH). To prepare spheroid images for further analysis, we performed maximum intensity projection on each channel separately. For automated nuclei instance detection and segmentation in spheroids, a deep learning object detection and instance segmentation workflow incorporating Mask R-CNN [86] was implemented. The neural network was initialized using pretrained models trained on the 'Microsoft COCO: Common Objects in Context' dataset [87] and fine-tuned using images of nuclei acquired from various unrelated sources. Maximum intensity projections of DAPI images were used as inputs for the neural network to produce segmentation for each individual nucleus as outputs. Nuclei sizes were calculated using these segmented DAPI masks, and objects smaller than 350 pixels were filtered out and excluded from subsequent analysis.

For quantification and analysis of transcript abundance of marker mRNAs specific for stem-like (*LGR5*), Paneth-like (*DEFA5*), and Tdiff-like (*FABP1*) cells, maximum intensity projections of RNA-FISH channels were binarized using 'Maximum Entropy' thresholding (FIJI/ImageJ; https://imagej.nih.gov/ij/). Transcript abundance was estimated by overlaying nuclei masks on maximum projected probe channels and calculating number of pixels lying within each mask. To account for cytoplasmic fluorescence signals localized outside of nuclei masks, we expanded nuclei before quantification by morphological dilation (two iterations) as implemented in scikit-image (Python). To quantify mitochondrial abundance per cell, Mitotracker signals were quantified similarly, but binarization of fluorescence signal was based on 'Moments' thresholding (FIJI/ImageJ). We then performed k-means clustering on frequency distributions of pixel counts per cell to identify and separate cells into two distinct positive 'ON' (high expression/abundance) and negative 'OFF' (low expression/abundance) states. k = 2 was used for mRNA probes, while k = 3 was used for

Mitotracker signals to better capture gradual differences between cells. Finally, the fraction of stem-like, Paneth-like, and Tdiff-like cells that are Mitotracker^{high} at the same time was calculated by dividing the number of Mitotracker^{high} *LGR5$^+$*, *DEFA5$^+$*, or *FABP1$^+$* cells by the total number of *LGR5$^+$*, *DEFA5$^+$*, or *FABP1$^+$* cells.

4.10. Flow Cytometry and Sorting of Metabolic Subpopulations

Spheroid cultures were dissociated into single-cell suspensions as described above. Sorted cells were collected in culture medium supplemented with 100 µg/mL streptomycin and 100 U/mL penicillin.

For MMP staining with Mitotracker, cells were resuspended in 25 nM Mitotracker Red CMXRos in PBS (1 mL per 10^6 cells). Staining was performed for 30 min at 37 °C. For dead cell exclusion, cells were stained with 200 nM TOTO-3 in PBS. Cells were resuspended in PBS, filtered through a 35 µm cell strainer (Becton Dickinson, Franklin Lakes, NJ, USA), and analyzed on a cell sorter (AriaII and FACS Diva). Mitotracker signal was detected in the PE-CF594 channel (561 nm laser; 600 LP, 610/20 filter). TOTO-3 signal was detected in the APC channel (633 nm laser; 670/30 filter). Samples were gated for cells (FSC-A vs. SSC-A), singlets (FSC-A vs. FSC-H, SSC-H vs. SSC-W), and living cells (FSC-A vs. APC-A). Sorting was performed based on Mitotracker signal intensity (FSC-A vs. PE-CF594-A; Figure S6a–f).

For MMP staining with JC-1, cells were counted and resuspended in 1 µg/mL JC-1 (ThermoFisher, Waltham, MA, USA) in PBS (1 mL per 10^6 cells). Staining was performed for 10 min at 37 °C. For dead cell exclusion, cells were stained with 200 nM TOTO-3 in PBS. Cells were resuspended in PBS, filtered through a 35 µm cell strainer, and analyzed on a cell sorter (AriaII and FACS Diva). JC-1 monomer signal was detected in the FITC channel (488 nm laser; 505 LP, 525/50 filter). JC-1 aggregate signal was detected in the PE channel (561 nm laser; 575/25 filter). TOTO-3 signal was detected in the APC channel (633 nm laser; 670/30 filter). Samples were gated for cells (FSC-A vs. SSC-A), singlets (FSC-A vs. FSC-H, SSC-H vs. SSC-W), and living cells (FSC-A vs. APC-A). Sorting was performed based on JC-1 aggregate/monomer ratio (PE-A vs. FITC-A). As negative control, 50 µM CCCP (Selleckchem, Houston, TX, USA) was added during the staining.

For ROS staining, cells were resuspended in 5 µM CellROX Deep Red Reagent (ThermoFisher, Waltham, MA, USA) in PBS (500 µL per 10^6 cells). Staining was performed for 45 min at 37 °C. For dead cell exclusion, cells were stained with 1 µg/mL propidium iodide (Sigma-Aldrich, St. Louis, MO, USA) in PBS. Cells were resuspended in PBS, filtered through a 35 µm cell strainer, and analyzed on a cell sorter (AriaII and FACS Diva). CellROX signal was detected in the APC channel (633 nm laser; 670/30 filter). Propidium iodide signal was detected in the PE-CF594 channel (561 nm laser; 600 LP, 610/20 filter). Samples were gated for cells (FSC-A vs. SSC-A), singlets (FSC-A vs. FSC-H, SSC-H vs. SSC-W), and living cells (FSC-A vs. PE-CF594-A). Sorting was performed based on CellROX signal intensity (FSC-A vs. APC-A).

4.11. Assessment of SFC Frequency

For each sorted cell population (OXPHOS^{low}, OXPHOS^{high}), 48 wells with 10 cells, 24 wells with 100 cells, and 16 wells with 1000 cells per well were sorted into 96-well ultra-low attachment plates (Corning, Corning, NY, USA) containing 100 µL of culture medium (50% fresh, 50% conditioned (filtered medium of the bulk culture harvested during collection of cells)) supplemented with 100 µg/mL streptomycin and 100 U/mL penicillin per well. Fresh cytokines and medium were added every four days. Spheroid formation was analyzed 5–7 days after sorting using conventional light microscopy (Axiovert 40C, Zeiss, Oberkochen, Germany). Based on the fraction of spheroid-containing wells for each dilution, SFC frequencies were calculated using Poisson statistics and maximum likelihood (L-Calc, StemCell Technologies, Vancouver, BC, Canada). In vitro limiting dilution assays upon Mitotracker staining were performed three times for MMP^{low} and MMP^{high} subpopulations of P1 and P4, twice for P5 as well as bulk (all living, i.e., TOTO3⁻ cells) populations of P1 and P4, and once for P2 and P10.

4.12. Assessment of TIC Frequency

Mitotracker stained cells were sorted as described above, pelleted, resuspended in medium, and counted. Different cell counts were pelleted, resuspended in medium, mixed with matrigel (Corning, Corning, NY, USA), and injected subcutaneously into the flanks of immunodeficient NSG mice. For MMP^{low} as well as MMP^{high} fractions of P1, four mice with 10^6 cells, five mice with 10^5 cells, six mice with 10^4 cells, and six mice with 10^3 cells were transplanted. For MMP^{low} as well as MMP^{high} fractions of P4, three mice with 3×10^5 cells, four mice with 3×10^4 cells, 5–6 mice with 3×10^3 cells (six mice for MMP^{low}, five mice for MMP^{high}), 5–6 mice with 3×10^2 cells (five mice for MMP^{low}, six mice for MMP^{high}), and six mice with 3×10^1 cells were transplanted. The experiments were performed blindly until observable tumor development.

Mice were monitored daily for tumor formation and sacrificed when tumors reached the maximum tolerable size or when experiments were ended (P1: seven weeks; P4: five weeks after transplantation). Based on the fraction of tumor formation for each dilution, TIC frequencies were calculated using Poisson statistics and maximum likelihood (L-Calc).

4.13. Co-Cultivation Experiments

Spheroid cultures ($n = 3$) were transduced with a lentiviral vector encoding for EGFP under control of the human PGK promoter at multiplicities of infection of 0.5 (P1, P4) or 1 (P5), yielding transduction efficiencies of ~40–50%. After expansion, cells were stained with Mitotracker and prepared for flow cytometry as described above. In addition to Mitotracker and TOTO-3, EGFP fluorescence was detected (488 nm laser; 505 LP, 525/50 filter). Cells were gated for low and high Mitotracker signal (MMP^{low}, MMP^{high}) as well as for negative or positive EGFP signal (EGFP⁻, EGFP⁺). For each culture, a set of 5×10^4 MMP^{high}EGFP⁺ and 5×10^4 MMP^{low}EGFP⁻ cells as well as a set of 5×10^4 MMP^{high}EGFP⁻ and 5×10^4 MMP^{low}EGFP⁺ cells were sorted simultaneously. To assess sorting efficiency, sorted samples were reanalyzed by recording 1000 living cells. Sorted cells were cultivated in 24-well ultra-low attachment plates (Corning, Corning, NY, USA). Spheroid formation and EGFP signal for each sample set were observed by fluorescence microscopy (Axiovert 200, Zeiss, Oberkochen, Germany). After 21 days in culture, cells were dissociated, stained with Mitotracker, and reanalyzed by flow cytometry as described.

4.14. Inhibitor Treatments

To assess SFC frequencies upon pretreatment with OXPHOS inhibitors, 5×10^5 tumor spheroid cells (P1, P4, P5) were seeded into two wells of 6-well ultra-low attachment plates (Corning, Corning, NY, USA). After seven days, 25 µM CCCP or DMSO (Sigma-Aldrich, St. Louis, MO, USA) were added and cells were incubated for 4 h at 37 °C. Cells were dissociated, stained with 200 nM TOTO-3 in PBS, and prepared for cell sorting as described. Living (i.e., TOTO-3⁻) cells were sorted into 96-well ultra-low attachment plates containing 100 µL of fresh culture medium supplemented with 100 µg/mL streptomycin and 100 U/mL penicillin per well. Limiting dilution and determination of SFC frequency were performed as described.

4.15. Immunohistochemistry

Formalin-fixed and paraffin-embedded tumor specimens of primary colorectal adeno-carcinomas ($n = 25$ patients) and liver metastases ($n = 25$ patients) resected between 2013 and 2016 at the University Hospital Heidelberg were extracted from the archive of the Institute of Pathology, Heidelberg University, with the support of the tissue bank of the NCT (#2831). Tissues were used in accordance with the ethical regulations of the tissue bank of the NCT defined by the local ethics committee. Diagnoses were made according to the recommendations of the World Health Organization classification of tumors of the digestive system [88].

Immunohistochemical staining was performed as previously described [89]. In brief, tissue sections were cut, pretreated with an antigen retrieval buffer, and stained for Ki-67, CA9, and LDH-A using an automatic staining device (Ventana Benchmark Ultra, Roche, Basel, Switzerland; Table S3).

4.16. Mass Spectrometry

Mass spectrometry was performed for $LGR5^+$ (i.e., $LGR5$ score > 1) patient-derived spheroid cultures (n = 4). Tumor spheroid cells were stained with Mitotracker, prepared for sorting as described, and 5×10^5 cells of MMP^{low} and MMP^{high} subfractions were sorted (n = 3 independent experiments). Cell pellets were reconstituted in 100 μL 0.1% RapiGest SF Surfactant (Waters, Milford, MA, USA) in 100 mM triethylammonium bicarbonate (Sigma-Aldrich, St. Louis, MO, USA) and 1× protease inhibitor cocktail (cOmplete, Sigma-Aldrich, St. Louis, MO, USA). Cells were lysed by probe-sonication twice for 15 s at 10% frequency, followed by centrifugation for 30 min at 15,000× g and 4 °C. 10 μg of protein per sample were denatured for 5 min at 95 °C, reduced with dithiothreitol (Biomol, Hamburg, Germany; 5 mM final concentration) for 30 min at 60 °C, and alkylated with chloroacetamide (Sigma-Aldrich, St. Louis, MO, USA; 15 mM final concentration) for 30 min at 23 °C. Proteins were digested overnight at 750 rpm and 37 °C, at an enzyme/protein ratio of 1:20 with sequencing-grade modified trypsin (Promega, Madison, WI, USA) in double-distilled water (ddH$_2$O). Samples were acidified by adding trifluoroacetic acid (Biosolve Chimie, Dieuze, France; 0.5% final concentration), incubated for another 30 min at 750 rpm and 37 °C, and centrifuged for 30 min at 15,000× g and 23 °C.

Peptides were separated using the Easy NanoLC 1200 fitted with a trapping column (Acclaim PepMap C18, ThermoFisher, Waltham, MA, USA; 5 μm, 100 Å, 100 μm × 2 cm) and an analytical column (nanoEase MZ BEH C18, Waters, Milford, MA, USA; 1.7 μm, 130 Å, 75 μm × 25 cm). The outlet of the analytical column was coupled directly to a Q-Exactive HF Orbitrap mass spectrometer (ThermoFisher, Waltham, MA, USA). Solvent A was ddH$_2$O (Biosolve Chimie, Dieuze, France), 0.1% (v/v) formic acid (Biosolve Chimie, Dieuze, France) and solvent B was 80% acetonitrile (ThermoFisher, Waltham, MA, USA) in ddH$_2$O, 0.1% (v/v) formic acid. Samples were loaded and peptides eluted with a 105 min gradient via the analytical column as described [90].

Raw files were processed using MaxQuant (https://www.maxquant.org; version 1.5.1.2) [91] against the human Uniprot database (20170801_Uniprot_homo-sapiens_ canonical_reviewed; 20,214 entries) using the Andromeda search engine with the default search criteria: enzyme was set to trypsin/P with up to two missed cleavages. Carbamidomethylation (C) and oxidation (M)/acetylation (protein N-term) were selected as fixed and variable modifications, respectively. Protein quantification was performed using the label-free quantification algorithm of MaxQuant. On top, intensity-based absolute quantification intensities were calculated with a log-fit enabled. Identification transfer between runs via the 'matching between runs' algorithm was allowed with a match time window of 0.3 min. Peptide and protein hits were filtered at a false discovery rate of 1% with a minimal peptide length of seven amino acids. The reversed sequences of the target database were used as a decoy database. Proteins only identified by a modification site, contaminants, as well as reversed sequences were removed from the dataset.

Differential expression analysis was performed using limma moderated t statistics (R package version 3.36.3; one-sample, two-sided) [92]. Here, data was first normalized based on median label-free quantification densities per sample. Next, ratios between MMP^{high} and MMP^{low} cells were calculated. Significantly differentially expressed proteins were defined to show a Benjamini–Hochberg adjusted p-value < 0.05 and an absolute log$_2$-fold change > 1.

4.17. Quantification and Statistical Analysis

Statistical tests and sample size used for individual experiments are described in the corresponding figure legends or methods. The threshold for statistical significance was defined as $p < 0.05$. Significance levels were denoted by asterisks: * $p < 0.05$, ** $p < 0.01$, **** $p < 0.0001$. The threshold for statistical significance in univariable and multivariable survival analyses was defined as $p < 0.1$.

4.18. Data and Code Availability

scRNA-seq data have been deposited at the European Genome-phenome Archive (EGA) which is hosted at the EBI and the CRG, under accession number EGAS00001004064.

The mass spectrometry proteomics data have been deposited to the ProteomeXchange Consortium via the PRIDE [93] partner repository with the dataset identifier PXD018230.

Codes for analysis of scRNA-seq and RNA-FISH data are available at the github repository (https://github.com/eilslabs/CRC_scRNAseq).

5. Conclusions

In this study, we show at single-cell resolution that transcriptional heterogeneity identifies functional states during tumor-initiating cell differentiation in colorectal cancer. Targeting specific transcriptional states associated with cancer cell differentiation unravels novel potential vulnerabilities in human colorectal cancer.

Supplementary Materials: The following are available online at https://www.mdpi.com/2072-669 4/13/5/1097/s1, Figure S1: Identification of heterogeneous gene expression programs, Figure S2: Cell cycle and proliferative activity of human colorectal cancer cells, Figure S3: Single-cell expression of representative stem and differentiation markers in patient-derived tumor models, Figure S4: Spatial localization in spheroid cultures in situ, Figure S5: Metabolic heterogeneity in colorectal tumors and patient-derived spheroids, Figure S6: Gating strategy for sorting according to mitochondrial membrane potential (MMP) using Mitotracker, Table S1: Signatures defined by non-negative matrix factorization (NNMF) of colorectal cancer spheroid single-cell RNA-sequencing data, Table S2: Univariable and multivariable survival models, Table S3: Antibodies for immunohistochemistry.

Author Contributions: Conceptualization, M.K.Z., S.M.T., S.M.D., R.E., H.G., C.C. and C.R.B.; formal analysis, M.K.Z., S.M.T., S.M.D., T.G.K., A.O., R.L.C., M.H., F.W.T., J.P., T.M., M.K. (Mathias Kalxdorf), M.K. (Mark Kriegsmann), K.K. and J.K.; investigation, M.K.Z., S.M.T., A.O., R.L.C., K.L., K.J. and T.M.; resources, M.S.; supervision, S.M.D., F.H., R.E., H.G., C.C. and C.R.B.; writing—original draft, M.K.Z., S.M.T., S.M.D., H.G., C.C. and C.R.B. All authors have read and agreed to the published version of the manuscript.

Funding: This research was funded by grants from the NCT Dresden (Translational Research Grants in Precision Oncology, TIC specific regulatory programs in GI cancers) to C.R.B., the NCT 3.0 Precision Oncology Program (NCT3.0_2015.4 TransOnco; NCT3.0_2015.54 DysregPT) to H.G., the EU Framework Program Horizon 2020 (TRANSCAN-2 ERA-NET, TACTIC consortium) to H.G., the Deutsche Krebshilfe Priority program 'Translational Oncology' (Colon-Resist-Net) to H.G. and C.R.B., the iMed Program (Helmholtz Association) to R.E., the BMBF-funded Heidelberg Center for Human Bioinformatics (HD-HuB) within the German Network for Bioinformatics Infrastructure (de.NBI; #031A537A, #031A537C) to R.E., and the Helmholtz Association (Incubator grant sparse2big; grant #ZT-I-0007) to C.C.. DKFZ-HIPO provided technical support and funding through Grant No. HIPO-H012. S.M.T. and K.L. were recipients of the stipend for the PhD program of the Helmholtz International Graduate School for Cancer Research (DKFZ, Heidelberg). S.M.D. and T.G.K. were supported by a DKTK postdoctoral fellowship from the Heidelberg School of Oncology.

Institutional Review Board Statement: The study was conducted according to the guidelines of the Declaration of Helsinki, and approved by the University Ethics Review Board of Heidelberg University Hospital (323/2004 (19 May 2009) and S-649/2012 (7 June 2013)). All animal experimentation performed in this study was conducted according to national guidelines and was reviewed and confirmed by an institutional review board/ethics committee headed by the responsible animal welfare officer. The Regional Authority of Karlsruhe, Germany finally approved the animal experiments as the responsible national authority (approval numbers G228/12 (29 January 2013), G49/14 (26 June 2014), G233/15 (17 November 2015)).

Informed Consent Statement: Informed consent was obtained from all subjects involved in the study.

Data Availability Statement: scRNA-seq data have been deposited at the European Genome-phenome Archive (EGA) which is hosted at the EBI and the CRG, under accession number EGAS00001004064. The mass spectrometry proteomics data have been deposited to the ProteomeXchange Consortium via the PRIDE [93] partner repository with the dataset identifier PXD018230. Codes for analysis of scRNA-seq and RNA-FISH data are available at the github repository (https://github.com/eilslabs/CRC_scRNAseq). Survival analysis was based on publicly available data generated by the TCGA Research Network: https://www.cancer.gov/tcga.

Acknowledgments: The authors thank Nina Hofmann, Tim Kindinger, Sylvia Martin, Celine Reifenberg, and Christiane Zgorzelski for technical assistance. The authors thank the Central Animal Laboratory unit of the Center for Preclinical Research, DKFZ Heidelberg, for providing excellent services. The authors thank David Ibberson (CellNetworks Deep Sequencing Core Facility, Heidelberg University) for NGS services, Daniel Liber and Marizela Kulisic (TakaraBio, formerly Wafergen) for technical support for the iCELL8 system, and Henrik Kaessmann and his group (ZMBH, Heidelberg University) for support and helpful discussions regarding the iCELL8 system and single-cell analysis. The results shown here are in part based on data generated by the TCGA Research Network: https://www.cancer.gov/tcga.

Conflicts of Interest: The authors declare no conflict of interest.

References

1. Jordan, C.T.; Guzman, M.L.; Noble, M. Cancer stem cells. *N. Engl. J. Med.* **2006**, *355*, 1253–1261. [CrossRef]
2. Reya, T.; Morrison, S.J.; Clarke, M.F.; Weissman, I.L. Stem cells, cancer, and cancer stem cells. *Nature* **2001**, *414*, 105–111. [CrossRef] [PubMed]
3. Visvader, J.E.; Lindeman, G.J. Cancer stem cells: Current status and evolving complexities. *Cell Stem Cell* **2012**, *10*, 717–728. [CrossRef]
4. Dieter, S.M.; Ball, C.R.; Hoffmann, C.M.; Nowrouzi, A.; Herbst, F.; Zavidij, O.; Abel, U.; Arens, A.; Weichert, W.; Brand, K.; et al. Distinct types of tumor-initiating cells form human colon cancer tumors and metastases. *Cell Stem Cell* **2011**, *9*, 357–365. [CrossRef]
5. Giessler, K.M.; Kleinheinz, K.; Huebschmann, D.; Balasubramanian, G.P.; Dubash, T.D.; Dieter, S.M.; Siegl, C.; Herbst, F.; Weber, S.; Hoffmann, C.M.; et al. Genetic subclone architecture of tumor clone-initiating cells in colorectal cancer. *J. Exp. Med.* **2017**, *214*, 2073–2088. [CrossRef]
6. Cortina, C.; Turon, G.; Stork, D.; Hernando-Momblona, X.; Sevillano, M.; Aguilera, M.; Tosi, S.; Merlos-Suárez, A.; Stephan-Otto Attolini, C.; Sancho, E.; et al. A genome editing approach to study cancer stem cells in human tumors. *EMBO Mol. Med.* **2017**, *9*, 869–879. [CrossRef]
7. Barker, N. Adult intestinal stem cells: Critical drivers of epithelial homeostasis and regeneration. *Nat. Rev. Mol. Cell Biol.* **2014**, *15*, 19–33. [CrossRef]
8. Merlos-Suárez, A.; Barriga, F.M.; Jung, P.; Iglesias, M.; Céspedes, M.V.; Rossell, D.; Sevillano, M.; Hernando-Momblona, X.; da Silva-Diz, V.; Muñoz, P.; et al. The intestinal stem cell signature identifies colorectal cancer stem cells and predicts disease relapse. *Cell Stem Cell* **2011**, *8*, 511–524. [CrossRef]
9. Bormann, F.; Rodríguez-Paredes, M.; Lasitschka, F.; Edelmann, D.; Musch, T.; Benner, A.; Bergman, Y.; Dieter, S.M.; Ball, C.R.; Glimm, H.; et al. Cell-of-Origin DNA Methylation Signatures Are Maintained during Colorectal Carcinogenesis. *Cell Rep.* **2018**, *23*, 3407–3418. [CrossRef]
10. Batlle, E.; Clevers, H. Cancer stem cells revisited. *Nat. Med.* **2017**, *23*, 1124–1134. [CrossRef]
11. Blanpain, C.; Fuchs, E. Stem cell plasticity. Plasticity of epithelial stem cells in tissue regeneration. *Science* **2014**, *344*, 1242281. [CrossRef]
12. Varga, J.; Greten, F.R. Cell plasticity in epithelial homeostasis and tumorigenesis. *Nat. Cell Biol.* **2017**, *19*, 1133–1141. [CrossRef]
13. De Sousa e Melo, F.; Kurtova, A.V.; Harnoss, J.M.; Kljavin, N.; Hoeck, J.D.; Hung, J.; Anderson, J.E.; Storm, E.E.; Modrusan, Z.; Koeppen, H.; et al. A distinct role for Lgr5$^+$ stem cells in primary and metastatic colon cancer. *Nature* **2017**, *543*, 676–680. [CrossRef]
14. Shimokawa, M.; Ohta, Y.; Nishikori, S.; Matano, M.; Takano, A.; Fujii, M.; Date, S.; Sugimoto, S.; Kanai, T.; Sato, T. Visualization and targeting of LGR5$^+$ human colon cancer stem cells. *Nature* **2017**, *545*, 187–192. [CrossRef]
15. Ball, C.R.; Oppel, F.; Ehrenberg, K.R.; Dubash, T.D.; Dieter, S.M.; Hoffmann, C.M.; Abel, U.; Herbst, F.; Koch, M.; Werner, J.; et al. Succession of transiently active tumor-initiating cell clones in human pancreatic cancer xenografts. *EMBO Mol. Med.* **2017**, *9*, 918–932. [CrossRef]

16. Dalerba, P.; Dylla, S.J.; Park, I.K.; Liu, R.; Wang, X.; Cho, R.W.; Hoey, T.; Gurney, A.; Huang, E.H.; Simeone, D.M.; et al. Phenotypic characterization of human colorectal cancer stem cells. *Proc. Natl. Acad. Sci. USA* **2007**, *104*, 10158–10163. [CrossRef]

17. Kreso, A.; O'Brien, C.A.; van Galen, P.; Gan, O.I.; Notta, F.; Brown, A.M.; Ng, K.; Ma, J.; Wienholds, E.; Dunant, C.; et al. Variable clonal repopulation dynamics influence chemotherapy response in colorectal cancer. *Science* **2013**, *339*, 543–548. [CrossRef] [PubMed]

18. Ricci-Vitiani, L.; Lombardi, D.G.; Pilozzi, E.; Biffoni, M.; Todaro, M.; Peschle, C.; De Maria, R. Identification and expansion of human colon-cancer-initiating cells. *Nature* **2007**, *445*, 111–115. [CrossRef]

19. Goldstein, L.D.; Chen, Y.J.; Dunne, J.; Mir, A.; Hubschle, H.; Guillory, J.; Yuan, W.; Zhang, J.; Stinson, J.; Jaiswal, B.; et al. Massively parallel nanowell-based single-cell gene expression profiling. *BMC Genom.* **2017**, *18*, 519. [CrossRef]

20. Butler, A.; Hoffman, P.; Smibert, P.; Papalexi, E.; Satija, R. Integrating single-cell transcriptomic data across different conditions, technologies, and species. *Nat. Biotechnol.* **2018**, *36*, 411–420. [CrossRef]

21. Tirosh, I.; Izar, B.; Prakadan, S.M.; Wadsworth, M.H., 2nd; Treacy, D.; Trombetta, J.J.; Rotem, A.; Rodman, C.; Lian, C.; Murphy, G.; et al. Dissecting the multicellular ecosystem of metastatic melanoma by single-cell RNA-seq. *Science* **2016**, *352*, 189–196. [CrossRef]

22. Tirosh, I.; Venteicher, A.S.; Hebert, C.; Escalante, L.E.; Patel, A.P.; Yizhak, K.; Fisher, J.M.; Rodman, C.; Mount, C.; Filbin, M.G.; et al. Single-cell RNA-seq supports a developmental hierarchy in human oligodendroglioma. *Nature* **2016**, *539*, 309–313. [CrossRef]

23. Rodríguez-Colman, M.J.; Schewe, M.; Meerlo, M.; Stigter, E.; Gerrits, J.; Pras-Raves, M.; Sacchetti, A.; Hornsveld, M.; Oost, K.C.; Snippert, H.J.; et al. Interplay between metabolic identities in the intestinal crypt supports stem cell function. *Nature* **2017**, *543*, 424–427. [CrossRef]

24. Lee, D.D.; Seung, H.S. Learning the parts of objects by non-negative matrix factorization. *Nature* **1999**, *401*, 788–791. [CrossRef] [PubMed]

25. Puram, S.V.; Tirosh, I.; Parikh, A.S.; Patel, A.P.; Yizhak, K.; Gillespie, S.; Rodman, C.; Luo, C.L.; Mroz, E.A.; Emerick, K.S.; et al. Single-Cell Transcriptomic Analysis of Primary and Metastatic Tumor Ecosystems in Head and Neck Cancer. *Cell* **2017**, *171*, 1611–1624.e1624. [CrossRef] [PubMed]

26. Barker, N.; Ridgway, R.A.; van Es, J.H.; van de Wetering, M.; Begthel, H.; van den Born, M.; Danenberg, E.; Clarke, A.R.; Sansom, O.J.; Clevers, H. Crypt stem cells as the cells-of-origin of intestinal cancer. *Nature* **2009**, *457*, 608–611. [CrossRef] [PubMed]

27. Barker, N.; van Es, J.H.; Kuipers, J.; Kujala, P.; van den Born, M.; Cozijnsen, M.; Haegebarth, A.; Korving, J.; Begthel, H.; Peters, P.J.; et al. Identification of stem cells in small intestine and colon by marker gene Lgr5. *Nature* **2007**, *449*, 1003–1007. [CrossRef]

28. Cancer Genome Atlas, N. Comprehensive molecular characterization of human colon and rectal cancer. *Nature* **2012**, *487*, 330–337. [CrossRef]

29. Eide, P.W.; Bruun, J.; Lothe, R.A.; Sveen, A. CMScaller: An R package for consensus molecular subtyping of colorectal cancer pre-clinical models. *Sci. Rep.* **2017**, *7*, 16618. [CrossRef]

30. Guinney, J.; Dienstmann, R.; Wang, X.; de Reyniès, A.; Schlicker, A.; Soneson, C.; Marisa, L.; Roepman, P.; Nyamundanda, G.; Angelino, P.; et al. The consensus molecular subtypes of colorectal cancer. *Nat. Med.* **2015**, *21*, 1350–1356. [CrossRef]

31. Li, H.; Courtois, E.T.; Sengupta, D.; Tan, Y.; Chen, K.H.; Goh, J.J.L.; Kong, S.L.; Chua, C.; Hon, L.K.; Tan, W.S.; et al. Reference component analysis of single-cell transcriptomes elucidates cellular heterogeneity in human colorectal tumors. *Nat. Genet.* **2017**, *49*, 708–718. [CrossRef]

32. Falkowska-Hansen, B.; Kollar, J.; Grüner, B.M.; Schanz, M.; Boukamp, P.; Siveke, J.; Rethwilm, A.; Kirschner, M. An inducible Tet-Off-H2B-GFP lentiviral reporter vector for detection and in vivo isolation of label-retaining cells. *Exp. Cell Res.* **2010**, *316*, 1885–1895. [CrossRef]

33. Haegebarth, A.; Clevers, H. Wnt signaling, lgr5, and stem cells in the intestine and skin. *Am. J. Pathol.* **2009**, *174*, 715–721. [CrossRef]

34. Oliver, P.L.; Finelli, M.J.; Edwards, B.; Bitoun, E.; Butts, D.L.; Becker, E.B.; Cheeseman, M.T.; Davies, B.; Davies, K.E. Oxr1 is essential for protection against oxidative stress-induced neurodegeneration. *PLoS Genet.* **2011**, *7*, e1002338. [CrossRef]

35. Li, W.; Kennedy, D.; Shao, Z.; Wang, X.; Kamdar, A.K.; Weber, M.; Mislick, K.; Kiefer, K.; Morales, R.; Agatisa-Boyle, B.; et al. Paraoxonase 2 prevents the development of heart failure. *Free Radic. Biol. Med.* **2018**, *121*, 117–126. [CrossRef] [PubMed]

36. Kaur, R.; Liu, X.; Gjoerup, O.; Zhang, A.; Yuan, X.; Balk, S.P.; Schneider, M.C.; Lu, M.L. Activation of p21-activated kinase 6 by MAP kinase kinase 6 and p38 MAP kinase. *J. Biol. Chem.* **2005**, *280*, 3323–3330. [CrossRef]

37. Klein, A.M.; Mazutis, L.; Akartuna, I.; Tallapragada, N.; Veres, A.; Li, V.; Peshkin, L.; Weitz, D.A.; Kirschner, M.W. Droplet barcoding for single-cell transcriptomics applied to embryonic stem cells. *Cell* **2015**, *161*, 1187–1201. [CrossRef]

38. Macosko, E.Z.; Basu, A.; Satija, R.; Nemesh, J.; Shekhar, K.; Goldman, M.; Tirosh, I.; Bialas, A.R.; Kamitaki, N.; Martersteck, E.M.; et al. Highly Parallel Genome-wide Expression Profiling of Individual Cells Using Nanoliter Droplets. *Cell* **2015**, *161*, 1202–1214. [CrossRef] [PubMed]

39. Sasaki, N.; Sachs, N.; Wiebrands, K.; Ellenbroek, S.I.; Fumagalli, A.; Lyubimova, A.; Begthel, H.; van den Born, M.; van Es, J.H.; Karthaus, W.R.; et al. Reg4+ deep crypt secretory cells function as epithelial niche for Lgr5+ stem cells in colon. *Proc. Natl. Acad. Sci. USA* **2016**, *113*, E5399–E5407. [CrossRef]

40. Baker, A.M.; Graham, T.A.; Elia, G.; Wright, N.A.; Rodriguez-Justo, M. Characterization of LGR5 stem cells in colorectal adenomas and carcinomas. *Sci. Rep.* **2015**, *5*, 8654. [CrossRef]

41. Sato, T.; van Es, J.H.; Snippert, H.J.; Stange, D.E.; Vries, R.G.; van den Born, M.; Barker, N.; Shroyer, N.F.; van de Wetering, M.; Clevers, H. Paneth cells constitute the niche for Lgr5 stem cells in intestinal crypts. *Nature* **2011**, *469*, 415–418. [CrossRef]
42. Dalerba, P.; Kalisky, T.; Sahoo, D.; Rajendran, P.S.; Rothenberg, M.E.; Leyrat, A.A.; Sim, S.; Okamoto, J.; Johnston, D.M.; Qian, D.; et al. Single-cell dissection of transcriptional heterogeneity in human colon tumors. *Nat. Biotechnol.* **2011**, *29*, 1120–1127. [CrossRef]
43. Tirier, S.M.; Park, J.; Preußer, F.; Amrhein, L.; Gu, Z.; Steiger, S.; Mallm, J.P.; Krieger, T.; Waschow, M.; Eismann, B.; et al. Pheno-seq-linking visual features and gene expression in 3D cell culture systems. *Sci. Rep.* **2019**, *9*, 12367. [CrossRef]
44. Heytler, P.G.; Prichard, W.W. A new class of uncoupling agents–carbonyl cyanide phenylhydrazones. *Biochem. Biophys. Res. Commun.* **1962**, *7*, 272–275. [CrossRef]
45. Joung, J.G.; Oh, B.Y.; Hong, H.K.; Al-Khalidi, H.; Al-Alem, F.; Lee, H.O.; Bae, J.S.; Kim, J.; Cha, H.U.; Alotaibi, M.; et al. Tumor Heterogeneity Predicts Metastatic Potential in Colorectal Cancer. *Clin. Cancer Res. Off. J. Am. Assoc. Cancer Res.* **2017**, *23*, 7209–7216. [CrossRef]
46. Yan, K.S.; Janda, C.Y.; Chang, J.; Zheng, G.X.Y.; Larkin, K.A.; Luca, V.C.; Chia, L.A.; Mah, A.T.; Han, A.; Terry, J.M.; et al. Non-equivalence of Wnt and R-spondin ligands during Lgr5$^+$ intestinal stem-cell self-renewal. *Nature* **2017**, *545*, 238–242. [CrossRef]
47. Dieter, S.M.; Glimm, H.; Ball, C.R. Colorectal cancer-initiating cells caught in the act. *EMBO Mol. Med.* **2017**, *9*, 856–858. [CrossRef]
48. Puig, I.; Tenbaum, S.P.; Chicote, I.; Arqués, O.; Martínez-Quintanilla, J.; Cuesta-Borrás, E.; Ramírez, L.; Gonzalo, P.; Soto, A.; Aguilar, S.; et al. TET2 controls chemoresistant slow-cycling cancer cell survival and tumor recurrence. *J. Clin. Investig.* **2018**, *128*, 3887–3905. [CrossRef] [PubMed]
49. Ayyaz, A.; Kumar, S.; Sangiorgi, B.; Ghoshal, B.; Gosio, J.; Ouladan, S.; Fink, M.; Barutcu, S.; Trcka, D.; Shen, J.; et al. Single-cell transcriptomes of the regenerating intestine reveal a revival stem cell. *Nature* **2019**, *569*, 121–125. [CrossRef]
50. Grün, D.; Lyubimova, A.; Kester, L.; Wiebrands, K.; Basak, O.; Sasaki, N.; Clevers, H.; van Oudenaarden, A. Single-cell messenger RNA sequencing reveals rare intestinal cell types. *Nature* **2015**, *525*, 251–255. [CrossRef]
51. Yan, K.S.; Gevaert, O.; Zheng, G.X.Y.; Anchang, B.; Probert, C.S.; Larkin, K.A.; Davies, P.S.; Cheng, Z.F.; Kaddis, J.S.; Han, A.; et al. Intestinal Enteroendocrine Lineage Cells Possess Homeostatic and Injury-Inducible Stem Cell Activity. *Cell Stem Cell* **2017**, *21*, 78–90.e76. [CrossRef]
52. Haber, A.L.; Biton, M.; Rogel, N.; Herbst, R.H.; Shekhar, K.; Smillie, C.; Burgin, G.; Delorey, T.M.; Howitt, M.R.; Katz, Y.; et al. A single-cell survey of the small intestinal epithelium. *Nature* **2017**, *551*, 333–339. [CrossRef]
53. Chen, E.C.; Karl, T.A.; Kalisky, T.; Gupta, S.K.; O'Brien, C.A.; Longacre, T.A.; van de Rijn, M.; Quake, S.R.; Clarke, M.F.; Rothenberg, M.E. KIT Signaling Promotes Growth of Colon Xenograft Tumors in Mice and Is Up-Regulated in a Subset of Human Colon Cancers. *Gastroenterology* **2015**, *149*, 705–717.e702. [CrossRef]
54. Patel, A.P.; Tirosh, I.; Trombetta, J.J.; Shalek, A.K.; Gillespie, S.M.; Wakimoto, H.; Cahill, D.P.; Nahed, B.V.; Curry, W.T.; Martuza, R.L.; et al. Single-cell RNA-seq highlights intratumoral heterogeneity in primary glioblastoma. *Science* **2014**, *344*, 1396–1401. [CrossRef]
55. Biton, M.; Haber, A.L.; Rogel, N.; Burgin, G.; Beyaz, S.; Schnell, A.; Ashenberg, O.; Su, C.W.; Smillie, C.; Shekhar, K.; et al. T Helper Cell Cytokines Modulate Intestinal Stem Cell Renewal and Differentiation. *Cell* **2018**, *175*, 1307–1320.e1322. [CrossRef]
56. Velten, L.; Haas, S.F.; Raffel, S.; Blaszkiewicz, S.; Islam, S.; Hennig, B.P.; Hirche, C.; Lutz, C.; Buss, E.C.; Nowak, D.; et al. Human haematopoietic stem cell lineage commitment is a continuous process. *Nat. Cell Biol.* **2017**, *19*, 271–281. [CrossRef] [PubMed]
57. Fumagalli, A.; Oost, K.C.; Kester, L.; Morgner, J.; Bornes, L.; Bruens, L.; Spaargaren, L.; Azkanaz, M.; Schelfhorst, T.; Beerling, E.; et al. Plasticity of Lgr5-Negative Cancer Cells Drives Metastasis in Colorectal Cancer. *Cell Stem Cell* **2020**, *26*, 569–578.e567. [CrossRef] [PubMed]
58. Morral, C.; Stanisavljevic, J.; Hernando-Momblona, X.; Mereu, E.; Álvarez-Varela, A.; Cortina, C.; Stork, D.; Slebe, F.; Turon, G.; Whissell, G.; et al. Zonation of Ribosomal DNA Transcription Defines a Stem Cell Hierarchy in Colorectal Cancer. *Cell Stem Cell* **2020**, *26*, 845–861 e812. [CrossRef]
59. Zheng, J. Energy metabolism of cancer: Glycolysis versus oxidative phosphorylation (Review). *Oncol. Lett.* **2012**, *4*, 1151–1157. [CrossRef] [PubMed]
60. Song, K.; Kwon, H.; Han, C.; Zhang, J.; Dash, S.; Lim, K.; Wu, T. Active glycolytic metabolism in CD133(+) hepatocellular cancer stem cells: Regulation by MIR-122. *Oncotarget* **2015**, *6*, 40822–40835. [CrossRef] [PubMed]
61. Feng, W.; Gentles, A.; Nair, R.V.; Huang, M.; Lin, Y.; Lee, C.Y.; Cai, S.; Scheeren, F.A.; Kuo, A.H.; Diehn, M. Targeting unique metabolic properties of breast tumor initiating cells. *Stem Cells* **2014**, *32*, 1734–1745. [CrossRef]
62. Palorini, R.; Votta, G.; Balestrieri, C.; Monestiroli, A.; Olivieri, S.; Vento, R.; Chiaradonna, F. Energy metabolism characterization of a novel cancer stem cell-like line 3AB-OS. *J. Cell. Biochem.* **2014**, *115*, 368–379. [CrossRef]
63. Shen, Y.A.; Lin, C.H.; Chi, W.H.; Wang, C.Y.; Hsieh, Y.T.; Wei, Y.H.; Chen, Y.J. Resveratrol Impedes the Stemness, Epithelial-Mesenchymal Transition, and Metabolic Reprogramming of Cancer Stem Cells in Nasopharyngeal Carcinoma through p53 Activation. *Evid.-Based Complementary Altern. Med. ECAM* **2013**, *2013*, 590393. [CrossRef]
64. Viale, A.; Pettazzoni, P.; Lyssiotis, C.A.; Ying, H.; Sánchez, N.; Marchesini, M.; Carugo, A.; Green, T.; Seth, S.; Giuliani, V.; et al. Oncogene ablation-resistant pancreatic cancer cells depend on mitochondrial function. *Nature* **2014**, *514*, 628–632. [CrossRef]

65. Janiszewska, M.; Suvà, M.L.; Riggi, N.; Houtkooper, R.H.; Auwerx, J.; Clément-Schatlo, V.; Radovanovic, I.; Rheinbay, E.; Provero, P.; Stamenkovic, I. Imp2 controls oxidative phosphorylation and is crucial for preserving glioblastoma cancer stem cells. *Genes Dev.* **2012**, *26*, 1926–1944. [CrossRef]

66. Lagadinou, E.D.; Sach, A.; Callahan, K.; Rossi, R.M.; Neering, S.J.; Minhajuddin, M.; Ashton, J.M.; Pei, S.; Grose, V.; O'Dwyer, K.M.; et al. BCL-2 inhibition targets oxidative phosphorylation and selectively eradicates quiescent human leukemia stem cells. *Cell Stem Cell* **2013**, *12*, 329–341. [CrossRef]

67. Elgendy, M.; Cirò, M.; Hosseini, A.; Weiszmann, J.; Mazzarella, L.; Ferrari, E.; Cazzoli, R.; Curigliano, G.; DeCensi, A.; Bonanni, B.; et al. Combination of Hypoglycemia and Metformin Impairs Tumor Metabolic Plasticity and Growth by Modulating the PP2A-GSK3β-MCL-1 Axis. *Cancer Cell* **2019**, *35*, 798–815.e795. [CrossRef]

68. Wiener, Z.; Högström, J.; Hyvönen, V.; Band, A.M.; Kallio, P.; Holopainen, T.; Dufva, O.; Haglund, C.; Kruuna, O.; Oliver, G.; et al. Prox1 promotes expansion of the colorectal cancer stem cell population to fuel tumor growth and ischemia resistance. *Cell Rep.* **2014**, *8*, 1943–1956. [CrossRef] [PubMed]

69. Ragusa, S.; Cheng, J.; Ivanov, K.I.; Zangger, N.; Ceteci, F.; Bernier-Latmani, J.; Milatos, S.; Joseph, J.M.; Tercier, S.; Bouzourene, H.; et al. PROX1 promotes metabolic adaptation and fuels outgrowth of Wnt(high) metastatic colon cancer cells. *Cell Rep.* **2014**, *8*, 1957–1973. [CrossRef] [PubMed]

70. Jeong, D.; Kim, H.; Kim, D.; Ban, S.; Oh, S.; Ji, S.; Kang, D.; Lee, H.; Ahn, T.S.; Kim, H.J.; et al. Defensin alpha 6 (DEFA6) is a prognostic marker in colorectal cancer. *Cancer Biomark. Sect. A Dis. Markers* **2019**, *24*, 485–495. [CrossRef] [PubMed]

71. Rothenberg, M.E.; Nusse, Y.; Kalisky, T.; Lee, J.J.; Dalerba, P.; Scheeren, F.; Lobo, N.; Kulkarni, S.; Sim, S.; Qian, D.; et al. Identification of a cKit$^+$ colonic crypt base secretory cell that supports Lgr5$^+$ stem cells in mice. *Gastroenterology* **2012**, *142*, 1195–1205 e1196. [CrossRef] [PubMed]

72. Altmann, G.G. Morphological observations on mucus-secreting nongoblet cells in the deep crypts of the rat ascending colon. *Am. J. Anat.* **1983**, *167*, 95–117. [CrossRef]

73. Dieter, S.M.; Giessler, K.M.; Kriegsmann, M.; Dubash, T.D.; Möhrmann, L.; Schulz, E.R.; Siegl, C.; Weber, S.; Strakerjahn, H.; Oberlack, A.; et al. Patient-derived xenografts of gastrointestinal cancers are susceptible to rapid and delayed B-lymphoproliferation. *Int. J. Cancer* **2017**, *140*, 1356–1363. [CrossRef]

74. Dubash, T.D.; Hoffmann, C.M.; Oppel, F.; Giessler, K.M.; Weber, S.; Dieter, S.M.; Hüllein, J.; Zenz, T.; Herbst, F.; Scholl, C.; et al. Phenotypic differentiation does not affect tumorigenicity of primary human colon cancer initiating cells. *Cancer Lett.* **2016**, *371*, 326–333. [CrossRef] [PubMed]

75. Riffle, S.; Hegde, R.S. Modeling tumor cell adaptations to hypoxia in multicellular tumor spheroids. *J. Exp. Clin. Cancer Res.* **2017**, *36*, 102. [CrossRef] [PubMed]

76. Fujii, M.; Shimokawa, M.; Date, S.; Takano, A.; Matano, M.; Nanki, K.; Ohta, Y.; Toshimitsu, K.; Nakazato, Y.; Kawasaki, K.; et al. A Colorectal Tumor Organoid Library Demonstrates Progressive Loss of Niche Factor Requirements during Tumorigenesis. *Cell Stem Cell* **2016**, *18*, 827–838. [CrossRef]

77. Van de Wetering, M.; Francies, H.E.; Francis, J.M.; Bounova, G.; Iorio, F.; Pronk, A.; van Houdt, W.; van Gorp, J.; Taylor-Weiner, A.; Kester, L.; et al. Prospective derivation of a living organoid biobank of colorectal cancer patients. *Cell* **2015**, *161*, 933–945. [CrossRef]

78. Sato, T.; Stange, D.E.; Ferrante, M.; Vries, R.G.; Van Es, J.H.; Van den Brink, S.; Van Houdt, W.J.; Pronk, A.; Van Gorp, J.; Siersema, P.D.; et al. Long-term expansion of epithelial organoids from human colon, adenoma, adenocarcinoma, and Barrett's epithelium. *Gastroenterology* **2011**, *141*, 1762–1772. [CrossRef]

79. Li, Y.; Liu, Y.; Liu, B.; Wang, J.; Wei, S.; Qi, Z.; Wang, S.; Fu, W.; Chen, Y.G. A growth factor-free culture system underscores the coordination between Wnt and BMP signaling in Lgr5$^+$ intestinal stem cell maintenance. *Cell Discov.* **2018**, *4*, 49. [CrossRef] [PubMed]

80. Castro, F.; Dirks, W.G.; Fähnrich, S.; Hotz-Wagenblatt, A.; Pawlita, M.; Schmitt, M. High-throughput SNP-based authentication of human cell lines. *Int. J. Cancer* **2013**, *132*, 308–314. [CrossRef]

81. Schmitt, M.; Pawlita, M. High-throughput detection and multiplex identification of cell contaminations. *Nucleic Acids Res.* **2009**, *37*, e119. [CrossRef]

82. Girardot, C.; Scholtalbers, J.; Sauer, S.; Su, S.Y.; Furlong, E.E. Je, a versatile suite to handle multiplexed NGS libraries with unique molecular identifiers. *BMC Bioinform.* **2016**, *17*, 419. [CrossRef]

83. Subramanian, A.; Tamayo, P.; Mootha, V.K.; Mukherjee, S.; Ebert, B.L.; Gillette, M.A.; Paulovich, A.; Pomeroy, S.L.; Golub, T.R.; Lander, E.S.; et al. Gene set enrichment analysis: A knowledge-based approach for interpreting genome-wide expression profiles. *Proc. Natl. Acad. Sci. USA* **2005**, *102*, 15545–15550. [CrossRef]

84. Wu, Z.; Wu, Z. Exploration, visualization, and preprocessing of high-dimensional data. *Methods Mol. Biol.* **2010**, *620*, 267–284. [CrossRef] [PubMed]

85. Stuart, T.; Butler, A.; Hoffman, P.; Hafemeister, C.; Papalexi, E.; Mauck, W.M., 3rd; Hao, Y.; Stoeckius, M.; Smibert, P.; Satija, R. Comprehensive Integration of Single-Cell Data. *Cell* **2019**, *177*, 1888–1902 e1821. [CrossRef]

86. He, K.; Gkioxari, G.; Dollár, P.; Girshick, R. Mask r-cnn. In Proceedings of the IEEE International Conference on Computer Vision, Venice, Italy, 22–29 October 2017; pp. 2980–2988.

87. Lin, T.-Y.; Maire, M.; Belongie, S.; Hays, J.; Perona, P.; Ramanan, D.; Dollár, P.; Zitnick, C.L. Microsoft coco: Common objects in context. In Proceedings of the European Conference on Computer Vision, Zurich, Switzerland, 6–12 September 2014; pp. 740–755.

88. Baretton, G. Pathology of the lower gastrointestinal tract. *Der Pathol.* **2011**, *32*, 273–274. [CrossRef]
89. Kriegsmann, K.; Cremer, M.; Zgorzelski, C.; Harms, A.; Muley, T.; Winter, H.; Kazdal, D.; Warth, A.; Kriegsmann, M. Agreement of CK5/6, p40, and p63 immunoreactivity in non-small cell lung cancer. *Pathology* **2019**, *51*, 240–245. [CrossRef]
90. Erich, K.; Reinle, K.; Müller, T.; Munteanu, B.; Sammour, D.A.; Hinsenkamp, I.; Gutting, T.; Burgermeister, E.; Findeisen, P.; Ebert, M.P.; et al. Spatial Distribution of Endogenous Tissue Protease Activity in Gastric Carcinoma Mapped by MALDI Mass Spectrometry Imaging. *Mol. Cell. Proteom. MCP* **2019**, *18*, 151–161. [CrossRef] [PubMed]
91. Cox, J.; Mann, M. MaxQuant enables high peptide identification rates, individualized p.p.b.-range mass accuracies and proteome-wide protein quantification. *Nat. Biotechnol.* **2008**, *26*, 1367–1372. [CrossRef] [PubMed]
92. Ritchie, M.E.; Phipson, B.; Wu, D.; Hu, Y.; Law, C.W.; Shi, W.; Smyth, G.K. limma powers differential expression analyses for RNA-sequencing and microarray studies. *Nucleic Acids Res.* **2015**, *43*, e47. [CrossRef] [PubMed]
93. Perez-Riverol, Y.; Csordas, A.; Bai, J.; Bernal-Llinares, M.; Hewapathirana, S.; Kundu, D.J.; Inuganti, A.; Griss, J.; Mayer, G.; Eisenacher, M.; et al. The PRIDE database and related tools and resources in 2019: Improving support for quantification data. *Nucleic Acids Res.* **2019**, *47*, D442–D450. [CrossRef] [PubMed]

 *cancers*

Article

The HROC-Xenobank—A High Quality Assured PDX Biobank of >100 Individual Colorectal Cancer Models

Stephanie Matschos [1], Florian Bürtin [1], Said Kdimati [1], Mandy Radefeldt [2], Susann Krake [2], Friedrich Prall [3], Nadja Engel [4], Mathias Krohn [1], Bianca Micheel [1], Michael Kreutzer [5], Christina Susanne Mullins [1] and Michael Linnebacher [1,*]

1 Clinic of General Surgery, Molecular Oncology and Immunotherapy, University Medicine Rostock, Schillingallee 69, 18057 Rostock, Germany; stephanie.matschos@med.uni-rostock.de (S.M.); florian.buertin@med.uni-rostock.de (F.B.); said.kdimati@med.uni-rostock.de (S.K.); mathias.krohn@med.uni-rostock.de (M.K.); b.micheel@keh-berlin.de (B.M.); christina.mullins@med.uni-rostock.de (C.S.M.)
2 CENTOGENE GmbH, 18055 Rostock, Germany; mandy.radefeldt@centogene.com (M.R.); susann.krake@centogene.com (S.K.)
3 Institute of Pathology, University Medicine Rostock, Strempelstraße 10, 18057 Rostock, Germany; friedrich.prall@med.uni-rostock.de
4 Department of Oral, Maxillofacial and Plastic Surgery, University Medicine Rostock, Schillingallee 69, 18057 Rostock, Germany; nadja.engel@med.uni-rostock.de
5 Medical Research Center, University Medicine Rostock, Schillingallee 69, 18057 Rostock, Germany; michael.kreutzer@med.uni-rostock.de
* Correspondence: michael.linnebacher@med.uni-rostock.de; Tel.: +49-381-494-6043

Citation: Matschos, S.; Bürtin, F.; Kdimati, S.; Radefeldt, M.; Krake, S.; Prall, F.; Engel, N.; Krohn, M.; Micheel, B.; Kreutzer, M.; et al. The HROC-Xenobank—A High Quality Assured PDX Biobank of >100 Individual Colorectal Cancer Models. *Cancers* **2021**, *13*, 5882. https://doi.org/10.3390/cancers13235882

Academic Editors: Marta Baiocchi and Ann Zeuner

Received: 28 October 2021
Accepted: 19 November 2021
Published: 23 November 2021

Publisher's Note: MDPI stays neutral with regard to jurisdictional claims in published maps and institutional affiliations.

Simple Summary: Considering recent research, it was established that the best experimental models to conserve biological features of human tumors and to predict individual clinical treatment success are patient-derived xenografts (PDX). Their recognized and growing importance for translational research, especially for late-stage preclinical testing of novel therapeutics, necessitates a high number of well-defined PDX models from individual patients' tumors. The starting platform for the Hansestadt Rostock colorectal cancer (HROC)-Xenobank was the assortment of colorectal tumor and normal tissue samples from patients stored in our university biobank.

Abstract: Based on our research group's large biobank of colorectal cancers (CRC), we here describe the ongoing activity of establishing a high quality assured PDX biobank for more than 100 individual CRC cases. This includes sufficient numbers of vitally frozen ($n > 30$ aliquots) and snap frozen ($n > 5$) backups, "ready to use". Additionally, PDX tumor pieces were paraffin embedded. At the current time, we have completed 125 cases. This resource allows histopathological examinations, molecular characterizations, and gene expression analysis. Due to its size, different issues of interest can be addressed. Most importantly, the application of low-passage, cryopreserved, and well-characterized PDX for in vivo studies guarantees the reliability of results due to the largely preserved tumor microenvironment. All cases described were molecularly subtyped and genetic identity, in comparison to the original tumor tissue, was confirmed by fingerprint analysis. The latter excludes ambiguity errors between the PDX and the original patient tumor. A cancer hot spot mutation analysis was performed for $n = 113$ of the 125 cases entities. All relevant CRC molecular subtypes identified so far are represented in the Hansestadt Rostock CRC (HROC)-Xenobank. Notably, all models are available for cooperative research approaches.

Keywords: PDX model; CRC; mutation analysis; histological examination

1. Introduction

Despite early diagnostic options and improved treatment, colorectal cancer (CRC) is still one of the leading causes of cancer-related deaths worldwide [1]. In particular, the fact that some patients do not respond even to targeted therapies underlines the necessity of further patient-derived models to promote the development of personalized treatments [1,2].

Currently the best model to reflect the characteristics of the original tumor is the patient-derived xenograft (PDX) model because of its conservation of the original tumor's biological features, including microarchitecture, pathomorphology, and genetic alterations [2,3]. Tentler et al. have stated that CRC PDX tumors retain the intratumoral clonal heterogeneity, chromosomal instability, and histology of the parent tumor for up to 14 passages [2,4]. Furthermore, the possibility of precisely predicting individual clinical treatment success, especially for the late-stage preclinical testing of novel therapeutics, implies a clear exigency for more academically run PDX-biobanks, containing large numbers of individual tumors [2,3,5,6]. Inspired by this notion, we used our large collection of patient material, which included matching tumor and normal epithelial tissue, as a starting platform to establish a high number of individual PDX models. This resulted in a quality assured PDX biobank containing more than 100 individual CRC cases and encompassing all specific CRC molecular subtypes. Thus, this PDX biobank represents an ideal platform to study new agents for adjuvant therapy. As such, it is feasible to target specific molecular subtypes or alterations, in combination with investigations concerning different molecular pathways within the tumor cells as compared to the normal epithelial tissue. Such an approach has recently been described by Medico and colleagues. Here the authors identified tumor specific changes that consist of clinically actionable kinase targets for which approved drugs are already available [7]. Moreover, omics data from both the PDX model and the original patient tumor could, on the one hand, accelerate the entry of novel drugs into the clinic, and, on the other hand, such paired data sets would facilitate the identification and validation of predictive biomarkers [2].

Finally, as has been described by us and other groups, the PDX-derived tissue is an ideal source for repetitive cell line establishment [8,9] and also patient-derived organoid (PDO) generation attempts [9,10]. This can significantly boost the overall success rate, from 10–13% for primary patient material derived cell lines [8,11] to about 30% for secondary, i.e., PDX-derived, cell lines [8]. Our vision is an integrated biobank collection, consisting of deeply characterized primary patient material, 2D cell lines, PDX, and PDO. With this vision we would like to support orchestrated research strategies, from more basic mechanistic approaches to translational drug development and tests to end-stage preclinical studies. Besides dogmatic animal welfare policies, the establishment and proper long-term maintenance of platforms such as the integrated Hansestadt Rostock CRC (HROC) biobank are essential for minimizing the overall number of animals involved in oncological in vivo studies.

2. Materials and Methods

Surgically resected tissues were collected from consenting patients at the UMR from 2006 to 2019. The study was approved by the ethics committee of the UMR (II HV 43/2004, A45/2007, A2018-0054, and A2019-0187) [8,12].

2.1. PDX Generation

Tumor engraftment was performed according to the guidelines of the local animal use and care committee, Landesamt für Landwirtschaft, Lebensmittelsicherheit und Fischerei Mecklenburg-Vorpommern with the permit numbers: LALLF M-V/TSD/7221.3-1.1-071/10; 7221.3-1-015/14; and 7221.3-2-020/17. The mice strains used were bred in the animal facility of the Rostock University Medical Center and maintained in specified, pathogen-free conditions, exposed to 12 h light/12 h darkness cycles. The mice received standard pellet food and water ad libitum.

Prior cryopreserved matching patient-derived tumor and normal tissue samples in our biobank of CRC patients served as the starting platform for all established PDX models.

Detailed information on the patients' tumors, as well as clinico-pathological information, is given in Supplementary Table S1.

Pieces of the patients' tumors were implanted subcutaneously into the animals' left and right flanks, under anesthesia (ketamine/xylazine, 90/6 mg/kg bw). Due to the engraftment rate of up to 80%, the preferred mouse strain for this first passage is NOD.Cg-Prkdc^{scid} Il2rg^{tm1Wjl}/SzJ (NSG). Further passaging can be performed either with NSG or with NMRI-*Foxn1*^{nu} (NMRI nude mice). The PDX tumors were named with anonymized patient information followed by the abbreviation Tx, standing for the passage of the PDX tumor, and then followed by the abbreviation Mx, standing for consecutively numbered mice. All tumor engraftments were performed on 6–12 week-old mice, both male and female, weighing 18–30 g. Prior to xenografting, four vital tumor aliquots (3 × 3 × 3 mm) were soaked in 100 μL Matrigel (Corning, Kaiserslautern, Germany) for >10 min at 4 °C. After 30 days of antibiotic treatment (drinking water containing cotrimoxazole: dosage 8 mg trimethoprim and 40 mg sulfamethoxazole per kg BW), tumor growth was monitored weekly until tumor establishment and growth to a maximal diameter of 14.2 mm. When the maximum tumor volume of 1500 mm^3 was reached or the mice became moribund, the tumors were explanted. The time of tumor growth until explantation was defined as tumor harvesting time.

After the explantation of the PDX tumors, the tumors were stored in Tissue Storage Solution (Miltenyi, Bergisch-Gladbach, Germany) until further processing. Snap frozen aliquots were made as soon as possible by immediately submerging tumor pieces in liquid nitrogen to ensure high quality, particularly for RNA molecules. Vital aliquots were made by transferring four tumor pieces of 3 × 3 × 3 mm in 1.5 mL freezer medium (fetal bovine serum with 10% DMSO) and cooling them down in CoolCell® LX—freezing containers (CryoShop, München, Germany) by −1 °C per minute to −80 °C [13].

While processing the PDX tumors, the degree of necrosis was assessed and documented, allowing a classification into not necrotic, barely necrotic, intermediately necrotic, and highly necrotic.

The process describing our PDX biobank establishment approach was recently published [14].

2.2. Histopathology

For each PDX model, one representative cross section of a subcutaneous PDX tumor or half of the PDX tumor was fixed immediately upon explantation in formalin and embedded in paraffin by routine procedures. H&E-stained sections (4–5 μm) were analyzed in light-microscopic studies to assess the morphologic features of each individual PDX model [15]. A comparison with the respective original patient tumor was performed by a board-certified pathologist (FP).

2.3. Quality Control via Short Tandem Repeat (STR) Analysis

The fluorescence-labeled, PCR-amplified DNA fragments of D5S818, D7S820, D16S539, D13S317, vWA, TPOX, THO1, CSF1PO, and Amelogenin were injected along with an appropriately sized standard GeneScan™ LIZ500 (appliedbiosystems Thermo Fisher Scientific, Waltham, MA, USA) into the capillary for electrophoresis size separation, using ABI instrumentation. Size-separated PCR fragments were detected by reading their fluorescence intensity at different emission wavelengths and were recorded as FSA after their migration through the capillary from cathode to anode, in which smaller fragments migrate faster than larger fragments [16]. The application of primer pairs labeled with three different fluorescence dyes—FAM (blue), HEX (green), and TAMRA (red)— enabled the fragment size determination of all markers mentioned (primers listed in detail in Table 1) in a single analysis.

Table 1. Sequences of STR primers.

Primer	Sequence
D5S818 for	5′-HEX-GGT GAT TTT CCT CTT TGG TAT CC-3′
D5S818 rev	5′-AGC CAC AGT TTA CAA CAT TTG TAT CT-3′
D7S820 for	5′-HEX-ATG TTG GTC AGG CTG ACT ATG-3′
D7S820 rev	5′-GAT TCC ACA TTT ATC CTC ATT GAC-3′
D16S539 for	5′-HEX-GGG GGT CTA AGA GCT TGT AAA AAG-3′
D16S539 rev	5′-GTT TGT GTG TGC ATC TGT AAG CAT GTA TC-3′
D13S317 for	5′-HEX-ATT ACA GAA GTC TGG GAT GTG GAG GA-3′
D13S317 rev	5′-GGC AGC CCA AAA AGA CAG A-3′
vWA for	5′-6-FAM-GCC CTA GTG GAT GAT AAG AAT AAT CAG TAT GTG-3′
vWA rev	5′-GGA CAG ATG ATA AAT ACA TAG GAT GGA TGG-3′
TPOX for	5′-6-FAM-ACT GGC ACA GAA CAG GCA CTT AGG-3′
TPOX rev	5′-GGA GGA ACT GGG AAC CAC ACA GGT TA-3′
THO1 for	5′-6-FAM-ATT CAA AGG GTA TCT GGG CTC TGG-3′
THO1 rev	5′-GTG GGC TGA AAA GCT CCC GAT TAT-3′
CSF1PO for	5′-6-FAM-AAC CTG AGT CTG CCA AGG ACT AGC-3′
CSF1PO rev	5′-TTC CAC ACA CCA CTG GCC ATC TTC-3′
Amelogenin for	5′-ACC TCA TCC TGG GCA CCC TGG TT-3′
Amelogenin rev	5′-TAMRA-AGG CTT GAG GCC AAC CAT CAG-3′

2.4. Molecular Classification Analyses

The microsatellite instability (MSI) and methylation status of CpG islands [17–19] were determined for all cases included in this study. The classification was MSI-H if two or more microsatellite markers of either the Bethesda panel or the "six mononucleotide repeat" panel, consisting of BAT25, BAT26, CAT25, NR21, NR24, and NR27, showed band shifts [8,17]. Classification concerning the CpG island methylator phenotype (CIMP) was carried out as follows: if the analysis was performed according to Ogino et al., the subtype was divided into CIMP-H, non MSI, when ≥4 loci, and CIMP-L, non MSI, if 1–3 CIMP loci out of 5 loci analyzed were methylated [17,18]. When analyzed according to Weisenberger et al., ≥3 methylated CIMP Loci out of 5 loci analyzed defined CIMP-H, non MSI; no further distinction of CIMP-L, non MSI took place [19].

2.5. Next Generation Sequencing (NGS) Analyses

In total, 121 datasets were either generated by Centogene (Rostock, Germany) or extracted from a previous dataset [20]. This dataset consisted of Whole Exom Sequencing (WES) analyses of 20 PDX cases and 12 primary tumors. The remaining analyses were performed using a Solid Tumor Panel from Centogene consisting of 105 fully sequenced genes, plus mutational hot spots from an additional 146 genes. Library preparation was performed with the Twist Library Preparation Enzymatic Fragmentation Kit (Twist Bioscience, San Francisco, CA, USA). Exome enrichment was carried out, using either the TWIST Human Core Exome Plus probes (covering 36.5 Mb of the human coding exome) or custom designed probes, in the case of the Solid Tumor panel. Sequencing was performed using the NextSeq500 (Solid Tumor panel) or the HiSeq4000 and NovaSeq (WES) systems (Illumina, Inc., San Diego, CA, USA) to produce 2 × 150 bp reads. Raw sequencing reads were converted to standard fastq format using bcl2fastq software 2.17.1.14 (Illumina, Inc., San Diego, CA, USA). The short-reads were aligned to the GRCh37(hg19) build of the human reference genome using Bowtie version 2.4.2 [21]. The alignments were sorted (samtools v. 1.11) [22] and de-duplicated (PicardTools v. 2.23.8) [23]. Variant calling was performed with Strelka Somatic pipeline (v. 2.9.2) [24]. The variant table was filtered with vcftools v. 0.1.16 [25] and annotated with snpEff [26]. The filters applied were set to protein-coding mutations, filter "Pass", allele frequency > 5%, quality > 50, and at least 20 reads for the tumor.

For one PDX case, tissues obtained from two different mice were analyzed, and for a second PDX case, tissues obtained from two different passages were analyzed.

Concerning the NGS data for this study we focused on mutations being pathogenic or likely pathogenic, but also included mutations of uncertain significance. Excluded were all benign mutations, as well as mutations classified as risk factor and influencing drug response. Moreover, only the mutations from the raw data which passed the following quality criteria, including the filter "Pass", ≠coding synonym, a quality ≥30, and a variant allele frequency of at least 15, were listed.

2.6. Statistical Analyses

Statistical analyses were performed using either the statistical program prism 8 or IBM SPSS Statistics. Heatmap and mutation frequency analyses were performed in Prism. In SPSS, a nonparametric bivariate correlation analysis according to Kendall-Tau and Spearman's rank correlation coefficient and Fisher's exact test were performed. The cluster analysis was performed with Origin Pro 2017G (parameters: cluster method = Furthest Neighbor; distance type = Euclidean).

3. Results

In order to achieve maximal quality and traceability, the data from the CRC PDX cases included in the end, were collected and were mostly presented according to the PDX-Minimal Information standard (PDX-MI) recommended by Meehan et al. [27]. The PDX-MI suggests four modules reflecting the process of generating and validating a PDX model: (1) clinical data, (2) model creation data, (3) model quality assurance data, and (4) model study and associated metadata. Detailed information on our analyzed PDX cases was arranged accordingly and listed in Supplementary Table S1.

3.1. Patient, Clinical and Molecular Tumor Data

In total, 261 CRC patients were included in this study in the time span of October 2006 to May 2019. From these cases, 167 individual PDX models could be generated (64.0%). The present study focuses on 125 of these cases, which have been selected according to the following criteria: (I) enduring growth in immunodeficient mice and (II) storing sufficient quantities of PDX tissues with (III) adequate quality. The latter criteria in particular, led to the exclusion of 20 cases (12.0%) due to very high proportions of necrotic areas reproducibly observed in the harvested PDX tissues. Analyses of 22 PDX are not yet finalized, and thus these were consequently excluded from the present study.

To anonymize the patient information, each case was assigned an alias consisting of: HRO for Hansestadt Rostock, C for colon cancer, and a consecutive number. Metastases included were given the identifier Met as an abbreviation of metastasis. This was added directly after the HROC number. In case of multiple tumors, an additional tumor numeration was included.

All available patient information following surgical removal of the tumor, e.g., further treatments, disease recurrence, progression free and overall survival, were collected as described before [8] and updated in May 2020. These data are listed in Supplementary Table S1. The patient tumor samples consist of 100 primary adenocarcinomas, including one neuroendocrine tumor. Twenty-five samples are of metastatic origin, largely of the liver (80.0%). Metastases also manifested in the abdominal wall, brain, lung, peritoneum, and multivisceral ($n = 1$, each).

The gender distribution of the 125 patient cases included was 56.8% male and 43.2% female. The mean age was 69.7 years (ranging from 30 to 98). Tumor UICC staging was 11% stage I, 29% stage II, 27% stage III, and 33% stage IV. T stages were 30% T4, 59% T3, 10% T2, and 1% T1; M stages 1% M2 and 32% M1; 67% had no metastases identified (M0). Tumor grading (G) was 2% G1, 55% G2, and 43% G3.

Due to the fact that the integrated biobanking activities started in 2006 and are an ongoing process, the included cases cover the time period of 2006 to 2019. Thus, it was not possible to calculate the 5 year survival rate for all patients. Accordingly, setting a cutoff for the calculation of survival time was necessary. At the cutoff of May 2020, 54 patients were still alive, and 71 patients were dead. Three patients died perioperative, within 30 days after surgery (congruent with Clavien-Dindo classification). For the remaining 68 deceased patients, progression free survival averaged 13.0 months (ranging from 0 to 119) and the mean overall survival was 32.6 months (ranging from 1 to 133).

Each patient's individual cancer history and therapy regiment was listed in detail if applicable, including type and duration of therapy, as well as applied chemotherapeutic agents (Supplementary Table S1). Because most of our in-house therapeutic studies with different agents compared to the standard of care are ongoing or will be published soon, the therapeutics which showed a reduction in the PDX tumor growth compared to the standard of care are simply listed in Supplementary Table S1.

Since an ideal CRC PDX collection should approximate the molecular heterogeneity of clinical cases, our 125 PDX were classified according to the following molecular subtypes [17]: chromosomal instable (CIN), sporadic microsatellite instable (spMSI), having the CpG island methylator phenotype (CIMP, sub classified into high level (CIMP-H) and low-level (CIMP-L)), and Lynch syndrome (LS) (Table 2).

Table 2. Molecular subclasses of the 125 investigated PDX, listed with total amount and percentage.

Molecular Subclass Determination (n = 125):		
CIN	65	52%
spMSI-H	29	23.2%
CIMP-H, non MSI	10	8%
CIMP-L, non MSI	10	8%
Lynch syndrome	10	8%
Neuroendocrine tumor	1	0.8%

The distribution of the molecular subtypes mostly corresponds to the general clinical distribution [28]. Only spMSI-H and LS are overrepresented, which is most likely attributable to the high engraftment rates of these molecular subtypes [29].

3.2. Biobanking of Established HROC PDX Models

Besides mirroring the clinical characteristics of patient cohorts, another major goal was to generate and biobank ample amounts of PDX tissue for subsequent analyses and future preclinical studies. In particular, $n \geq 30$ vital PDX tissue backups (consisting of four small cubes of approximately 3 mm side-length, to allow for a total of at least 120 implantations), plus a minimum of n = 5 snap frozen samples, ideally suited for molecular analyses, were generated and stored in the gas phase above liquid nitrogen for each case. It is notable that the backups of all 125 cases were generated within less than 10 passages, and usually in less than 5 passages. This ensured closest achievable proximity to the tissue of origin. Moreover, representative cross-sections and halves of the PDX tumors were fixed in formalin and paraffin-embedded (in the following, this is termed FFPE tissue) for histopathological assessment.

The HROC Xenobank contains eight sets of primary tumor and metastases derived tissues from the very same patients; namely: HROC72 and HROC72Met1; HROC147 and HROC147Met1; HROC277, HROC277Met1 (synchronous), and HROC277Met2 (metachronous); HROC278 and HROC278Met1; HROC300 and HROC300Met1; HROC348 and HROC348Met1; HROC362 and HROC362Met1; and HROC405 and HROC405Met1. Additionally, three sets of two metastases from the very same patient are included: HROC103Met1 and Met2, HROC230Met1 and Met2, as well as HROC313Met1 and Met2. Furthermore, two sets of different primary tumors from the very same patients are included as well: HROC252Tu1, Tu2, and Tu3, plus HROC386Tu1 and Tu2.

The mean duration from implantation to harvest for all the included PDX models and overall passages was 105 days (range 38 to 287). No significant differences in duration until harvest were observed for the initial passages with 120 days (range 36 to 329) compared to 106 days (range 35 to 324) for the last performed passages of the included PDX models. Mice presenting health conditions leading to premature harvest were excluded from calculation of duration to harvest. These cases are indicated with >x days in Supplementary Table S1. Harvesting times were compared between the first and the last passage of each individual patient case, and increased or decreased growth is indicated with arrows in Supplementary Table S1 column AG and column AH. A trend for shorter duration until harvest for the last passage was observed in the majority of cases: 70/125 (56%). No correlations with molecular subtype or features of the PDX became apparent.

Moreover, a direct comparison of the utilization of fresh vs. vitally frozen tissues for subsequent passaging was possible for 28 cases (Supplementary Table S2). In 23 of those cases (82.1%), a shorter time to harvest was observed when PDX tissues were passaged fresh (unpaired t-test $p = 0.0003$).

A correlation analysis revealed, beside the expected positive correlations between the UICC stage and patient's progression free and overall survival, correlations between the PDX model features and the properties of the patient tumor (Supplementary Table S3). Further, 62.0% of the PDX established from male patients originated from patients in the age group 50–69 years old, whereas 53.7% of the tumors from female patients originated from patients > 70 years. Concerning only primary tumors, the distribution of models derived from male or female patients is balanced at 50.5% vs. 49.5%, but concerning metastases, more models derived from male patients are represented, with a division of 80.0% vs. 20.0%. Moreover, within the molecular subtypes, male patients' tumors dominated within the CIN cases (70.8%). Cases of the subtype spMSI-H were predominantly from female patients (79.3%).

Further, the molecular subtype and the time to harvest of PDX tumors were correlated (correlation $p < 0.001$, exact fisher test $p = 0.008$). Moreover, the number of mice necessary to generate sufficient backups correlated with the molecular subtype (correlation $p = 0.002$, exact fisher test $p = 0.048$). We observed again a correlation between the localization of the tumor and the duration until harvest (correlation $p = 0.002$, exact fisher test $p = 0.047$). Tumors of the right colon generally needed less than 90 days. For the other localizations, the duration was 90–180 days until outgrowth. Concerning the molecular subtype, it can be pointed out that MSI-H tumors, both sporadic as well as Lynch-associated, for the most part grew out in less than 90 days. The remaining molecular subtypes took 90–180 days to reach the designated size. Ample backups could, for most PDX cases of the spMSI-H type, be generated using fewer than 3 mice, whereas the highest numbers of mice, frequently > 4.5 mice, were necessary for CIMP-L, non MSI-type tumors. Furthermore, it is worth mentioning that there is a positive correlation between the time to harvest of the last passage performed and patients' overall survival (correlation $p = 0.033$, but not significant in exact fisher test).

3.3. Identity Testing

The genetic identity of all established PDX models in comparison to the original patient tissue was confirmed by fingerprint analysis as described before [30]. The PDX were found to be either genetically identical to or descended from the respective patient (data not shown) with two exceptions: no evaluable signals could be generated by fingerprint analysis for PDX HROC32 T3 M7 and HROC223 T2 M1. Allelic imbalances were regularly observed as well as small shifts in allele length for PDX of the MSI molecular subtype. These phenomena are both well-known. However, the possibility of comparing different patient tissues, as well as primary and secondary cell lines in many of the MSI-cases, still allowed valid identity verification.

3.4. Histological Examination

H&E-stained sections from FFPE blocks were used to assess morphologic features of each individual PDX model. Moreover, PDX tumors were compared to the respective original patient tumor by a board-certified pathologist (FP). Figure 1 depicts three selected cases: HROC172, HROC260, and HROC386Tu1. Details of the histological investigations are listed in Supplementary Table S4.

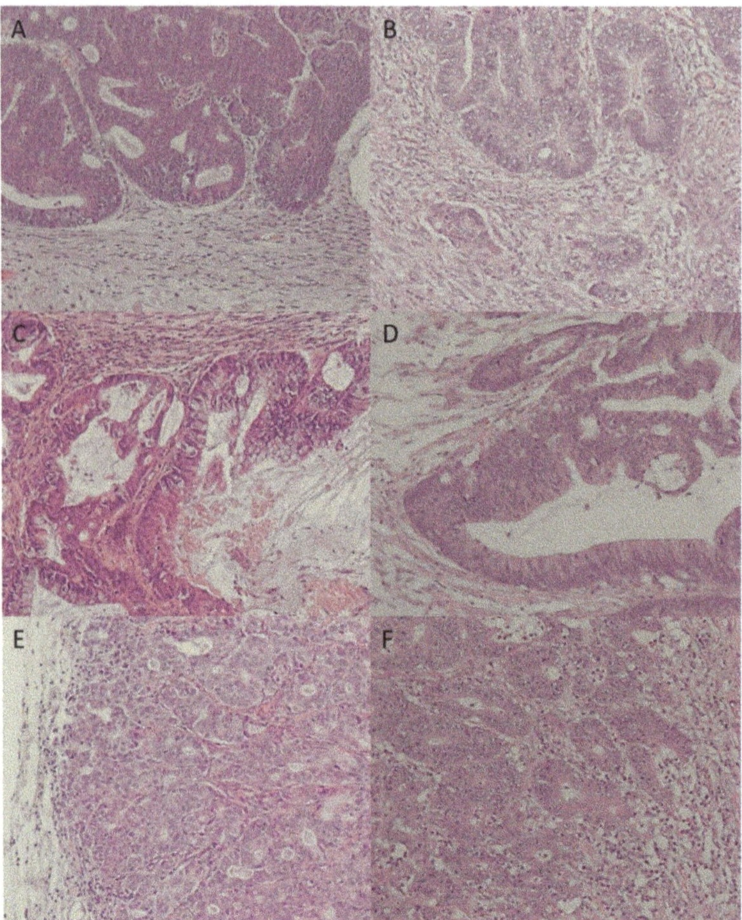

Figure 1. Comparison of primary tumor vs. PDX tumor in 20-fold magnification: (**A**) = HROC172 primary tumor, (**B**) = HROC172 T2 M2; (**C**) = HROC260 primary tumor, (**D**) = HROC260 T2 M5; (**E**) = HROC386Tu1 primary tumor, (**F**) = HROC386Tu1 T1 M1. In the case of HROC172, PDX cytomorphology and architecture match the primary tumor—stroma desmoplasia and tumor budding were markedly reduced; in the case of HROC260, PDX cytomorphology and architecture match the primary tumor—villous-mucinous structure was also reproduced; and, in the case of PDX HROC386Tu1, cytomorphology and architecture of the primary tumor was reproduced precisely.

A comparison was not possible for 17 out of the 125 cases due to a lack of patient tumor FFPE material, thus allowing a direct comparison for a total of 108 cases. Concordance of patient and PDX tumor structure was found in 92 cases (85.2%), minor differences were noticed 11 times (10.2%) and marked differences occurred in 5 cases (4.6%). Here, a fingerprint analysis performed with gDNA, isolated from sections of the very same FFPE tissue blocks used for the histological examination, confirmed genetic identity with their respective patients of origin for three out of the five cases (HROC251, HROC370, and HROC447). For the cases HROC32 and HROC223, the pathologist suspected heavy contamination of the PDX tissues with murine or human lymphatic cells. Since the fingerprint analysis failed for these two PDX tissues, as mentioned above, species-specific PCR analyses were performed. The results allowed the conclusion that murine thymoma cells predominated in the PDX tissues. Of note, identity tests from HROC32 PDX tissue after the initial mouse passage, as well as from two PDX-derived cell lines generated from the same passage as the FFPE tissues, matched the patients' identities.

A side-by-side comparison of PDX tumors derived from the same passage, but different animals, for six cases with different molecular subtypes (HROC92, HROC111Met1, HROC131, HROC169, HROC324, and HROC430) delivered exactly matching pathomorphological results (Supplementary Table S4). Thus, we would conclude that the conservation of the original tumor's biological features such as microarchitecture and pathomorphology are an intrinsic feature of the individual model and we expect these to be stably maintained for several passages as shown before by Tentler et al. [2].

Finally, when PDX tumors were explanted, the degree of necrosis was assessed. Although this is not a very precise method, it allowed the classification of the PDX cases into four categories: not, barely, intermediately, and highly necrotic (Supplementary Table S1, column AJ). Here, we observed no significant correlations between the degree of necrosis and patient data. The MSI cases, both sporadic and LS, the PDX models which needed fewer mice for complete asservation, and the PDX cases with shorter duration until harvest were rarely highly necrotic: there were only 4/39 cases (10.3%) vs. 27/86 cases (31.4%) for the remaining molecular subtypes. Highly necrotic PDX tumors maintained this characteristic also in later passages. The paired PDX cases of primary and metastasis derived tumors from the very same patients ($n = 8$) always had very similar necrosis categories (Supplementary Table S1).

3.5. Mutation Analysis

Selected cases ($n = 113$) were analyzed using a Solid Tumor Panel NGS approach. The NGS data are presented in detail in Supplementary Table S5. For two cases (HROC277Met2 and HROC405), two individual PDX tumors were analyzed. Here, the same pathogenic or likely pathogenic mutations were observed. The PDX tumor with the lower passage for patient HROC405 presented additional mutations of uncertain significance. Furthermore, for four cases, NGS analyses were conducted with patient and PDX tumor tissue (HROC285, HROC404, HROC415Met1, and HROC419). Pathogenic or likely pathogenic mutations detected in the original patient tumors were also detected in the PDX tumors, with one exception: a mutation in RHOA was only found in the primary patient tumor of HROC419. However, the PDX tumors of HROC285, HROC404, and HROC415Met1 displayed additional pathogenic or likely pathogenic mutations. In case of HROC285, four additional mutations in the genes *ABCB4*, *KRAS*, *MSH2*, and *NF1* were observed, whereas HROC404 and HROC415Met1 displayed two additional mutations in *AXIN2* and *HRAS*, as well as in *KMT2D* and *KRAS*, respectively. Besides, the PDX tumors displayed more additional mutations of uncertain significance than the original tumors.

Next, an unsupervised cluster analysis including only the pathogenic or likely pathogenic mutations, was performed (Figure 2).

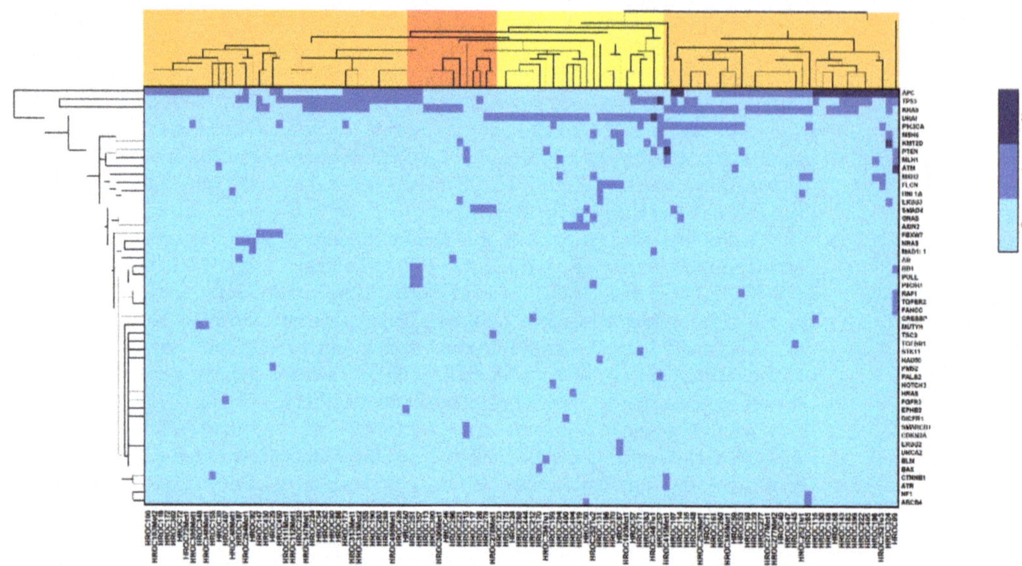

Figure 2. Unsupervised cluster analysis for all investigated tumors concerning the pathogenic or likely pathogenic mutations, with the following parameters: cluster method = Furthest Neighbor; distance type = Euclidean.

The clusters highlighted in orange contain mostly CIN cases (70.0 and 65.7%), whereas the cluster highlighted in yellow consists almost exclusively of sporadic MSI tumors (88%). However, the cluster highlighted in red could not be linked to a specific molecular subtype.

Furthermore, the number of pathogenic and likely pathogenic mutations detected per gene and case are illustrated in a heat map (Figure 3A). The mutation frequency of each gene can be found in Figure 3B. The most frequently mutated genes in the HROC-Xenobank are *APC* (50.4%), *KRAS* (39.8%), *TP53* (37.2%), *BRAF* (23.0%) and *PIK3CA* (17.7%).

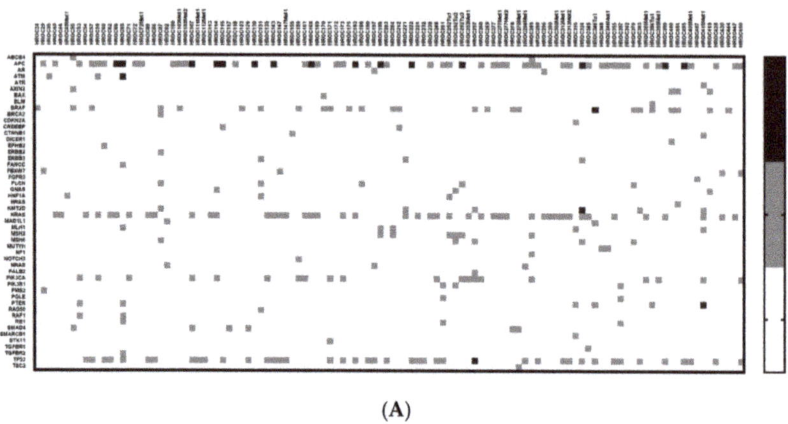

(A)

Figure 3. *Cont.*

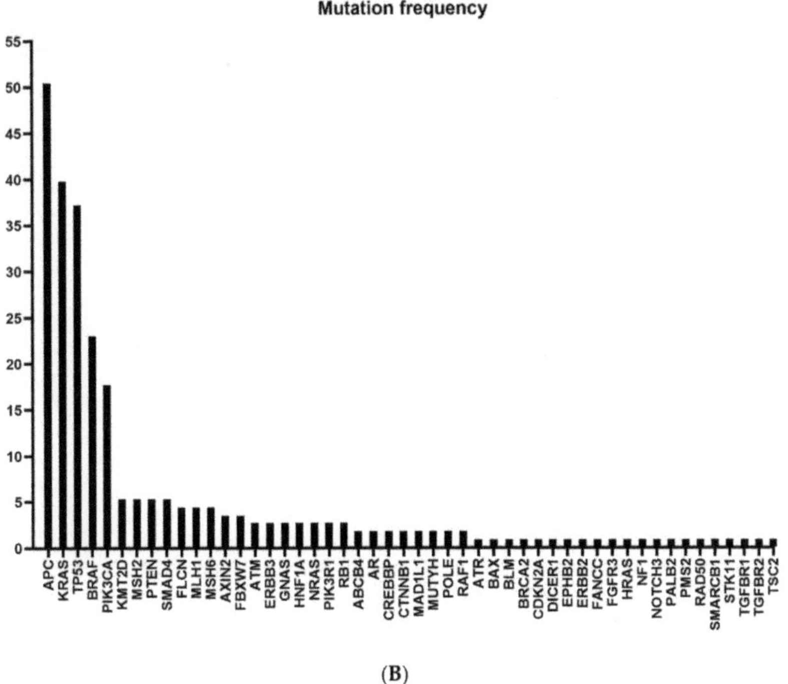

(B)

Figure 3. (**A**) = heat map. (**B**) = mutation frequency; (**A**) illustrates the number of pathogenic and likely pathogenic mutations per gene and case in a heat map, and the mutation frequency for each gene calculated out of this data is illustrated in (**B**).

4. Discussion

In summary, in this study we succeeded in establishing a CRC xenobank containing 125 individual PDX models with sufficient numbers of vital backups, snap frozen aliquots for molecular analysis, and FFPE material.

Mattar et al. described the challenges of creating a PDX biobank and proposed requirements for sample characterization and validation [31]. These included the collection of the above-mentioned sample types, namely, vital tissue, snap frozen samples, and FFPE specimens. In addition, they urged for genomic profiling and comparative histological reviewing. All of these recommendations were followed in our study. Comparative pathomorphological analysis and genetic identity testing confirmed the close proximity of the HROC-Xenobank models to the original patient tumors. Moreover, we could confirm previous findings that PDX models, in the majority of cases, maintain the original tumor's biological features [2,3,32]. For each individual HROC PDX model we described in detail, the pathomorphological structures were reproduced from the original patient tumor. A side-by-side comparison of PDX tumors derived from the same passage but different animals selected randomly revealed exactly matching pathomorphological results. Thus, we concluded that the original tumor's biological features, such as microarchitecture and pathomorphology, are intrinsic features of the individual tumor; thus, they are also conserved in the derived models and are most likely stable for several passages. It has been shown before that CRC PDX retain the histology as well as other features of the parental tumor, including intratumoral clonal heterogeneity and chromosomal instability, for up to 14 passages [2].

A surprising observation of the present study was the comparable duration until harvest for PDX tumors of the first and the last performed passage, at least with regard to mean values. Others have described a significantly accelerated growth rate with increasing passages [33]. One possible explanation for this discrepancy might come from the fact that we used different mouse strains. Because the engraftment efficacy for CRC ranges between ~60 and 70% in NMRI nude mice and up to ~80–90% in NSG mice [34], the preferred mouse strain for the first passage was NSG. Subsequent passaging was performed either with NSG or NMRI nude mice. The latter strain was preferred, since the risk of murine and/or human lymphoma development is reduced [35–37]. When considering this, we cannot formally exclude the possibility that this may have biased our results and this might explain why similar harvesting times between different passages were observed. Another factor affecting the time to harvest is the tissue condition at the time of engraftment, i.e., fresh vs. vitally frozen samples. For 23/28 (82.1%) cases, passaging with fresh material resulted in a significantly diminished duration until harvest.

Abdirahman et al. reported an engraftment success of 22/33 (67%) cases for their CRC PDX series, but because four cases turned out to be human lymphomas, their rate dropped to 18/33 (55%) [32]. Their lymphoma rate was surprisingly high (4/22; 18.2%). When comparing this with the 1.6% (2/125) rate of murine lymphoma in our study, and considering that Abdirahman and colleagues performed all passaging in NSG mice, it becomes clear why switching from NSG for initial engraftment to passaging in NMRI nude mice is preferable.

Further, the HROC PDX models precisely recapitulated the mutation profiles of the original patient tumors, thereby confirming previous data [32,38,39]. In particular, our study pointed out that pathogenic or likely pathogenic mutations detected in the original tumors were maintained in the PDX tumors. The current gold standard of NGS data analysis to compare are the data contained in the Cancer Genome Atlas Network (TCGA). In comparison to the TCGA results, the HROC-Xenobank mutational landscape differed in parts. The frequency for the most commonly mutated gene, *APC*, was 72.5% in the TCGA data set, compared to our frequency of only 50.4%. The *PIK3CA* gene had a frequency of 27.5% in TCGA and of 17.7% in the HROC-Xenobank. Frequencies for *BRAF* mutations were 11.6% (TCGA) and 23% (HROC-Xenobank). The most striking difference was noticed for *TP53* mutation frequency. Here 58.8% were reported for TCGA, and we observed merely 37.2% in our cohort. However, similar frequencies were observed for KRAS with 40.8% (compared to 39.8%) [40]. Compared to mutation frequencies reported by Lee et al. for *APC* (60%) and *KRAS* (49%), no apparent differences to our results were seen [41]. Similarly, the mutation frequency published by Burgenske et al. for *PIK3CA*, 15–25%, did not differ from our observation (17.7% in the HROC-Xenobank) [42].

When focusing on hypermutated tumors, the mutation frequency in TCGA increased to 57.5% for *BRAF*. The overrepresentation of hypermutated tumors in our cohort most likely explains the higher frequency of *BRAF* mutations observed in our cohort.

Mutation patterns of CRC from adolescent and young adults are also different. Tricoli et al. stated that in tumors of younger patients, genes were mutated significantly more frequently, particularly genes associated with DNA repair pathways BRCA2 (39% vs. 3%) and RAD9B (22% vs. 0%), as well as the cell-cycle checkpoint kinases ATM (35% vs. 7%) and ATR (48% vs. 13%). Despite the limited number of mutations associated with DNA repair pathway genes in our cohort, some of the HROC-Xenobank models might yet be interesting for functional analyses in that particular area of research.

In addition, many of our PDX models carry a higher number of mutations with uncertain significance than the original tumors, whereas Abdirahman et al. did not observe such additional mutations in their serially transplanted tumors [32]. Brown et al. suggested that mutations detected in PDX but not in original patient tumor samples could reflect hard to detect low-frequency clones in the original tumors [43].

The data from the German cancer registry ZfKD (Zentrum für Krebsregisterdaten) pointed out that 56% of patients with advanced CRC were male. The mean age among male patients ranged from 67.6 to 68.3 years and among female patients from 70.6 to 71.0 years [44]. The mean age of our study population is 69.7 (range 30 to 98) and thus lies precisely within the ZfKD ranges. Moreover, the male percentage of our study population is, with 56.8% of patients being male, the same as reported by the ZfKD. Tumor UICC staging was 11% (stage I), 29% (stage II), 27% (stage III), and 33% (stage IV) in our study, compared to 19.5% (stage I), 29% (stage II), 30% (stage III), and 21.5% (stage IV) in the general German CRC population [45]. Thus, the proportion of CRC stage II exactly matches, and the percentage of stage III nearly matches, the general German CRC population. However, CRC stage IV was considerably overrepresented and stage I considerably underrepresented in our study, as compared to the normal distribution of CRC stages in Germany. Partly, this is simply attributable to the fact, that our biobank collection is restricted to cases with sufficient tumor material available upon diagnosis in the pathology; therefore smaller, lower-staged cases frequently must be excluded. Additionally, all cases were collected at a university center, which is typically also biased towards more advanced, higher staged cases by the referring doctors.

The molecular subtype analyses revealed that our HROC-Xenobank cohort represents the common CRC subtypes [28]. However, spMSI-H and LS cases are overrepresented. This can best be explained by the fact that MSI tumors engraft significantly better than MSS tumors. Such a discrepancy in engraftment rates linked to the MS status was also reported for gastric cancer with 55.93% vs. 23.64%; $p < 0.0001$ [29]. This improved biological fitness in the xenograft environment was also highlighted by the fact that these PDX grew out faster, and fewer mice were necessary to generate ample amounts of backups. Our group previously described the molecular subtype ($p = 0.003$), especially the MS status ($p = 0.001$), as a potent parameter likely to influence the success rate of PDX establishment from CRC resection specimens [8]. It is also of relevance, albeit to a lesser extent, that MSI-H PDX tumors were not often highly necrotic and were, thus, rarely excluded from our final cohort due to this undesirable characteristic.

We want to emphasize the fact that the HROC-Xenobank has, already in its establishment phase, supported "standard" in vivo studies [46,47], detailed molecular pathway investigations [48], biomarker studies, and more basic studies which took advantage of snap-frozen samples [49]. A first pre-clinical PDX trial has been started in-house, preceded by a dose finding study [34]. Due to the fact that all samples have been collected exclusively after informed consent of all patients and the irrevocable anonymization of any personal or clinical data, the use of the HROC-Xenobank models is not restricted by the General Data Protection Regulation in the EU.

5. Conclusions

In summary, this study succeeded in generating 125 individual PDX models with sufficient numbers of vital and snap frozen aliquots as well as FFPE material. The preservation of the original tumor's biological features such as microarchitecture and pathomorphology as intrinsic features of the individual models remain stable for several passages. Moreover, we were able to confirm the high concordance of pathomorphological as well as mutation patterns of the HROC-Xenobank to the underlying clinical case series. This enables the selection of individual models according to desired features and allows future investigations, such as pre-clinical PDX trials and detailed molecular pathway investigations, with well-characterized samples. Notably, the models of the HROC-Xenobank are available upon reasonable request.

Supplementary Materials: The following are available online at https://www.mdpi.com/article/10.3390/cancers13235882/s1, Table S1: Patient data and PDX detail, The patient clinical data, including patient information, tumor information, patient treatment information, and PDX model information as well as PDX model validation and sharing and contact, Table S2: Comparison of fresh vs. vitally frozen transfer, Data for the direct comparison of fresh vs. vitally frozen tissues for subsequent passaging, presented for 28 cases listed with the individual time to harvest of each individual case and the calculated mean values, Table S3: Correlation analysis, Data for nonparametric bivariate correlation analysis according to Kendall-Tau and Spearman's rank correlation coefficient, as well as data of the consequential Fisher exact tests, Table S4: FFPE tissue Comparison PDX vs. primary tumor, Data of PDX tumors concerning detailed histological investigation and comparison to original patient tumor Table S5: NGS data, The included datasets are derived according to the quality criteria described in materials and methods. Three further datasets of WES data (HROC405Tu, HROC405Met1, and HROC425Tu) failed these quality criteria and are, therefore, not listed. The NGS data include mutations classified as pathogenic or likely pathogenic as well as mutations of uncertain significance. Excluded are all benign mutations as well as mutations classified as risk factor and drug response.

Author Contributions: Conceptualization, S.M. and M.L.; methodology, S.M., F.B. and M.K. (Mathias Krohn); investigation, S.M., F.B., S.K. (Said Kdimati), F.P., N.E. and M.K. (Mathias Krohn); data curation, S.M., B.M., M.R., S.K. (Susann Krake) and M.K. (Michael Kreutzer); writing—original draft preparation, S.M.; writing—review and editing, C.S.M.; writing—review and editing supervision, M.L.; visualization, S.M. and M.K. (Michael Kreutzer); project administration, S.M. and M.L. All authors have read and agreed to the published version of the manuscript.

Funding: This research was, in part, funded by grant number TBI-V-1-241-VBW-084 from the state of Mecklenburg-Vorpommern for the project "Peptide-Based Immunization for Colon- and Pancreas-Carcinoma (PiCoP)".

Institutional Review Board Statement: The study was conducted according to the guidelines of the Declaration of Helsinki. All procedures were approved by the Ethics Committee of the University of Rostock University Medical Centre (Reference numbers: II HV 43/2004, A 45/2007, A 2018-0054, and A 2019-0187) in accordance with generally accepted guidelines for the use of human material. All animal experiments were performed according to the guidelines of the local animal use and care committee (Landesamt für Landwirtschaft, Lebensmittelsicherheit und Fischerei Mecklenburg-Vorpommern, permit numbers: LALLF M-V/TSD/7221.3-1.1-071/10; 7221.3-1-015/14; and 7221.3-2-020/17).

Informed Consent Statement: Informed consent was obtained from all subjects involved in the study.

Data Availability Statement: The data and materials are available from the corresponding author upon reasonable request.

Acknowledgments: We thank Daniel Wolter and Mohamed Elhensheri (University Medicine Rostock, Department of oral and maxillofacial surgery, Rostock University Medical Center) for their excellent technical assistance.

Conflicts of Interest: The authors declare no conflict of interest.

References

1. Oh, B.Y.; Lee, W.Y.; Jung, S.; Hong, H.K.; Nam, D.H.; Park, Y.A.; Huh, J.W.; Yun, S.H.; Kim, H.C.; Chun, H.K.; et al. Correla-tion between tumor engraftment in patient-derived xenograft models and clinical outcomes in colorectal cancer patients. *Oncotarget* **2015**, *6*, 16059–16068. [CrossRef] [PubMed]

2. Tentler, J.J.; Tan, A.C.; Weekes, C.D.; Jimeno, A.; Leong, S.; Pitts, T.M.; Arcaroli, J.J.; Messersmith, W.A.; Eckhardt, S.G. Patient-derived tumour xenografts as models for oncology drug development. *Nat. Rev. Clin. Oncol.* **2012**, *9*, 338–350. [CrossRef]

3. Hidalgo, M.; Amant, F.; Biankin, A.V.; Budinská, E.; Byrne, A.T.; Caldas, C.; Clarke, R.B.; De Jong, S.; Jonkers, J.; Mari, G. Europe PMC Funders Group Patient Derived Xenograft Models: An Emerging Platform for Translational Cancer Research. *Cancer Discov.* **2015**, *4*, 998–1013. [CrossRef]

4. Bleijs, M.; Van De Wetering, M.; Clevers, H.; Drost, J. Xenograft and organoid model systems in cancer research. *EMBO J.* **2019**, *38*, e101654. [CrossRef]

5. Yoshida, G.J. Applications of patient-derived tumor xenograft models and tumor organoids. *J. Hematol. Oncol.* **2020**, *13*, 4. [CrossRef]

6. Inoue, A.; Deem, A.K.; Kopetz, S.; Heffernan, T.P.; Draetta, G.F.; Carugo, A. Deem Current and Future Horizons of Patient-Derived Xenograft Models in Colorectal Cancer Translational Research. *Cancers* **2019**, *11*, 1321. [CrossRef] [PubMed]

7. Medico, E.; Russo, M.; Picco, G.; Cancelliere, C.; Valtorta, E.; Corti, G.; Buscarino, M.; Isella, C.; Lamba, S.E.; Martinoglio, B.; et al. The molecular landscape of colorectal cancer cell lines unveils clinically actionable kinase targets. *Nat. Commun.* **2015**, *6*, 7002. [CrossRef]

8. Mullins, C.S.; Micheel, B.; Matschos, S.; Leuchter, M.; Bürtin, F.; Krohn, M.; Hühns, M.; Klar, E.; Prall, F.; Linnebacher, M. Integrated biobanking and tumor model establishment of human colorectal carcinoma provides excellent tools for preclinical research. *Cancers* **2019**, *11*, 1520. [CrossRef] [PubMed]

9.   Namekawa, T.; Ikeda, K.; Horie-Inoue, K.; Inoue, S. Application of Prostate Cancer Models for Preclinical Study: Advantages and Limitations of Cell Lines, Patient-Derived Xenografts, and Three-Dimensional Culture of Patient-Derived Cells. *Cells* **2019**, *8*, 74. [CrossRef] [PubMed]

10. Nelson, S.; Zhang, C.; Roche, S.; O'Neill, F.; Swan, N.; Luo, Y.; Larkin, A.; Crown, J.; Walsh, N. Modelling of pancreatic cancer biology: Transcriptomic signature for 3D PDX-derived organoids and primary cell line organoid development. *Sci. Rep.* **2020**, *10*, 2778. [CrossRef] [PubMed]

11. Dangles-Marie, V.; Pocard, M.; Richon, S.; Weiswald, L.-B.; Assayag, F.; Saulnier, P.; Judde, J.-G.; Janneau, J.-L.; Auger, N.; Validire, P.; et al. Establishment of Human Colon Cancer Cell Lines from Fresh Tumors versus Xenografts: Comparison of Success Rate and Cell Line Features. *Cancer Res.* **2007**, *67*, 398–407. [CrossRef] [PubMed]

12. Oberländer, M.; Linnebacher, M.; König, A.; Bogoevska, V.; Brodersen, C.; Kaatz, R.; Krohn, M.; Hackmann, M.; Ingenerf, J.; et al.; On behalf of the ColoNet consortium. The "North German Tumor Bank of Colorectal Cancer": Status report after the first 2 years of support by the German Cancer Aid Foundation. *Lagenbeck's Arch. Surg.* **2013**, *398*, 251–258. [CrossRef] [PubMed]

13. Linnebacher, M.; Maletzki, C.; Ostwald, C.; Klier, U.; Krohn, M.; Klar, E.; Prall, F. Cryopreservation of human colorectal carcinomas prior to xenografting. *BMC Cancer* **2010**, *10*, 362. [CrossRef] [PubMed]

14. Bürtin, F.; Matschos, S.; Prall, F.; Mullins, C.S.; Krohn, M.; Linnebacher, M. Creation and maintenance of a living bi-obank-how we do it. *J. Vis. Exp.* **2021**, *2021*, 62065.

15. Prall, F.; Maletzki, C.; Hühns, M.; Krohn, M.; Linnebacher, M. Colorectal carcinoma tumour budding and podia formation in the xenograft microenvironment. *PLoS ONE* **2017**, *12*, e0186271. [CrossRef]

16. Covarrubias-Pazaran, G.; Diaz-Garcia, L.; Schlautman, B.; Salazar, W.; Zalapa, J. Fragman: An R package for fragment analysis. *BMC Genet.* **2016**, *17*, 62. [CrossRef] [PubMed]

17. Prall, F.; Ostwald, C.; Linnebacher, M.; Weirich, V. Chromosomally and microsatellite stable colorectal carcinomas without the CpG island methylator phenotype in a molecular classification. *Int. J. Oncol.* **2009**, *35*, 321–327. [CrossRef]

18. Ogino, S.; Cantor, M.; Kawasaki, T.; Brahmandam, M.; Kirkner, G.J.; Weisenberger, D.J.; Campan, M.; Laird, P.W.; Loda, M.; Fuchs, C.S. CpG island methylator phenotype (CIMP) of colorectal cancer is best characterised by quantitative DNA methylation analysis and prospective cohort studies. *Gut* **2006**, *55*, 1000–1006. [CrossRef]

19. Weisenberger, D.J.; Siegmund, K.D.; Campan, M.; Young, J.; Long, T.I.; Faasse, M.A.; Kang, G.H.; Widschwendter, M.; Weener, D.; Buchanan, D.; et al. CpG island methylator phenotype underlies sporadic microsatellite instability and is tightly associated with BRAF mutation in colorectal cancer. *Nat. Genet.* **2006**, *38*, 787–793. [CrossRef]

20. Lazzari, L.; Corti, G.; Picco, G.; Isella, C.; Montone, M.; Arcella, P.; Durinikova, E.; Zanella, E.R.; Novara, L.; Barbosa, F.; et al. Patient-Derived Xenografts and Matched Cell Lines Identify Pharmacogenomic Vulnerabilities in Colorectal Cancer. *Clin. Cancer Res.* **2019**, *25*, 6243–6259. [CrossRef]

21. Langmead, B.; Salzberg, S.L. Fast gapped-read alignment with Bowtie 2. *Nat. Methods* **2012**, *9*, 357–359. [CrossRef] [PubMed]

22. Li, H.; Handsaker, B.; Wysoker, A.; Fennell, T.; Ruan, J.; Homer, N.; Marth, G.; Abecasis, G.; Durbin, R. Subgroup, 1000 Genome Project Data Processing The Sequence Alignment/Map format and SAMtools. *Bioinformatics* **2009**, *25*, 2078.

23. Broad Institute. Picard Toolkit. Available online: http://broadinstitute.github.io/picard/ (accessed on 1 September 2020).

24. Kim, S.; Scheffler, K.; Halpern, A.L.; Bekritsky, M.A.; Noh, E.; Källberg, M.; Chen, X.; Kim, Y.; Beyter, D.; Krusche, P.; et al. Strelka2: Fast and accurate calling of germline and somatic variants. *Nat. Methods* **2018**, *15*, 591–594. [CrossRef] [PubMed]

25. Danecek, P.; Auton, A.; Abecasis, G.; Albers, C.A.; Banks, E.; DePristo, M.A.; Handsaker, R.E.; Lunter, G.; Marth, G.T.; Sherry, S.T.; et al. The variant call format and VCFtools. *Bioinformatics* **2011**, *27*, 2156–2158. [CrossRef]

26. Cingolani, P.; Platts, A.; Wang, L.L.; Coon, M.; Nguyen, T.; Wang, L.; Land, S.J.; Lu, X.; Ruden, D.M. A program for annotating and predicting the effects of single nucleotide polymorphisms, SnpEff: SNPs in the genome of Drosophila melanogaster strain w1118; iso-2; iso-3. *Fly* **2012**, *6*, 80–92. [CrossRef]

27. Meehan, T.F.; Conte, N.; Goldstein, T.C.; Inghirami, G.; Murakami, M.A.; Brabetz, S.; Gu, Z.; Wiser, J.A.; Dunn, P.; Begley, D.A.; et al. PDX-MI: Minimal Information for Patient-Derived Tumor Xenograft Models. *Cancer Res.* **2017**, *77*, e62–e66. [CrossRef] [PubMed]

28. Wang, W.; Kandimalla, R.; Huang, H.; Zhu, L.; Li, Y.; Gao, F.; Goel, A.; Wang, X. Molecular subtyping of colorectal cancer: Recent progress, new challenges and emerging opportunities. *Semin. Cancer Biol.* **2018**, *55*, 37–52. [CrossRef]

29. Corso, S.; Isella, C.; Bellomo, S.E.; Apicella, M.; Durando, S.; Migliore, C.; Ughetto, S.; D'Errico, L.; Menegon, S.; Rull, D.M.; et al. A Comprehensive PDX Gastric Cancer Collection Captures Cancer Cell–Intrinsic Transcriptional MSI Traits. *Cancer Res.* **2019**, *79*, 5884–5896. [CrossRef]

30. Maletzki, C.; Gock, M.; Randow, M.; Klar, E.; Huehns, M.; Prall, F.; Linnebacher, M. Establishment and characterization of cell lines from chromosomal instable colorectal cancer. *World J. Gastroenterol.* **2015**, *21*, 164–176. [CrossRef]

31. Mattar, M.; McCarthy, C.; Kulick, A.R.; Qeriqi, B.; Guzman, S.; De Stanchina, E. Establishing and Maintaining an Extensive Library of Patient-Derived Xenograft Models. *Front. Oncol.* **2018**, *8*, 19. [CrossRef]

32. Abdirahman, S.M.; Christie, M.; Preaudet, A.; Burstroem, M.C.U.; Mouradov, D.; Lee, B.; Sieber, O.M.; Putoczki, T.L. A biobank of colorectal cancer patient-derived xenografts. *Cancers* **2020**, *12*, 2340. [CrossRef]

33. Houghton, J.A.; Taylor, D.M. Growth characteristics of human colorectal tumours during serial passage in immune-deprived mice. *Br. J. Cancer* **1978**, *37*, 213.

34. Maletzki, C.; Bock, S.; Fruh, P.; Macius, K.; Witt, A.; Prall, F.; Linnebacher, M. NSG mice as hosts for oncological precision medicine. *Lab. Investig.* **2019**, *100*, 27–37. [CrossRef]
35. Tillman, H.; Janke, L.J.; Funk, A.; Vogel, P.; Rehg, J.E. Morphologic and Immunohistochemical Characterization of Spontaneous Lymphoma/Leukemia in NSG Mice. *Veter. Pathol.* **2019**, *57*, 160–171. [CrossRef]
36. Bondarenko, G.; Ugolkov, A.; Rohan, S.; Kulesza, P.; Dubrovskyi, O.; Gursel, D.; Mathews, J.; O'Halloran, T.V.; Wei, J.J.; Mazar, A.P. Patient-Derived Tumor Xenografts Are Susceptible to Formation of Human Lymphocytic Tumors. *Neoplasia* **2015**, *17*, 735–741. [CrossRef]
37. Chateau-Joubert, S.; Hopfe, M.; Richon, S.; Decaudin, D.; Roman-Roman, S.; Reyes-Gomez, E.; Bieche, I.; Nemati, F.; Dangles-Marie, V. Spontaneous mouse lymphoma in patient-derived tumor xenografts: The importance of systematic analysis of xenografted human tumor tissues in preclinical efficacy trials. *Transl. Oncol.* **2021**, *14*, 101133. [CrossRef] [PubMed]
38. Woo, X.Y.; Giordano, J.; Srivastava, A.; Zhao, Z.-M.; Lloyd, M.W.; de Bruijn, R.; Suh, Y.-S.; Patidar, R.; Chen, L.; Scherer, S.; et al. Conservation of copy number profiles during engraftment and passaging of patient-derived cancer xenografts. *Nat. Genet.* **2021**, *53*, 86.
39. Rizzo, G.; Bertotti, A.; Leto, S.M.; Vetrano, S. Patient-derived tumor models: A more suitable tool for pre-clinical studies in colorectal cancer. *J. Exp. Clin. Cancer Res.* **2021**, *40*, 178. [CrossRef] [PubMed]
40. cBioPortal for Cancer Genomics. Available online: http://www.cbioportal.org/study/summary?id=coadread_tcga_pan_can_atlas_2018 (accessed on 7 October 2021).
41. Lee, C.S.; Song, I.H.; Lee, A.; Kang, J.; Lee, Y.S.; Lee, I.K.; Song, Y.S.; Lee, S.H. Enhancing the landscape of colorectal cancer using targeted deep sequencing. *Sci. Rep.* **2021**, *11*, 8154. [CrossRef] [PubMed]
42. Burgenske, D.M.; Monsma, D.J.; Dylewski, D.; Scott, S.B.; Sayfie, A.D.; Kim, D.G.; Luchtefeld, M.; Martin, K.; Stephenson, P.; Hostetter, G.; et al. Establishment of genetically diverse patient-derived xenografts of colorectal cancer. *Am. J. Cancer Res.* **2014**, *4*, 824–837. [PubMed]
43. Brown, K.M.; Xue, A.; Mittal, A.; Samra, J.S.; Smith, R.; Hugh, T.J. Patient-derived xenograft models of colorectal cancer in pre-clinical research: A systematic review. *Oncotarget* **2016**, *7*, 66212–66225. [CrossRef]
44. Oppelt, K.A.; Luttmann, S.; Kraywinkel, K.; Haug, U. Incidence of advanced colorectal cancer in Germany: Comparing claims data and cancer registry data. *BMC Med Res. Methodol.* **2019**, *19*, 142. [CrossRef] [PubMed]
45. Robert Koch Institut. Krebs in Deutschland 2015/2016. 2015. Available online: https://www.krebsdaten.de/Krebs/DE/Content/Publikationen/Krebs_in_Deutschland/kid_2019/krebs_in_deutschland_2019.pdf?__blob=publicationFile (accessed on 5 October 2021).
46. Shang, Y.; Zhang, X.; Lu, L.; Jiang, K.; Krohn, M.; Matschos, S.; Mullins, C.S.; Vollmar, B.; Zechner, D.; Gong, P.; et al. Pharmaceutical immunoglobulin G impairs anti-carcinoma activity of oxaliplatin in colon cancer cells. *Br. J. Cancer* **2021**, *124*, 1411–1420. [CrossRef] [PubMed]
47. Nörz, D.; Mullins, C.S.; Smit, D.J.; Linnebacher, M.; Hagel, G.; Mirdogan, A.; Siekiera, J.; Ehm, P.; Izbicki, J.R.; Block, A.; et al. Combined Targeting of AKT and mTOR Synergistically Inhibits Formation of Primary Colorectal Carcinoma Tumouroids In Vitro: A 3D Tumour Model for Pre-therapeutic Drug Screening. *Anticancer. Res.* **2021**, *41*, 2257–2275. [CrossRef]
48. Marx, C.; Sonnemann, J.; Beyer, M.; Maddocks, O.D.K.; Lilla, S.; Hauzenberger, I.; Piée-Staffa, A.; Siniuk, K.; Nunna, S.; Marx-Blümel, L.; et al. Mechanistic insights into p53-regulated cytotoxicity of combined entinostat and irinotecan against colorectal cancer cells. *Mol. Oncol.* **2021**. [CrossRef] [PubMed]
49. Bock, S.; Mullins, C.S.; Klar, E.; Pérot, P.; Maletzki, C.; Linnebacher, M. Murine Endogenous Retroviruses Are Detectable in Patient-Derived Xenografts but Not in Patient-Individual Cell Lines of Human Colorectal Cancer. *Front. Microbiol.* **2018**, *9*, 789. [CrossRef]

Review

Patient-Derived In Vitro Models for Drug Discovery in Colorectal Carcinoma

George M. Ramzy [1,2], Thibaud Koessler [3], Eloise Ducrey [1,2], Thomas McKee [4], Frédéric Ris [5], Nicolas Buchs [5], Laura Rubbia-Brandt [4], Pierre-Yves Dietrich [3] and Patrycja Nowak-Sliwinska [1,2,*]

1. Molecular Pharmacology Group, School of Pharmaceutical Sciences, Institute of Pharmaceutical Sciences of Western Switzerland, University of Geneva, 1211 Geneva, Switzerland; George.Ramzy@unige.ch (G.M.R.); Eloise.Ducrey@unige.ch (E.D.)
2. Translational Research Center in Oncohaematology, University of Geneva, 1211 Geneva, Switzerland
3. Department of Oncology, Geneva University Hospitals, 1211 Geneva, Switzerland; Thibaud.Kossler@hcuge.ch (T.K.); Pierre-Yves.Dietrich@unige.ch (P.-Y.D.)
4. Division of Clinical Pathology, Diagnostic Department, University Hospitals of Geneva (HUG), 1211 Geneva, Switzerland; Thomas.A.McKee@hcuge.ch (T.M.); Laura.Rubbia-Brandt@unige.ch (L.R.-B.)
5. Translational Department of Digestive and Transplant Surgery, Faculty of Medicine, Geneva University Hospitals, 1211 Geneva, Switzerland; Frederic.Ris@unige.ch (F.R.); Nicolas.Buchs@unige.ch (N.B.)
* Correspondence: Patrycja.Nowak-Sliwinska@unige.ch; Tel.: +41-22-379-3352

Received: 3 May 2020; Accepted: 28 May 2020; Published: 31 May 2020

Abstract: Lack of relevant preclinical models that reliably recapitulate the complexity and heterogeneity of human cancer has slowed down the development and approval of new anti-cancer therapies. Even though two-dimensional in vitro culture models remain widely used, they allow only partial cell-to-cell and cell-to-matrix interactions and therefore do not represent the complex nature of the tumor microenvironment. Therefore, better models reflecting intra-tumor heterogeneity need to be incorporated in the drug screening process to more reliably predict the efficacy of drug candidates. Classic methods of modelling colorectal carcinoma (CRC), while useful for many applications, carry numerous limitations. In this review, we address the recent advances in in vitro CRC model systems, ranging from conventional CRC patient-derived models, such as conditional reprogramming-based cell cultures, to more experimental and state-of-the-art models, such as cancer-on-chip platforms or liquid biopsy.

Keywords: colorectal cancer; organoids; 3D bioprinting; patient-derived xenograft; cancer-on-chip; drug combination

---

1. Introduction

Colorectal carcinoma (CRC) is the third most commonly diagnosed form of cancer in the world, with an estimated incidence of 1.8 million cases in 2018, and is expected to reach 2.2 million by 2030 [1,2]. Although cancer treatment in general has improved over the past few decades, the need for more personalized targeted therapies remains present, specifically for late-stage metastatic CRC (mCRC) patients for whom treatment options—and consequently overall survival rates—are limited [3].

The attrition rate of anticancer drugs candidates is very high, and only approximately 5% of drugs successfully complete phase III clinical trials [4,5]. One of the problems that might impair the development and approval of new anticancer therapies is the lack of relevant models that recapitulate the complexity of human cancer nature.

The main traits of an "ideal" CRC model for testing new treatment options reside in its capacity to resemble the in vivo conditions. This includes characteristics such as the genetic-, functional- and histological features of the patient's tumor, along with sequential mutagenesis (i.e., loss of adenomatous polyposis coli, APC), followed by activating Kirsten rat sarcoma viral oncogene homolog (*KRAS*) mutations and loss of *TP53*, the presence of stromal- and immune cells, as well as the presence and composition of tumor stroma.

In patient-derived xenograft (PDX) models, small pieces of surgical patient tumor tissue are used for implantation into an immunodeficient mouse. Detailed protocols of the engraftment and propagation procedure for CRC PDX were described by several groups [6–8]. Different studies have demonstrated the potential use of PDX as a preclinical model in the drug screening cascade, as it can reliably predict and recapitulate CRC patient drug responses in colorectal cancer [9–15]. Although the tumors grow in a biologically more relevant microenvironment than can be provided in vitro, the mice are immunocompromised, therefore some of the complex interactions between the host and the tumor might not be preserved. Furthermore, some genetic and epigenetic changes may occur in the tumor cells during manipulations, such as resection, culture or implantation. Among several factors, the quality of the patient specimen, tumor type and stage, technique and time to implantation may affect the engraftment rates [16–21].

Multiple reports have shown that, after engraftment, the human tumor stroma is preserved, but is slowly replaced by murine stroma over time throughout the consequent passages [22]. In addition, in PDX models the microenvironment of the subcutaneous space differs greatly from that of the colon. The latter led scientists to develop orthotopic mouse models, where the tumor is directly implanted into the caecum. The main goal was to create an in vivo model that would allow tumor development locally in the colon, allowing all stages progression for CRC [23,24].

Genetically engineered mice (e.g., germline APC mutant models [25] or models presenting mismatch repair-deficiency [26]) and carcinogen-induced models [27] are widely used to investigate CRC and its treatment screenings. These models remain the most developed CRC in vivo models, due to their genetic controllability. Several of them are elegantly reviewed by others [28,29] and they are not discussed in this review.

Since the use of laboratory animals is laborious, time-consuming and expensive, in vitro models would greatly contribute to higher efficiency in drug screening. In addition, given that animal experimentation [30] is being widely discussed, the development of alternative in vitro models is needed to support the idea of reduction, refinement and replacement of laboratory animals. Even though drug discovery cannot be based purely on in vitro models, they can deliver important results that may, in turn, help in the reduction of further in vivo experiments.

In this review we discuss a broad spectrum of in vitro CRC model systems, ranging from recent advances in conventional CRC patient-derived models, to more experimental and state-of-the-art technology models (Figure 1). We also suggest potentially attractive models used in other cancer types that would need further validation for CRC.

Recently developed CRC patient-derived models are usually established from freshly resected tumor tissue that undergoes enzymatic and mechanical digestion. Patient-derived models have emerged from liquid biopsies i.e., blood containing circulating tumor cells. Current advances in tissue engineering allowed patient-derived cells to be incorporated into a bioink and bioprinted into 3D constructs.

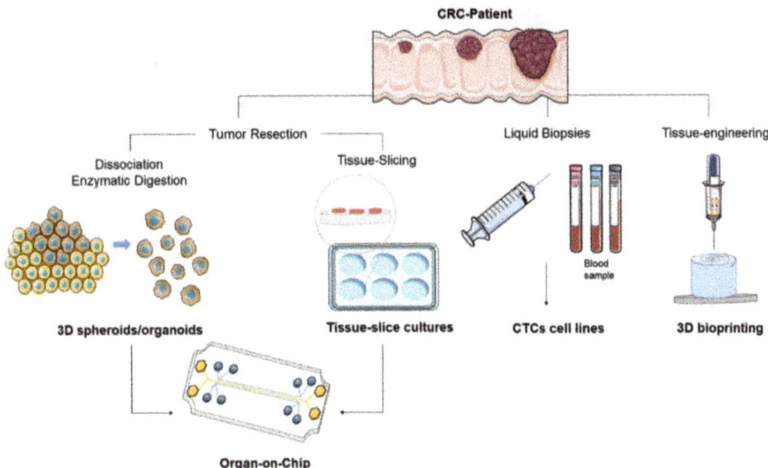

Figure 1. Overview of colorectal carcinoma (CRC) patient-derived preclinical models.

2. Conventional CRC In Vitro Models

2.1. Patient-Derived Tumor Cell Lines Cultivated in Two Dimensions

For decades, preclinical cancer research has been widely based on the use of cancer cell lines cultured in vitro and xenografts derived from these, grown in vivo. The culture of cells in vitro lead to the acquisition of multiple genetic and epigenetic alterations that diverge drastically from the original tumor they were derived from. Expanding and maintaining tissue-derived cell lines in culture often implicates the use of exogenous immortalization methods to keep cells in culture. These cells exhibit a similar gene expression and epigenetic profile, and can be propagated in vitro, into several germ layers, providing great potential for disease modelling such as cancer [31–34].

Recently, conditional reprogramming became widely used as a preclinical model in cancer research [35]. It is a co-culture based technology that makes it possible to efficiently expand patient-derived cells in culture medium supplemented with Rho kinase inhibitor (ROCK inhibitor, Y27632) and irradiated feeder fibroblasts [36]. Y27632 has been shown to induce the proliferative capacity of primary tumor cells resulting in efficient immortalization of cells into stratified epithelium without any DNA damage [37,38]. Several groups have elaborated specific protocols for cell isolation from various tumor types, including hepatocellular carcinoma [39], prostate cancer [40], tongue squamous cell carcinoma [41] and non-small lung cancers [42]. Liu et al. described a detailed protocol of CRC patient-derived cell cultures establishment from both cancerous and non-cancerous tissue biopsies that had the capacity to grow indefinitely in vitro, while maintaining phenotypic and genotypic features of the original tissue [36]. The study included freshly resected CRC tumors that generated approximately 10,000 cells after four weeks being in culture. This was done using conditional reprogramming, i.e., a novel next generation tool for long-term culture of primary epithelium cells derived from almost all origins without alteration of genetic background of primary cells. Moreover, Kodack et al. reported on a platform for functional testing on tumor-biopsy-derived cultures [43]. The criteria of a successful generation of colon cancer cell lines were defined as the point where the cells no longer required feeder cells to grow, could be cryopreserved and thawed and regrown, while maintaining the genotypic and phenotypic features. The authors also elaborated an immunofluorescence-based assay using a cocktail of monoclonal antibodies targeting cytokeratin (CK) 8 and cytokeratin 18 to specifically identify cancer cells, since both CK8/CK18 are nearly present in all tumors of epithelial origin.

The induction of conditional reprogramming in cancer cells is fast, and, unlike in the case of conventional cell lines, results in the generation of whole cell populations without clone selection.

In addition, this technology makes it possible to maintain the phenotypic features of the primary tumor in culture. Future studies are still needed to confirm the genetic diversity within the isolated tumor cells.

In general, 2D cell cultures lack in vivo characteristics, such as tissue specific architecture, which can affect the proliferation of the cells and their response to external stimuli. This lack of complex cellular interactions fails to replicate the aggressiveness and heterogeneity of the disease, making 2D cultures poor models to predict drug response for complex diseases such as cancer. Two-dimensional models are used in a relatively high-throughput in vitro screening, but are constrained by the limited viability and the low and/or short proliferative capacity of the cultivated cells [20]. Moreover, the result highly depends on the isolation and culturing conditions, e.g., composition of the cell culture medium, seeding density and the addition of supplements or cellular matrixes [44–47].

Thus, traditional cell lines cultured in monolayers are not perfectly suited for complex CRC research and its further development led to creation of three-dimensional cell culture systems. To better recapitulate the organ and tumor complexity, researchers expanded technology of cell cultivation using the spheroid and organoid technology [46].

2.2. Patient-Derived Cells Cultivated in Three Dimensions

Spheroids are spherical cellular constructs, consisting of an external proliferating zone surrounding an internal quiescent zone [48]. The co-existence of these multilayers makes it possible to mimic the cellular heterogeneity observed in solid tumors [49,50]. We have recently developed a robust short-term 3D spheroid model, where human CRC cells were simultaneously co-cultured with human fibroblasts and human endothelial cells in a clinical relevant ratio [51].

Jeppesen et al. established short-term spheroids cultures obtained with a high success rate of over 80% from freshly-derived primary CRC tumors [52]. They show that the initial tumor fragment size does not affect the success rate of spheroid formation or the cellular characteristics of the spheroid, which preserve both the molecular and histological characteristics of CRC, while maintaining inter-patient sensitivity towards a given treatment.

The cell culture media composition has a major impact on the success rate of maintaining high viability of tumor-derived spheroids in culture. Available protocols remain inconsistent, as the report on various combinations of cell medium supplements and their positive effect on cell viability [47,53–56].

To date, the 3D CRC spheroids rarely contain immune cells [57,58]. Integrating a potential immune response in the patient-derived spheroid to a treatment might represent an important factor in the treatment design. Recently, CRC patient-derived spheroids were co-cultured with tumor infiltrating lymphocytes from the same patient [59] to study the infiltration, activation and function of immune cells in tumors. This study proved that both activated natural killer cells and activated T cells infiltrated the spheroids induced the death of cancer cells and disrupted the 3D spheroid structure. Heterotypic co-cultures of tumor spheroids with other immune cells types could further expand our knowledge of human anti-tumor immune responses.

Each of the above-mentioned models has its own advantages and drawbacks in terms of replicating the in vivo physiology of original tumor architecture, TME, cellular composition, as well as response to different exogenous stimuli. This is very often highly dependent on the initial patient specimen. These models are being constantly improved, to better recapitulate complex cancer biology. Further research regarding the automation, miniaturization and adaptation of spheroid co-culture models to human tumor types will make it possible to dynamically study the anti-tumor immune response. Several critical aspects of cell isolation and culture conditions need to be carefully considered when handling patient-derived in vitro material. The type of dissociation method (mechanical vs. enzymatic) used might influence the number and quality of isolated cells. In addition, the choice of an adequate antibiotic or a mixture of thereof, as well as the concentration of this in the culture medium, is critical to avoiding microbial contamination during transport and culture of the CRC cells [60]. Furthermore, evaluation of the purity and tumoral nature of the isolated cells, e.g., by flow cytometric analysis or immunohistochemistry, is essential for preserving a representative ratio of the different cell populations in the tumor of origin [60,61]. Patient-derived intestinal crypts (see Section 3.4) require the presence of Matrigel, and a defined mixture of the Lgr-5-ligand R-spondin, epidermal growth factors, and Noggin, that are known to be the most essential stem cell maintenance factors [62].

3. CRC Patient-Derived In Vitro Models

3.1. Patient-Derived Tissue Slice Culture

Three-dimensional culture systems have been developed to mimic in vivo tumors as closely as possible by considering two aspects of cancer: heterogeneity and stromal interactions. It is important to note that most of the models take only partially into account tumor complexity, and most of the components of the stroma are absent [63]. Patient-derived tissue slice culture model is a tumor slice of approximately 200–300 μm thick, which is enough to preserve histological features of the original tumor, as well as important cellular components such as immune, vascular and mesenchymal cells [64]. The latter are important key features of this culture model [65]. Patient-derived tissue slice culture models have already been established for various cancer types, such as prostate [66,67], breast [68] and lung [64]. Until now, only a few reports included this model to study CRC.

Sönnichsen et al. showed that slice culture from patient-derived colorectal tumor tissue represented similar morphological features to the original tumor over the observed cultivation period of 3 days. Persisting tumor cell proliferation in tissue slice culture under treatment with 5-FU, as highlighted in the study, can help identify a non-responding patients to a treatment, and therefore may help in preventing administration of ineffective treatment in clinically applicable timeframe [69]. Unlike 3D culture models, the initial step for this technique does not include an enzymatic digestion step of the tumor-tissue before the cells are stimulated to grow in 3D, which, in turn, makes it possible to maintain the complexity of the tumor without extra manipulation of the tissue [63]. Ironically, a key advantage of this model can also be a limiting factor, as the normalization of tumor cell fractions is a major inconvenience. The exact number of tumor cells in the tissue slice culture is not determined prior to the initiation step, which can greatly impact the reproducibility of the results [65]. While this model represents a promising technology to assess drug sensitivity in individual colorectal tumors, further correlations with clinical outcomes in larger cohorts of patients to validate the clinical application of the technique, are to be considered [69].

Tumor tissue slices of hepatic CRC metastases were used for the first time to evaluate the response to oxaliplatin, cetuximab and pembrolizumab and measure anti-proliferative and pro-apoptotic features of the tumor and morphometric changes [70]. Moreover, the RAS mutation status, as well as the immunohistochemical evaluation of microsatellite stability and checkpoint protein (PD1) expression, was assessed. This study identified non-responders and responding patients. Moreover, the original tumor sections showed moderate to high infiltrates of PD1 positive tumor-associated immune cells, indicating susceptibility to selected treatments.

One of the obstacles of the tissue-slice technique is the lack of long-term tissue preservation methods. In order to address this issue, Zhang et al. developed a new method of vitrification-based cryopreservation of tumor biopsies [71]. The patient-derived xenograft models were then successfully established. The observed drug responses in the xenograft model were consistent with those in tissue slice cultures performed in vitro. Importantly, the cultivation retained the heterogeneous architecture of the original tumor giving opportunity to further analysis of tumor biology.

3.2. Liquid Biopsy and Circulating Tumor Cells

Liquid biopsy refers to the analysis of biomarkers in any body fluid [72]. In oncology, liquid biopsy represents a non-invasive test using blood to analyze tumor-derived genetic materials (DNA, RNA and miRNA) and proteins that either can be circulating freely in the blood or incorporated in circulating tumor cells (CTCs) [73]. CTCs play an important key role in the understanding of the biology of metastasis in patients with cancer, as recent studies have shown that they are found in the blood of cancer patients, as single CTC or CTC clusters with a strong ability to seed metastasis [74]. CTCs are new potential biomarkers that have been recently employed as diagnostic, prognostic and predictive to a wide range of cancer type including breast, lung, prostate and colorectal cancers [75,76]. The detection and study of CTCs in peripheral blood have caught the attention of scientists over the past decade, mainly for their promising clinical implication. A recent study showed that the disruption of CTC clusters, which have been linked to high metastatic potential [77], increases the proportions of single CTCs in the blood stream, but suppresses overall metastasis formation, highlighting the importance of CTC clusters as potential therapeutic targets in cancer treatment [74].

The CellSearch® platform, approved by the Food and Drug Administration (FDA), is currently used to identify, isolate and enumerate the CTC in the blood samples [78,79]. This technique consists mainly of the antibodies attached to magnetic beads against epithelial cell-adhesion molecule that are present on the surface of the CTC, and not on the healthy blood cells, separating magnetically the CTC from the bulk of other cells in the blood sample [80]. The low number of CTCs in peripheral blood makes it very challenging to establish their in vitro cultures. Recently, however, several groups managed to obtain CTC cell lines from patients with prostate [81] and breast cancer [82], two tumor types known to have a higher number of CTCs.

Cayrefourcq et al. reported for the first time the establishment of a permanent cell line from approximately 300 CTCs of one CRC patient, using the CellSearch® platform [83]. This cell line has been cultured for more than one year. Thorough analysis of the cells at the genomic, transcriptomic, proteomic and secretomic levels showed high similarity to the tumor of the patient with colon cancer that they were derived from. This approach opened a new avenue for a potential platform for novel drug screening in CRC, by eventually generating single CTC or CTC spheres from patient-derived blood samples.

Another protocol that can be used to establish a CTC colorectal cancer patient-derived cell line was described by Grillet et al. The authors generated sufficient cellular material (5 million cells) within three weeks after sample collection, and then used it to perform cytotoxicity assays. The study offered preliminary clinical data suggesting that toxicity assays on CTC might predict patient response to drugs. A patient, from whom a CTC line was established, died after being treated with FOLFIRI (FOL = Folinic acid (Leucovorin) + F = 5-fluorouracil + IRI = irinotecan), a first line treatment in patients diagnosed with mCRC [84]. The CTC line was shown in this study to be resistant to this chemotherapy combination in vitro [85].

A potential future application of CTC in personalized medicine would be to develop CTC-derived organoids and CTC-derived xenografts that could be used in drug screening for CRC treatment, using minimally invasive methods, while reflecting to a high extent tumor heterogeneity ex vivo.

A major inconvenience of the use of liquid biopsy and CTC is the low concentration and yield of CTC extraction, especially in the blood of patients with CRC [83]. Moreover, major discrepancies have been highlighted, depending on the technique used for CTC detection, i.e., EpCAM antigen detection-based (CellSearch®) or cell size-based (ISET assay, based on the use of specific designed filtration membranes that allow size based exclusion of blood components, in different tumor types) [86]. The limited number of FDA approved technologies available for CTC detection and extraction makes the technology less accessible. Its wide application in clinical practice is also limited by its high costs. Lately, the development of microfluidic technology (see Section 3.3) is considered a potential alternative solution for CTC isolation [87]. Nevertheless, liquid biopsy holds great promise for revolutionizing cancer diagnostics in the future, to enable early detection.

3.3. Organ-on-Chip

The development of organ-on-chip (OOC) technology has made it possible to bridge the gap between conventional cell cultures, preclinical animal models and clinical trials in patients. In addition to recapitulating to a high extent the biology and physiology of the organ of origin, the OOC allow high-resolution-real-time imaging of living human cells in a functional human tissue and organ context [88].

OOC are microfluidic culture devices consisting essentially of flexible polymers, such as polydimethylsiloxane, containing perfused micro-channels harboring living cells that mimic in vivo organ architecture and physiology. The viability of cells can be maintained over an extended period (weeks to months) by flowing medium through the micro-channels. When the system is stabilized, medium can be replaced by whole blood perfusion for a couple of hours [89].

Multiple research groups managed to establish OOC platforms by replacing healthy cells and associated extracellular matrixes with those of cancers [90]. The concept behind the technology is to model cancer cell behavior within human-relevant tissue and organ microenvironments in vitro. OOC enable researchers to vary local cellular, molecular, chemical and biophysical parameters in a controlled manner, both individually and in precise combinations, while analyzing how they contribute to human cancer formation, progression and response to therapy. OOC have been developed for a wide range of solid tumors like brain [91], bladder [92], breast [93] and non-small lung cancer [94]. Not only has this technology been used to create organs and solid cancer-on-chip, but some research groups, like Zhou et al., have also employed it to isolate CTCs in cancer patients [87], or to model bone marrow angiogenesis [95]. Traditional static intestine models containing human epithelial cells (e.g., Cako-2 or HT-29 cell lines) cultured on extra-cellular matrix-coated membranes within the trans-well devices, could not support several intestine properties or its functions. Over the last few years, the intestine chips have been developed with increased complexity that include channels lined to human microvascular endothelium, immune cells or pathogenic bacteria, and allow interaction between them [96]. That, in turn, enabled studying physiology, as well as pathology of the intestine. A good example of such a device is the microfluidic two-channel gut chip, which contains human epithelial, endothelial, immune and microbial cells, co-existing on ECM-coated transparent silicon polymer [97]. In this model, the pneumatic application of suctions applied in cycles enabled device deformations, resembling the movements of intestine during the peristalsis. Importantly, under these conditions, epithelial cells that, in conventional 2D conditions, grow in monolayers, spontaneously undergo villus morphogenesis. This is an important improvement, as compared to organoids that do not experience physical stretching resulting from peristaltic contractions. Those villi are lined in a columnar manner similar to that in a real intestine [97]. Whether the human gut chip might be a potentially important application in CRC treatment remains to be demonstrated, but it is highly possible given its successful use in other complex disorders [96].

Along with an understanding of colonic epithelial cell behavior in the presence of microfabrication substrates, improved crypt isolation and 3D culture was an important step in the development of 'organ-on-chip' approaches for studies using primary colonic epithelium. Ahmad et al. reported on a protocol to standardize the isolation of intact murine colonic crypts by optimizing matrix concentrations on different surfaces i.e., PDMSs. The authors presented a reproducible low-cost crypt culture protocol, which may pave the way for further intestinal studies on patient-derived material using "organ-on-chip" platforms [98].

Concerning tumor-on-chip platforms, only a few studies are available, leaving great possibilities for further development. Carvalho et al. employed OOC technology to recapitulate the human colorectal tumor microenvironment, and assess the efficacy of gemcitabine loaded nanoparticles for the treatment of CRC using a microfluidic gradient [99]. In this device, the human CRC-like core, containing HCT-116 and HCoMECS cells, is surrounded by a vascularized microtissue, and serves as a tool to study radial drug penetration and efficacy of the microvascular network into a cancer-mimicking tissue. Although the oxygen gradient is not present in this device, its potential application is vast in CRC, or in solid tumors in general. This 3D microfluidic cell culture seems to be an extremely useful tool in the study of various phenomena, such as vascularization and oncogenesis under dynamic conditions. In the development of CRC, or its dissemination during the metastasis, the organ-on-chip-like microfluidic device has been developed [100]. This device, which merges microfluidics and photoconvertible protein technology, enables tracing the velocity of the circulating cells in the selectin-regulated process of adhesion and metastasis in a spatiotemporal manner.

Effectively, the OOC can be considered as a reliable platform to evaluate drug toxicity, given their high capacity to mimic in vivo like structures and functions. However, with these models, only the tissue or organ responses are considered without taking into account multi-organ interactions, which is crucial to evaluating the pharmacodynamics and the pharmacokinetics parameters of a drug [101]. To overcome this challenge, so called multiple-organs-on-chip (body-on-chip) were developed [102]. The latter technology recapitulates numerous organ interactions on a limited surface, while maintaining the highest degree of similarity to the in vivo situation. These systems usually do not require the use of pumps, using gravity to drive fluid flow to better replicate the physiologic flow. Oleaga et al. developed a system consisting of four different 2D tissue cultures (i.e., liver, cardiac, skeletal muscle and neuronal components), which were integrated within a single device to evaluate the toxicity of doxorubicin, valproic acid, acetaminophen and atorvastatin [103]. This phenotypic culture model exhibited a multi-organ toxicity response, representing the next generation of in vitro systems.

Esch et al. integrated liver and gastro-intestinal tract tissues within their device to evaluate intestinal barrier functions and metabolic rates of orally administered drugs and nutrients [104]. Fluidic flow through the organ chips was maintained via gravity and controlled passively via hydraulic resistances of the microfluidic channel network.

Another improvement of OOC was reported by Kassendra et al. on a generation of a "hybrid model" of OOC that integrated intestinal organoids, resulting in a higher similarity to the in vivo situation. They were able to recapitulate "normal" intestinal functions by integrating fluid flow and peristalsis-like motions, along with immune cells to a vascular compartment, which are all key factors to the normal intestinal physiology. In these culture conditions, biopsy-derived epithelial cells used were differentiated into villus-like epithelium (thin brush border of the colon epithelium). The primary human intestine chip model can be adapted for a wide range of applications, particularly in personalized medicine, by establishing a platform that could help investigate patient-specific disease mechanisms, as well as novel drug screening and anti-cancer therapy response [105].

3.4. Patient-Derived Organoids

Another recent development in human 3D in vitro technologies are the constructs derived from self-organizing stem cells that mimic the architecture, functionality and genetic feature of their corresponding organ [106]. The introduction of human patient-derived organoids (PDO) has enabled disease modelling, highlighting their great potential in biomedical applications, translational medicine and personalized therapy [107–109].

Moreover, PDO established from metastases taken by sequential biopsies at multiple time points, and multiple regions of heavily pre-treated CRC patients were used as pre-clinical models in co-clinical trials [110–112]. Those organoids were exposed to anti-cancer drugs, and the results were compared to patients' responses in clinical trials. The findings revealed the capacity of PDO to mimic in vivo tumor organization, at histopathological, molecular and functional levels, and to predict patient's treatment response [111].

Organoids can also be used to analyze mechanisms of drug resistance. Cancer stem cells expressing specific surface markers, such as CD44 and LGR5, known to be strongly associated with therapeutic drug resistance were co-cultured with epithelial and stromal cells to simulate the in vivo TME, with the use of an air-liquid interface (ALI) method [112,113]. ALI does not require exogenous growth factor supplementation and enables multilineage differentiation and sustained growth for over 60 days [112,114]. ALI patient-derived organoids presented higher resistance than CRC cell lines exposed to 5-fluorouracil and irinotecan (standard-of-care treatment of advanced CRC) [115].

Despite their advantages, PDO possess also shortcomings. Their self-organizing structure leads to different phenotypes between organoids, and might induce high background noise during drug-screening. The use of scaffolds and other bioengineering methods could help standardize cancer stem cell development into a specific and robust organoid morphology [116]. Lab-grown organoids showed some abnormalities, such as lack of cellular diversity and altered gene expression patterns [117]. Another important drawback of organoid development is time, as it takes several weeks to form a relevant organoid [111,118]. Moreover, the lack of some epithelial components, tumor stroma and microbiome remain major limitations of the PDO model [119]. Only very recent studies reported the incorporation of immune cells derived from CRC patient biopsies. The infiltration of immune cells was found to correlate with tumor growth and drug response. The ALI organoids could therefore be a promising approach to better understanding the tumor immune microenvironment and its impact on treatment response [120].

The PDO is an interesting in vitro model for preclinical drug development, due to its ability to mimic human physiopathology. As this still needs further technical and cost-effective improvements, it is unlikely that PDOs will fully replace existing drug development models.

3.5. Biomarkers-Based Drug Discovery

One other way to leverage tissue to predict drug sensitivity is to interrogate the tissue directly. For certain tumors, this approach has been a routine part of pathological assessment of patient samples. Breast cancers that express the receptor for estrogens and for progesterone are therefore rapidly and cheaply detected by immunohistochemistry, and can be treated effectively with one of the range of anti-estrogens available [121]. Similarly, B-cell lymphomas that express CD20 have had their prognosis transformed by the introduction of anti-CD20 therapeutic antibodies [122].

A recent extension of this approach has been an FDA-approved tumor treatment based on the tumor molecular characteristics, and not on the tissue origin or pathohistological type. This was based on the observation that patients, whose tumors have lost their mismatch repair machinery respond better to immunotherapy than patients with tumors, in which the machinery is intact [123]. The identification of this phenotype is routinely detected by immunohistochemistry. Other approaches, such as the evaluation of the immune response, tumor mutation burden and expression of programmed death-ligand (PD-L1) are also aimed at identifying patients whose tumors are likely to respond to immunotherapy [124].

With the democratization of molecular analysis of tumors many other anomalies have been and are being identified that can be targeted by specific therapies. The most established are the *EGFR* mutations in lung cancer [125] and *BRAF* mutations in melanoma with loss of the homology directed repair mechanism through loss of *BRCA1/2* or other associated genes being a more recent example [126]. However, at the moment, we are experiencing an explosion of new molecules that target different molecular abnormalities, a detailed review of which is beyond the scope of this review.

It is expected that, in the near future, we will witness a further expansion of our ability to characterize the phenotype of tumors, probably in the domain of expression analysis and proteomics through tissue analysis by mass cytometry [127]. These will allow the characterization of the pathways activated in different tumors, allowing the development of pathway instead of mutation directed therapies [128]. However, the high cost of the mass spectrometry remains the major constraint.

Moreover, Coppe et al. reported on a high-throughput kinase-activity mapping (HT-KAM) assay, which makes it possible to reveal the phosphor-catalytic signatures of tumors [129]. The HT-KAM is based on identifying catalytically hyper-active kinases in cell models or tissue, in order to highlight drug resistance and identify potential new drug targets. The authors synthesized a 228-peptide library that include 151 biological substrate protein regions phosphorylated by oncogenic kinases, and serve as combinatorial sensors of kinases phosphor-catalytic activity. Peptide phosphorylation signatures can be converted in kinase activity profiles, which will make it possible to identify the activity of druggable proto-oncogenic kinases in these models. This platform was tested to determine the new mechanism/targets of drug resistance in $BRAF^{V600E}$ CRC. In CRC cells (WiDr), which were exposed to treatment with a *BRAF* inhibitor vemurafenib [130], downregulation of the phospho-proteins MEK1/2 and ERK1/2 and upregulation in phospho-EGFR were observed, which was in line with the previously

The authors further created the 3D bioprinted pancreatic tumor model containing a PDX cell line surrounded by endothelial cells and pancreatic stellate cells (PSCs). The tumor tissue was treated for 6 days with gemcitabine (standard of care chemotherapy for pancreatic cancer), and results showed a dose-dependent response of cancer cell death. Furthermore, a patient-derived pancreatic tumor tissue, after enzymatic digestion, into the bioink surrounded by endothelial cells and PSCs. Bioprinted patient-derived cells maintained their tumorigenic properties, as confirmed by proliferation marker Ki67 staining, and showed similar morphological properties when compared to matching in vivo PDX or primary patient tumor [139].

Among the most common challenges faced by 3D bioprinting reside maintaining high cell viability and functionality, establishing constructs harboring in vivo like vascularization, obtaining resolution and reproducibility. Therefore, in order to improve cell viability during the bioprinting procedure, Colosi et al., employed a bioink that consists of a blend of alginate and gelatin methacroyl [140]. The authors optimized the formulation of a low viscosity bioink, which matches the physiological pH and osmolarity, and promotes cells adhesion as well as cellular migration, resulting in 80% of cell viability.

Furthermore, scientists from Rice University have recently developed a new bioprinting model that allows to create highly complex vascular networks, which recapitulate the body's natural passageways for blood, air, lymph and other vital fluids. Grigoryan et al. underlined the fact that their bioprint of a tissue was not only phenotypically similar to its healthy counterpart in the organism but also able to "breath" like the organism, through the oxygenation and flow of red blood cells through a complex distensible vascular network model [141].

3.7. Clinical Point of View

The clinical management of CRC has not majorly changed over the last two decades, as compared to other tumor types. The need for more representative preclinical models in the drug screening cascade is essential, as the attrition rate for anti-cancer drugs is very high especially for CRC. Most therapeutic agents developed mainly target VEGF (bevacizumab, aflibercept) or its receptors (regorafenib) or EGFR (cetuximab, panitumab). These discoveries have been made using cell lines and xenografts [142].

Each of the presented platforms possesses its own strengths and drawbacks in terms of study design and expected outcome (see Figure 2). Patient-derived cell lines cultured in 2D monolayers are simple to manipulate, and usually allow for large high-throughput screenings. The lack of tumor microenvironment (TME) is improved in 3D organoids/spheroids, and they often retain the characteristics of the original tumor including tumor heterogeneity and complexity. In contrast to in vitro platforms, patient-derived xenograft models tumor microenvironment, but ethical limitations and host background represent main drawbacks of these models. Cancer-on-chip models have been recently developed as more physiologically relevant platforms [93,119,143,144].

Interestingly, in the case of the $BRAF^{V600E}$, single or double inhibition (BRAF inhibitor and/or MEK inhibitor) has been unsuccessful in CRC treatment [130]. It was later shown that insensitivity to the double inhibition was due to a feedback activation of EGFR [145]. Current standard of care for $BRAF^{V600E}$-mutated CRC involve the triple inhibition of BRAF, MEK and EGFR [146]. This remains a major consideration that needs to be integrated through patient-derived models for drug discovery in CRC. This said, it is fair to assume that the potential treatment or combination of treatment options have been missed, due to the model selection. The determinants of such paucity are the choice of the model and lack of integration into the model of tumor heterogeneity.

There are several expectations to be addressed from the clinician point of view regarding a patient-derived preclinical model in CRC. First, the treatment resulting from this process has to be more efficient than the current standard-of-care with similar or inferior toxicity. Second, when dealing with de novo CRC diagnosis or metastatic disease, two weeks is an acceptable turnaround time [147]. From a clinical point of view, from the diagnosis to the initial treatment, time should be as short as possible. However, preoperative workup and pre-habilitation are often time-consuming, but remain mandatory.

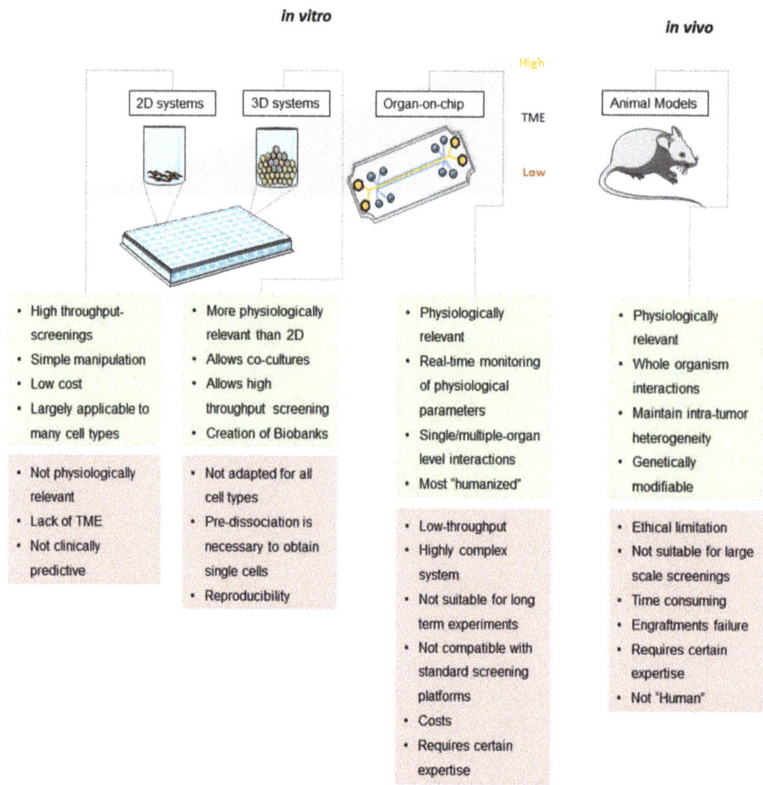

Figure 2. Advantages and weakness of CRC patient-derived models in preclinical studies.

An adequate preoperative staging is important when considering neoadjuvant treatment. Microscopic tissue assessment of CRC by a pathologist aims at describing the complex composition and architecture of the tumor. The recent development of computer-aided approaches helped in advances also in this discipline. A machine learning-based approaches for automated analysis of digitized microscopic images of CRC samples with the goal to improve prognostic stratification of patients are already available [148]. For colon cancer, preoperative chemotherapy or even radio-chemotherapy showed interesting results with less surgical complications, but no impact on disease relapsing [149]. On the other hand, the assessment of the resected surgical specimen is the cornerstone before starting any adjuvant treatment. The importance to evaluate precisely the tumor, node, metastasis (TNM) stage is obvious, as it determines whether the patient should receive adjuvant treatment or not. Tissue availability is thus limited in localized disease, compared to the metastatic setting. A good collaboration between the pathologist and the researcher, dealing with the presented models, is fundamental, in order to maintain a high level of quality for the tumor staging. The part devoted for the research should not alter this quality.

It is important to underline that we are entering a new era of data-driven medicine. This is what offers the next-generation DNA sequencing (NGS). NGS describes the high-throughput technology that allows the sequencing of the entire human genome within a single day [150]. NGS-based diagnostic assays have achieved clinical utility, on one hand, by being a solid platform for direct therapeutic decision-making [151]. Today, the NGS enables to cluster patients' tumor cases, based on their genomic profile, in virtual cohorts, in a way to determine whether the cancer of a specific patient is similar to that of another patient on the globe who received treatment A or B that saved them, matching genomic alterations with curated databases of evidence-based associations. Such platforms are being used in treatment of glioblastomas, lung, colorectal or gastrointestinal tumors, and can also be applied on liquid biopsy samples to help better diagnose, treat and monitor cancer in a less invasive manner.

Another important technological tool are the imaging techniques that allow 3D visualization of the patient-specific tumor phenotype with prognostic pre- and post-treatment relevance. New non-invasive imaging techniques, apart from structural evaluation, help in assessment of TME and certain hallmarks of cancer, as elegantly reviewed by García-Figueiras et al. [152].

Last but not least, the CRC in vitro model development might be supported by computer-aided approaches that facilitate experimental testing per se by guiding the researchers in the choice of tested conditions. It is extremely important especially in the context of CRC, where combinatory treatment approaches are mostly applied. Testing all combination options with multiple drugs is not trivial and requires a high time- and cost-effort. We have used the learning algorithm or statistical methods to lead experimental testing of multidrug combination candidates [153–156]. The generated data we modelled, made it possible to generate regression models that, in turn, enabled the elimination of ineffective and/or antagonistic compounds from the initial drug pool and led to identification of the most effective synergistic multidrug mixtures [155,156]. This approach brings the possibility of rapid patient-specific treatment optimization.

4. Conclusions

Despite the absence of an "ideal" preclinical model that would completely recapitulate the complexity of colorectal cancer with its stages and heterogeneity, as well as genomic signature, the choice of available models is wide-ranging. A careful decision on which model to employ should be taken, depending on a specific scientific question prior experimentation. A careful consideration of the advantages and shortcomings of each model, as presented above, should help in the most optimal model selection.

It is particularly important to mention that fundamental researchers should readily discuss the model choice with their clinical partners, i.e., oncologists, surgeons and pathologists, in order to secure the optimal conditions from tissue resection till its use in selected models. Already available in vitro models might provide very valuable information on several treatment aspects that can be further verified in more complex in vivo conditions. Model improvement should involve tumor phenotyping and genotyping (e.g., consensus molecular subtypes classifications), as well as a better representation of the TME [157]. Through the combination of multiple imaging along with biological and clinical information, the computer-aided-based platforms process the data using statistical models and the result is an accurate prediction of tumor growth and evolution. With the information gathered, eventual in vitro model can be further optimized through the characterization of a patient-specific tumor peculiarity.

Author Contributions: Conceptualization, P.N.-S.; Methodology, G.M.R.; Writing, G.M.R., T.K., P.N.-S.; Writing—Review & Editing, G.M.R., T.K., E.D., T.M., F.R., N.B., L.R.-B., P.-Y.D., P.N.-S.; Visualization, G.M.R.; Supervision, P.N.-S.; Project Administration, P.N.-S.; Funding Acquisition, P.N.-S. All authors have read and agreed to the published version of the manuscript.

Funding: We acknowledge the European Research Council (ERC-StG-680209 to P.N.-S.) and Foundation for the fight against cancer and for medico-biological research (to P.N.-S.) for funding.

Conflicts of Interest: The authors declare no competing interests.

References

1. Siegel, R.L.; A Torre, L.; Soerjomataram, I.; Hayes, R.B.; Bray, F.; Weber, T.K.; Jemal, A. Global patterns and trends in colorectal cancer incidence in young adults. *Gut* **2019**, *68*, 2179–2185. [CrossRef]

2. Bray, F.; Ferlay, J.; Soerjomataram, I.; Siegel, R.L.; Torre, L.A.; Jemal, A. Global cancer statistics 2018: GLOBOCAN estimates of incidence and mortality worldwide for 36 cancers in 185 countries. *CA Cancer J. Clin.* **2018**, *68*, 394–424. [CrossRef]

3. McQuade, R.; Stojanovska, V.; De Leiris, J.; Nurgali, K. Colorectal Cancer Chemotherapy: The Evolution of Treatment and New Approaches. *Curr. Med. Chem.* **2017**, *24*, 1537–1557. [CrossRef] [PubMed]

4. Centenera, M.M.; Raj, G.V.; Knudsen, K.E.; Tilley, W.D.; Butler, L.M. Ex vivo culture of human prostate tissue and drug development. *Nat. Rev. Urol.* **2013**, *10*, 483–487. [CrossRef] [PubMed]

5. Hutchinson, L.; Kirk, R. High drug attrition rates—Where are we going wrong? *Nat. Rev. Clin. Oncol.* **2011**, *8*, 189–190. [CrossRef] [PubMed]

6. Uronis, J.M.; Osada, T.; McCall, S.J.; Yang, X.Y.; Mantyh, C.; Morse, M.A.; Lyerly, H.K.; Clary, B.M.; Hsu, D.S. Histological and Molecular Evaluation of Patient-Derived Colorectal Cancer Explants. *PLoS ONE* **2012**, *7*, e38422. [CrossRef] [PubMed]

7. Burgenske, D.M.; Monsma, D.J.; MacKeigan, J. Patient-Derived Xenograft Models of Colorectal Cancer: Procedures for Engraftment and Propagation. In *Colorectal Cancer*; Humana Press: New York, NY, USA, 2018; pp. 307–314. [CrossRef]

8. Ji, X.; Chen, S.; Guo, Y.; Li, W.; Qi, X.; Yang, H.; Xiao, S.; Fang, G.; Hu, J.; Wen, C.; et al. Establishment and evaluation of four different types of patient-derived xenograft models. *Cancer Cell Int.* **2017**, *17*, 122. [CrossRef]

9. Bertotti, A.; Migliardi, G.; Galimi, F.; Sassi, F.; Torti, D.; Isella, C.; Corà, D.; Di Nicolantonio, F.; Buscarino, M.; Petti, C.; et al. A Molecularly Annotated Platform of Patient-Derived Xenografts ("Xenopatients") Identifies HER2 as an Effective Therapeutic Target in Cetuximab-Resistant Colorectal Cancer. *Cancer Discov.* **2011**, *1*, 508–523. [CrossRef]

10. Jung, J.; Kim, J.; Lim, H.K.; Kim, K.M.; Lee, Y.S.; Park, J.S.; Yoon, D.-S. Establishing a colorectal cancer liver metastasis patient-derived tumor xenograft model for the evaluation of personalized chemotherapy. *Ann. Surg. Treat. Res.* **2017**, *93*, 173–180. [CrossRef]

11. Maletzki, C.; Bock, S.; Fruh, P.; Macius, K.; Witt, A.; Prall, F.; Linnebacher, M. NSG mice as hosts for oncological precision medicine. *Lab. Investig.* **2019**, *100*, 27–37. [CrossRef] [PubMed]

12. Brown, K.M.; Xue, A.; Mittal, A.; Samra, J.S.; Smith, R.; Hugh, T.J. Patient-derived xenograft models of colorectal cancer in pre-clinical research: A systematic review. *Oncotarget* **2016**, *7*, 66212–66225. [CrossRef]

13. McIntyre, R.E.; Buczacki, S.J.; Arends, M.J.; Adams, D.J. Mouse models of colorectal cancer as preclinical models. *BioEssays* **2015**, *37*, 909–920. [CrossRef] [PubMed]

14. Gao, H.; Korn, J.M.; Ferretti, S.; Monahan, J.; Wang, Y.; Singh, M.; Zhang, C.; Schnell, C.; Yang, G.; Zhang, Y.; et al. High-throughput screening using patient-derived tumor xenografts to predict clinical trial drug response. *Nat. Med.* **2015**, *21*, 1318–1325. [CrossRef] [PubMed]

15. Okada, S.; Vaeteewoottacharn, K.; Kariya, R. Application of Highly Immunocompromised Mice for the Establishment of Patient-Derived Xenograft (PDX) Models. *Cells* **2019**, *8*, 889. [CrossRef] [PubMed]

16. Katsiampoura, A.; Raghav, K.; Jiang, Z.-Q.; Menter, D.G.; Varkaris, A.; Morelli, M.P.; Manuel, S.; Wu, J.; Sorokin, A.V.; Rizi, B.S.; et al. Modeling of Patient-Derived Xenografts in Colorectal Cancer. *Mol. Cancer Ther.* **2017**, *16*, 1435–1442. [CrossRef] [PubMed]

17. Rubio-Viqueira, B.; Jimeno, A.; Cusatis, G.; Zhang, X.; Iacobuzio-Donahue, C.; Karikari, C.; Shi, C.; Danenberg, K.; Danenberg, P.V.; Kuramochi, H.; et al. An In vivo Platform for Translational Drug Development in Pancreatic Cancer. *Clin. Cancer Res.* **2006**, *12*, 4652–4661. [CrossRef]

18. Hidalgo, M.; Bruckheimer, E.; RajeshKumar, N.V.; Garrido-Laguna, I.; De Oliveira, E.; Rubio-Viqueira, B.; Strawn, S.; Wick, M.J.; Martell, J.; Sidransky, D. A pilot clinical study of treatment guided by personalized tumorgrafts in patients with advanced cancer. *Mol. Cancer Ther.* **2011**, *10*, 1311–1316. [CrossRef]

19. Stebbing, J.; Paz, K.; Schwartz, G.K.; Wexler, L.H.; Maki, R.G.; Pollock, R.E.; Morris, R.; Cohen, R.; Shankar, A.; Blackman, G.; et al. Patient-derived xenografts for individualized care in advanced sarcoma. *Cancer* **2014**, *120*, 2006–2015. [CrossRef]

20. Fiore, D.; Di Giacomo, F.; Kyriakides, P.; Inghirami, G. Patient-Derived-Tumor-Xenograft: Modeling cancer for basic and translational cancer research. *Clin. Diagn. Pathol.* **2017**, *1*, 1. [CrossRef]

21. Pauli, C.; Hopkins, B.D.; Prandi, D.; Shaw, R.; Fedrizzi, T.; Sboner, A.; Sailer, V.; Augello, M.; Puca, L.; Rosati, R.; et al. Personalized In Vitro and In Vivo Cancer Models to Guide Precision Medicine. *Cancer Discov.* **2017**, *7*, 462–477. [CrossRef] [PubMed]

22. Chao, C.; Widen, S.G.; Wood, T.G.; Zatarain, J.R.; Johnson, P.; Gajjar, A.; Gomez, G.; Qiu, S.; Thompson, J.; Spratt, H.; et al. Patient-Derived Xenografts from Colorectal Carcinoma: A Temporal and Hierarchical Study of Murine Stromal Cell Replacement. *Anticancer. Res.* **2017**, *37*, 3405–3412. [CrossRef] [PubMed]

23. O'Rourke, K.P.; Loizou, E.; Livshits, G.; Schatoff, E.M.; Baslan, T.; Manchado, E.; Simon, J.; Romesser, P.B.; Leach, B.; Han, T.; et al. Transplantation of engineered organoids enables rapid generation of metastatic mouse models of colorectal cancer. *Nat. Biotechnol.* **2017**, *35*, 577–582. [CrossRef]

24. Céspedes, M.V.; Espina, C.; García-Cabezas, M.A.; Trias, M.; Boluda, A.; del Pulgar, M.T.G.; Sancho, F.J.; Nistal, M.; Lacal, J.C.; Mangues, R. Orthotopic Microinjection of Human Colon Cancer Cells in Nude Mice Induces Tumor Foci in All Clinically Relevant Metastatic Sites. *Am. J. Pathol.* **2007**, *170*, 1077–1085. [CrossRef] [PubMed]

25. Swamy, M.V.; Patlolla, J.M.; Steele, V.E.; Kopelovich, L.; Reddy, B.S.; Rao, C.V. Chemoprevention of Familial Adenomatous Polyposis by Low Doses of Atorvastatin and Celecoxib Given Individually and in Combination to APCMinMice. *Cancer Res.* **2006**, *66*, 7370–7377. [CrossRef] [PubMed]

26. Washington, M.K.; Powell, A.E.; Sullivan, R.; Sundberg, J.P.; A Wright, N.; Coffey, R.J.; Dove, W. Pathology of rodent models of intestinal cancer: Progress report and recommendations. *Gastroenterol.* **2013**, *144*, 705–717. [CrossRef]

27. Chen, J.; Huang, X.-F. The signal pathways in azoxymethane-induced colon cancer and preventive implications. *Cancer Boil. Ther.* **2009**, *8*, 1313–1317. [CrossRef]

28. Young, M.; Ordonez, L.; Clarke, A. What are the best routes to effectively model human colorectal cancer? *Mol. Oncol.* **2013**, *7*, 178–189. [CrossRef]

29. Golovko, D.; Kedrin, D.; Yilmaz, O.H.; Roper, J. Colorectal cancer models for novel drug discovery. *Expert Opin. Drug Discov.* **2015**, *10*, 1217–1229. [CrossRef]

30. Tannenbaum, J.; Bennett, B.T. Russell and Burch's 3Rs then and now: The need for clarity in definition and purpose. *J. Am. Assoc. Lab. Anim. Sci. JAALAS* **2015**, *54*, 120–132.

31. Navarro, A.M.; Susanto, E.; Falk, A.; Wilhelm, M. Modeling cancer using patient-derived induced pluripotent stem cells to understand development of childhood malignancies. *Cell Death Discov.* **2018**, *4*, 7. [CrossRef] [PubMed]

32. Curry, E.L.; Moad, M.; Robson, C.N.; Heer, R. Using induced pluripotent stem cells as a tool for modelling carcinogenesis. *World J. Stem Cells* **2015**, *7*, 461–469. [CrossRef] [PubMed]

33. Jesudoss, M.X.D.; Sachinidis, A. Current Challenges of iPSC-Based Disease Modeling and Therapeutic Implications. *Cells* **2019**, *8*, 403. [CrossRef]

34. Rowe, R.G.; Daley, G.Q. Induced pluripotent stem cells in disease modelling and drug discovery. *Nat. Rev. Genet.* **2019**, *20*, 377–388. [CrossRef]

35. Suprynowicz, F.A.; Upadhyay, G.; Krawczyk, E.; Kramer, S.C.; Hebert, J.D.; Liu, X.; Yuan, H.; Cheluvaraju, C.; Clapp, P.; Boucher, R.C.; et al. Conditionally reprogrammed cells represent a stem-like state of adult epithelial cells. *Proc. Natl. Acad. Sci. USA* **2012**, *109*, 20035–20040. [CrossRef]

36. Liu, X.; Krawczyk, E.; A Suprynowicz, F.; Palechor-Ceron, N.; Yuan, H.; Dakic, A.; Simic, V.; Zheng, Y.-L.; Sripadhan, P.; Chen, C.; et al. Conditional reprogramming and long-term expansion of normal and tumor cells from human biospecimens. *Nat. Protoc.* **2017**, *12*, 439–451. [CrossRef] [PubMed]

37. Chapman, S.; Liu, X.; Meyers, C.; Schlegel, R.; McBride, A.A. Human keratinocytes are efficiently immortalized by a Rho kinase inhibitor. *J. Clin. Investig.* **2010**, *120*, 2619–2626. [CrossRef] [PubMed]

38. Terunuma, A.; Limgala, R.P.; Park, C.J.; Choudhary, I.; Vogel, J.C. Efficient Procurement of Epithelial Stem Cells from Human Tissue Specimens Using a Rho-Associated Protein Kinase Inhibitor Y-27632. *Tissue Eng. Part A* **2010**, *16*, 1363–1368. [CrossRef] [PubMed]

39. Wang, Z.; Bi, B.; Song, H.; Liu, L.; Zheng, H.; Wang, S.; Shen, Z. Proliferation of human hepatocellular carcinoma cells from surgically resected specimens under conditionally reprogrammed culture. *Mol. Med. Rep.* **2019**, *19*, 4623–4630. [CrossRef] [PubMed]

40. Timofeeva, O.A.; Palechor-Ceron, N.; Li, G.; Yuan, H.; Krawczyk, E.; Zhong, X.; Liu, G.; Upadhyay, G.; Dakic, A.; Yu, S.; et al. Conditionally reprogrammed normal and primary tumor prostate epithelial cells: A novel patient-derived cell model for studies of human prostate cancer. *Oncotarget* **2016**, *8*, 22741–22758. [CrossRef]

41. Palechor-Ceron, N.; Krawczyk, E.; Dakic, A.; Simic, V.; Yuan, H.; Blancato, J.; Wang, W.; Hubbard, F.; Zheng, Y.L.; Dan, H.; et al. Conditional Reprogramming for Patient-Derived Cancer Models and Next-Generation Living Biobanks. *Cells* **2019**, *8*, 1327. [CrossRef]

42. Correa, B.R.S.; Hu, J.; Penalva, L.O.F.; Schlegel, R.; Rimm, D.L.; Galante, P.A.F.; Agarwal, S. Patient-derived conditionally reprogrammed cells maintain intra-tumor genetic heterogeneity. *Sci. Rep.* **2018**, *8*, 4097. [CrossRef]

43. Kodack, D.P.; Farago, A.F.; Dastur, A.; Held, M.A.; Dardaei, L.; Friboulet, L.; Von Flotow, F.; Damon, L.J.; Lee, D.-J.; Parks, M.; et al. Primary Patient-Derived Cancer Cells and Their Potential for Personalized Cancer Patient Care. *Cell Rep.* **2017**, *21*, 3298–3309. [CrossRef] [PubMed]

44. Dame, M.K.; Bhagavathula, N.; Mankey, C.; DaSilva, M.; Paruchuri, T.; Aslam, M.N.; Varani, J. Human colon tissue in organ culture: Preservation of normal and neoplastic characteristics. *Vitr. Cell. Dev. Boil. Anim.* **2010**, *46*, 114–122. [CrossRef] [PubMed]

45. Zirvi, K.A. Development of serum-free media for the growth of human gastrointestinal adenocarcinoma xenografts as primary tissue cultures. *J. Cancer Res. Clin. Oncol.* **1991**, *117*, 515–518. [CrossRef] [PubMed]

46. Sato, T.; Stange, D.E.; Ferrante, M.; Vries, R.G.; Van Es, J.H.; Brink, S.V.D.; Van Houdt, W.J.; Pronk, A.; Van Gorp, J.M.; Siersema, P.D.; et al. Long-term Expansion of Epithelial Organoids from Human Colon, Adenoma, Adenocarcinoma, and Barrett's Epithelium. *Gastroenterology* **2011**, *141*, 1762–1772. [CrossRef] [PubMed]

47. Miyoshi, H.; Maekawa, H.; Kakizaki, F.; Yamaura, T.; Kawada, K.; Sakai, Y.; Taketo, M.M. An improved method for culturing patient-derived colorectal cancer spheroids. *Oncotarget* **2018**, *9*, 21950–21964. [CrossRef]

48. Edmondson, R.; Broglie, J.J.; Adcock, A.F.; Yang, L. Three-Dimensional Cell Culture Systems and Their Applications in Drug Discovery and Cell-Based Biosensors. *ASSAY Drug Dev. Technol.* **2014**, *12*, 207–218. [CrossRef]

49. Zanoni, M.; Piccinini, F.; Arienti, C.; Zamagni, A.; Santi, S.; Polico, R.; Bevilacqua, A.; Tesei, A. 3D tumor spheroid models for in vitro therapeutic screening: A systematic approach to enhance the biological relevance of data obtained. *Sci. Rep.* **2016**, *6*, 19103. [CrossRef]

50. Cattin, S.; Ramont, L.; Ruegg, C. Characterization and In Vivo Validation of a Three-Dimensional Multi-Cellular Culture Model to Study Heterotypic Interactions in Colorectal Cancer Cell Growth, Invasion and Metastasis. *Front. Bioeng. Biotechnol.* **2018**, *6*, 97. [CrossRef]

51. Zoetemelk, M.; Rausch, M.; Colin, D.J.; Dormond, O.; Nowak-Sliwinska, P. Short-term 3D culture systems of various complexity for treatment optimization of colorectal carcinoma. *Sci. Rep.* **2019**, *9*, 7103. [CrossRef] [PubMed]

52. Jeppesen, M.; Hagel, G.; Glenthoj, A.; Vainer, B.; Ibsen, P.; Harling, H.; Thastrup, O.; Jørgensen, L.N.; Thastrup, J. Short-term spheroid culture of primary colorectal cancer cells as an in vitro model for personalizing cancer medicine. *PLoS ONE* **2017**, *12*, e0183074. [CrossRef] [PubMed]

53. Weiswald, L.-B.; Richon, S.; Validire, P.; Briffod, M.; Lai-Kuen, R.; Cordelières, F.P.; Bertrand, F.; Dargere, D.; Massonnet, G.; Marangoni, E.; et al. Newly characterised ex vivo colospheres as a three-dimensional colon cancer cell model of tumour aggressiveness. *Br. J. Cancer* **2009**, *101*, 473–482. [CrossRef] [PubMed]

54. Ashley, N.; Jones, M.; Ouaret, D.; Wilding, J.; Bodmer, W.F. Rapidly derived colorectal cancer cultures recapitulate parental cancer characteristics and enable personalized therapeutic assays. *J. Pathol.* **2014**, *234*, 34–45. [CrossRef]

55. Kondo, J.; Endo, H.; Okuyama, H.; Ishikawa, O.; Iishi, H.; Tsujii, M.; Ohue, M.; Inoue, M. Retaining Cell-Cell Contact Enables Preparation and Culture of Spheroids Composed of Pure Primary Cancer Cells from Colorectal Cancer. *Gastroenterology* **2011**, *140*, 6235–6240. [CrossRef]

56. Qureshi-Baig, K.; Ullmann, P.; Rodriguez, F.; Frasquilho, S.; Nazarov, P.V.; Haan, S.; Letellier, E. What Do We Learn from Spheroid Culture Systems? Insights from Tumorspheres Derived from Primary Colon Cancer Tissue. *PLoS ONE* **2016**, *11*, e0146052. [CrossRef]

57. Hoffmann, O.; Ilmberger, C.; Magosch, S.; Joka, M.; Jauch, K.-W.; Mayer, B. Impact of the spheroid model complexity on drug response. *J. Biotechnol.* **2015**, *205*, 14–23. [CrossRef]

58. Hirt, C.; Papadimitropoulos, A.; Mele, V.; Muraro, M.G.; Mengus, C.; Iezzi, G.; Terracciano, L.; Martin, I.; Spagnoli, G.C. "In vitro" 3D models of tumor-immune system interaction. *Adv. Drug Deliv. Rev.* **2014**, *79*, 145–154. [CrossRef]

59. Courau, T.; Bonnereau, J.; Chicoteau, J.; Bottois, H.; Remark, R.; Miranda, L.A.; Toubert, A.; Bléry, M.; Aparicio, T.; Allez, M.; et al. Cocultures of human colorectal tumor spheroids with immune cells reveal the therapeutic potential of MICA/B and NKG2A targeting for cancer treatment. *J. Immunother. Cancer* **2019**, *7*, 74. [CrossRef]

60. Failli, A.; Consolini, R.; Legitimo, A.; Spisni, R.; Castagna, M.; Romanini, A.; Crimaldi, G.; Miccoli, P. The challenge of culturing human colorectal tumor cells: Establishment of a cell culture model by the comparison of different methodological approaches. *Tumori J.* **2009**, *95*, 343–347. [CrossRef]

61. Fan, F.; Bellister, S.; Lü, J.; Ye, X.; Boulbès, D.R.; Tozzi, F.; Sceusi, E.; Kopetz, S.; Tian, F.; Xia, L.; et al. The requirement for freshly isolated human colorectal cancer (CRC) cells in isolating CRC stem cells. *Br. J. Cancer* **2014**, *112*, 539–546. [CrossRef] [PubMed]

62. Pereira, J.F.D.S.; Awatade, N.T.; Loureiro, C.; Matos, P.; Amaral, M.D.; Jordan, P. The third dimension: New developments in cell culture models for colorectal research. *Cell. Mol. Life Sci.* **2016**, *73*, 3971–3989. [CrossRef] [PubMed]

63. Meijer, T.G.; Naipal, K.A.; Jager, A.; Van Gent, D.C. Ex vivotumor culture systems for functional drug testing and therapy response prediction. *Future Sci. OA* **2017**, *3*, FSO190. [CrossRef] [PubMed]

64. Davies, E.J.; Dong, M.; Gutekunst, M.; Närhi, K.; Van Zoggel, H.J.A.A.; Blom, S.; Nagaraj, A.; Metsalu, T.; Oswald, E.; Erkens-Schulze, S.; et al. Capturing complex tumour biology in vitro: Histological and molecular characterisation of precision cut slices. *Sci. Rep.* **2015**, *5*, 17187. [CrossRef] [PubMed]

65. Brand, D.V.D.; Massuger, L.F.; Brock, R.; Verdurmen, W.P. Mimicking Tumors: Toward More Predictive In Vitro Models for Peptide- and Protein-Conjugated Drugs. *Bioconjug. Chem.* **2017**, *28*, 846–856. [CrossRef] [PubMed]

66. Bläuer, M.; Tammela, T.L.; Ylikomi, T. A novel tissue-slice culture model for non-malignant human prostate. *Cell Tissue Res.* **2008**, *332*, 489–498. [CrossRef]

67. Van De Merbel, M.; Van Der Horst, G.; Van Der Mark, M.H.; Van Uhm, J.I.M.; Van Gennep, E.J.; Kloen, P.; Beimers, L.; Pelger, R.C.M.; Van Der Pluijm, G. An ex vivo Tissue Culture Model for the Assessment of Individualized Drug Responses in Prostate and Bladder Cancer. *Front. Oncol.* **2018**, *8*. [CrossRef]

68. Carranza-Torres, I.E.; Guzmán-Delgado, N.E.; Coronado-Martínez, C.; Bañuelos-García, J.I.; Valdez, E.V.; Morán-Martínez, J.; Carranza-Rosales, P. Organotypic Culture of Breast Tumor Explants as a Multicellular System for the Screening of Natural Compounds with Antineoplastic Potential. *BioMed Res. Int.* **2015**, *2015*, 618021. [CrossRef]

69. Sönnichsen, R.; Hennig, L.; Blaschke, V.; Winter, K.; Körfer, J.; Hähnel, S.; Monecke, A.; Wittekind, C.; Jansen-Winkeln, B.; Thieme, R.; et al. Individual Susceptibility Analysis Using Patient-derived Slice Cultures of Colorectal Carcinoma. *Clin. Color. Cancer* **2018**, *17*, e189–e199. [CrossRef]

70. Martin, S.Z.; Wagner, D.C.; Hörner, N.; Horst, D.; Lang, H.; Tagscherer, K.E.; Roth, W. Ex vivo tissue slice culture system to measure drug-response rates of hepatic metastatic colorectal cancer. *BMC Cancer* **2019**, *19*, 1030. [CrossRef]

71. Zhang, Y.; Huang, W.; Yang, Q.; Zhang, H.; Zhu, X.; Zeng, M.; Zhou, X.; Wang, Z.; Li, W.; Jing, H.; et al. Cryopreserved biopsy tissues of rectal cancer liver metastasis for assessment of anticancer drug response in vitro and in vivo. *Oncol. Rep.* **2019**, *43*, 405–414. [CrossRef]

72. Siravegna, G.; Marsoni, S.; Siena, S.; Bardelli, A. Integrating liquid biopsies into the management of cancer. *Nat. Rev. Clin. Oncol.* **2017**, *14*, 531–548. [CrossRef] [PubMed]

73. Normanno, N.; Cervantes, A.; Ciardiello, F.; De Luca, A.; Pinto, C. The liquid biopsy in the management of colorectal cancer patients: Current applications and future scenarios. *Cancer Treat. Rev.* **2018**, *70*, 1–8. [CrossRef] [PubMed]

74. Gkountela, S.; Castro-Giner, F.; Szczerba, B.M.; Vetter, M.; Landin, J.; Scherrer, R.; Krol, I.; Scheidmann, M.C.; Beisel, C.; Stirnimann, C.U.; et al. Circulating Tumor Cell Clustering Shapes DNA Methylation to Enable Metastasis Seeding. *Cell* **2019**, *176*, 98–112.e14. [CrossRef] [PubMed]

75. Cabel, L.; Proudhon, C.; Gortais, H.; Loirat, D.; Coussy, F.; Pierga, J.-Y.; Bidard, F.-C. Circulating tumor cells: Clinical validity and utility. *Int. J. Clin. Oncol.* **2017**, *22*, 421–430. [CrossRef] [PubMed]

76. Bork, U.; Rahbari, N.N.; Schölch, S.; Reissfelder, C.; Kahlert, C.; Büchler, M.W.; Weitz, J.; Koch, M. Circulating tumour cells and outcome in non-metastatic colorectal cancer: A prospective study. *Br. J. Cancer* **2015**, *112*, 1306–1313. [CrossRef]

77. Aceto, N.; Bardia, A.; Miyamoto, D.T.; Donaldson, M.C.; Wittner, B.S.; Spencer, J.A.; Yu, M.; Pely, A.; Engstrom, A.; Zhu, H.; et al. Circulating tumor cell clusters are oligoclonal precursors of breast cancer metastasis. *Cell* **2014**, *158*, 1110–1122. [CrossRef]

78. Andree, K.C.; Van Dalum, G.; Terstappen, L.W.M.M. Challenges in circulating tumor cell detection by the CellSearch system. *Mol. Oncol.* **2015**, *10*, 395–407. [CrossRef]

79. Wang, L.; Balasubramanian, P.; Chen, A.P.; Kummar, S.; Evrard, Y.A.; Kinders, R.J. Promise and limits of the CellSearch platform for evaluating pharmacodynamics in circulating tumor cells. *Semin. Oncol.* **2016**, *43*, 464–475. [CrossRef]

80. Williams, S.C.P. Circulating tumor cells. *Proc. Natl. Acad. Sci. USA* **2013**, *110*, 4861. [CrossRef]

81. Gao, D.; Vela, I.; Sboner, A.; Iaquinta, P.J.; Karthaus, W.R.; Gopalan, A.; Dowling, C.; Wanjala, J.N.; Undvall, E.A.; Arora, V.K.; et al. Organoid cultures derived from patients with advanced prostate cancer. *Cell* **2014**, *159*, 176–187. [CrossRef]

82. Zhang, L.; Ridgway, L.D.; Wetzel, M.D.; Ngo, J.; Yin, W.; Kumar, D.; Goodman, J.C.; Groves, M.D.; Marchetti, D. The Identification and Characterization of Breast Cancer CTCs Competent for Brain Metastasis. *Sci. Transl. Med.* **2013**, *5*, 180ra48. [CrossRef] [PubMed]

83. Cayrefourcq, L.; Mazard, T.; Joosse, S.; Solassol, J.; Ramos, J.; Assenat, E.; Schumacher, U.; Costes-Martineau, V.; Maudelonde, T.; Pantel, K.; et al. Establishment and Characterization of a Cell Line from Human Circulating Colon Cancer Cells. *Cancer Res.* **2015**, *75*, 892–901. [CrossRef] [PubMed]

84. Souglakos, J.; Androulakis, N.; Syrigos, K.; Polyzos, A.; Ziras, N.; Athanasiadis, A.; Kakolyris, S.; Tsousis, S.; Kouroussis, C.; Vamvakas, L.; et al. FOLFOXIRI (folinic acid, 5-fluorouracil, oxaliplatin and irinotecan) vs FOLFIRI (folinic acid, 5-fluorouracil and irinotecan) as first-line treatment in metastatic colorectal cancer (MCC): A multicentre randomised phase III trial from the Hellenic Oncology Research Group (HORG). *Br. J. Cancer* **2006**, *94*, 798–805. [CrossRef] [PubMed]

85. Grillet, F.; Bayet, E.; Villeronce, O.; Zappia, L.; Lagerqvist, E.L.; Lunke, S.; Charafe-Jauffret, E.; Pham, K.; Molck, C.; Rolland, N.; et al. Circulating tumour cells from patients with colorectal cancer have cancer stem cell hallmarks inex vivoculture. *Gut* **2016**, *66*, 1802–1810. [CrossRef]

86. Farace, F.; Massard, C.; Vimond, N.; Drusch, F.; Jacques, N.; Billiot, F.; Laplanche, A.; Chauchereau, A.; Lacroix, L.; Planchard, D.; et al. A direct comparison of CellSearch and ISET for circulating tumour-cell detection in patients with metastatic carcinomas. *Br. J. Cancer* **2011**, *105*, 847–853. [CrossRef]

87. Zhou, J.; Kulasinghe, A.; Bogseth, A.; O'Byrne, K.; Punyadeera, C.; Papautsky, I. Isolation of circulating tumor cells in non-small-cell-lung-cancer patients using a multi-flow microfluidic channel. *Microsyst. Nanoeng.* **2019**, *5*, 8. [CrossRef]

88. Esch, E.W.; Bahinski, A.; Huh, D. Organs-on-chips at the frontiers of drug discovery. *Nat. Rev. Drug Discov.* **2015**, *14*, 248–260. [CrossRef]

89. Sontheimer-Phelps, A.; Hassell, B.A.; Ingber, D.E. Modelling cancer in microfluidic human organs-on-chips. *Nat. Rev. Cancer* **2019**, *19*, 65–81. [CrossRef]

90. Zhang, Y.S.; Zhang, Y.-N.; Zhang, W. Cancer-on-a-chip systems at the frontier of nanomedicine. *Drug Discov. Today* **2017**, *22*, 1392–1399. [CrossRef]

91. Akay, M.; Hite, J.; Avci, N.G.; Fan, Y.; Akay, Y.; Lu, G.; Zhu, J.-J. Drug Screening of Human GBM Spheroids in Brain Cancer Chip. *Sci. Rep.* **2018**, *8*, 15423. [CrossRef] [PubMed]

92. Liu, P.-F.; Cao, Y.-W.; Zhang, S.-D.; Zhao, Y.; Liu, X.-G.; Shi, H.-Q.; Hu, K.-Y.; Zhu, G.-Q.; Ma, B.; Niu, H.-T. A bladder cancer microenvironment simulation system based on a microfluidic co-culture model. *Oncotarget* **2015**, *6*, 37695–37705. [CrossRef] [PubMed]

93. Pradhan, S.; Smith, A.M.; Garson, C.J.; Hassani, I.; Seeto, W.J.; Pant, K.; Arnold, R.; Prabhakarpandian, B.; Lipke, E. A Microvascularized Tumor-mimetic Platform for Assessing Anti-cancer Drug Efficacy. *Sci. Rep.* **2018**, *8*, 3171. [CrossRef] [PubMed]

94. Yu, T.; Guo, Z.; Fan, H.; Song, J.; Liu, Y.; Gao, Z.; Wang, Q. Cancer-associated fibroblasts promote non-small cell lung cancer cell invasion by upregulation of glucose-regulated protein 78 (GRP78) expression in an integrated bionic microfluidic device. *Oncotarget* **2016**, *7*, 25593–25603. [CrossRef] [PubMed]

95. Zheng, Y.; Sun, Y.; Yu, X.; Shao, Y.; Zhang, P.; Dai, G.; Fu, J. Angiogenesis in Liquid Tumors: An In Vitro Assay for Leukemic-Cell-Induced Bone Marrow Angiogenesis. *Adv. Heal. Mater.* **2016**, *5*, 1014–1024. [CrossRef]

96. Bein, A.; Shin, W.; Jalili-Firoozinezhad, S.; Park, M.H.; Sontheimer-Phelps, A.; Tovaglieri, A.; Chalkiadaki, A.; Kim, H.J.; Ingber, D.E. Microfluidic Organ-on-a-Chip Models of Human Intestine. *Cell. Mol. Gastroenterol. Hepatol.* **2018**, *5*, 659–668. [CrossRef] [PubMed]

97. Kim, H.J.; Li, H.; Collins, J.J.; Ingber, D.E. Contributions of microbiome and mechanical deformation to intestinal bacterial overgrowth and inflammation in a human gut-on-a-chip. *Proc. Natl. Acad. Sci. USA* **2015**, *113*, E7–E15. [CrossRef]

98. Ahmad, A.A.; Wang, Y.; Gracz, A.D.; Sims, C.E.; Magness, S.T.; Allbritton, N.L. Optimization of 3-D organotypic primary colonic cultures for organ-on-chip applications. *J. Boil. Eng.* **2014**, *8*, 9. [CrossRef]

99. Carvalho, M.; Barata, D.; Teixeira, L.S.M.; Giselbrecht, S.; Reis, R.L.; Oliveira, J.M.; Truckenmüller, R.; Habibovic, P. Colorectal tumor-on-a-chip system: A 3D tool for precision onco-nanomedicine. *Sci. Adv.* **2019**, *5*, eaaw1317. [CrossRef]

100. Edwards, E.E.; Birmingham, K.G.; O'Melia, M.J.; Oh, J.; Thomas, S.N. Fluorometric Quantification of Single-Cell Velocities to Investigate Cancer Metastasis. *Cell Syst.* **2018**, *7*, 496–509.e6. [CrossRef]

101. Zhao, Y.; Kankala, R.K.; Wang, S.-B.; Chen, A.-Z. Multi-Organs-on-Chips: Towards Long-Term Biomedical Investigations. *Molecules* **2019**, *24*, 675. [CrossRef] [PubMed]

102. Skardal, A.; Shupe, T.; Atala, A. Organoid-on-a-chip and body-on-a-chip systems for drug screening and disease modeling. *Drug Discov. Today* **2016**, *21*, 1399–1411. [CrossRef]

103. Oleaga, C.; Bernabini, C.; Smith, A.S.; Srinivasan, B.; Jackson, M.; McLamb, W.; Platt, V.; Bridges, R.; Cai, Y.; Santhanam, N.; et al. Multi-Organ toxicity demonstration in a functional human in vitro system composed of four organs. *Sci. Rep.* **2016**, *6*, 20030. [CrossRef] [PubMed]

104. Esch, M.B.; Ueno, H.; Applegate, D.R.; Shuler, M.L. Modular, pumpless body-on-a-chip platform for the co-culture of GI tract epithelium and 3D primary liver tissue. *Lab Chip* **2016**, *16*, 2719–2729. [CrossRef] [PubMed]

105. Kasendra, M.; Tovaglieri, A.; Sontheimer-Phelps, A.; Jalili-Firoozinezhad, S.; Bein, A.; Chalkiadaki, A.; Scholl, W.; Zhang, C.; Rickner, H.; Richmond, C.A.; et al. Development of a primary human Small Intestine-on-a-Chip using biopsy-derived organoids. *Sci. Rep.* **2018**, *8*, 2871. [CrossRef]

106. Lou, Y.-R.; Leung, A. Next generation organoids for biomedical research and applications. *Biotechnol. Adv.* **2018**, *36*, 132–149. [CrossRef]

107. Dutta, D.; Heo, I.; Clevers, H. Disease Modeling in Stem Cell-Derived 3D Organoid Systems. *Trends Mol. Med.* **2017**, *23*, 393–410. [CrossRef]

108. Drost, J.; Clevers, H. Organoids in cancer research. *Nat. Rev. Cancer* **2018**, *18*, 407–418. [CrossRef]

109. Van De Wetering, M.; Francies, H.E.; Francis, J.M.; Bounova, G.; Iorio, F.; Pronk, A.; Van Houdt, W.; Van Gorp, J.; Taylor-Weiner, A.; Kester, L.; et al. Prospective derivation of a living organoid biobank of colorectal cancer patients. *Cell* **2015**, *161*, 933–945. [CrossRef]

110. Nardella, C.; Lunardi, A.; Patnaik, A.; Cantley, L.C.; Pandolfi, P.P. The APL Paradigm and the "Co-Clinical Trial" Project. *Cancer Discov.* **2011**, *2011*, 108–116. [CrossRef]

111. Vlachogiannis, G.; Hedayat, S.; Vatsiou, A.; Jamin, Y.; Fernández-Mateos, J.; Khan, K.; Lampis, A.; Eason, K.; Huntingford, I.; Burke, R.; et al. Patient-derived organoids model treatment response of metastatic gastrointestinal cancers. *Science* **2018**, *359*, 920–926. [CrossRef] [PubMed]

112. Li, X.; Ootani, A.; Kuo, C. An Air–Liquid Interface Culture System for 3D Organoid Culture of Diverse Primary Gastrointestinal Tissues. In *Gastrointestinal Physiology and Diseases: Methods and Protocols*; Ivanov, A.I., Ed.; Springer: New York, NY, USA, 2016; pp. 33–40. [CrossRef]

113. Katano, T.; Ootani, A.; Mizoshita, T.; Tanida, S.; Tsukamoto, H.; Ozeki, K.; Ebi, M.; Mori, Y.; Kataoka, H.; Kamiya, T.; et al. Establishment of a long-term three-dimensional primary culture of mouse glandular stomach epithelial cells within the stem cell niche. *Biochem. Biophys. Res. Commun.* **2013**, *432*, 558–563. [CrossRef] [PubMed]

114. Li, X.; Nadauld, L.; Ootani, A.; Corney, D.C.; Pai, R.K.; Gevaert, O.; Cantrell, M.A.; Rack, P.D.; Neal, J.; Chan, C.W.-M.; et al. Oncogenic transformation of diverse gastrointestinal tissues in primary organoid culture. *Nat. Med.* **2014**, *20*, 769–777. [CrossRef] [PubMed]

115. Elbadawy, M.; Usui, T.; Yamawaki, H.; Sasaki, K. Development of an Experimental Model for Analyzing Drug Resistance in Colorectal Cancer. *Cancers* **2018**, *10*, 164. [CrossRef]

116. Lancaster, M.A.; Corsini, N.S.; Wolfinger, S.; Gustafson, E.H.; Phillips, A.; Burkard, T.R.; Otani, T.; Livesey, F.J.; Knoblich, J.A. Guided self-organization and cortical plate formation in human brain organoids. *Nat. Biotechnol.* **2017**, *35*, 659–666. [CrossRef]

117. Urbischek, M.; Rannikmae, H.; Foets, T.; Ravn, K.; Hyvönen, M.; De La Roche, M. Organoid culture media formulated with growth factors of defined cellular activity. *Sci. Rep.* **2019**, *9*, 6193. [CrossRef]

118. Lancaster, M.A.; Huch, M. Disease modelling in human organoids. *Dis. Model. Mech.* **2019**, *12*, dmm039347. [CrossRef]

119. Sasaki, N.; Clevers, H. Studying cellular heterogeneity and drug sensitivity in colorectal cancer using organoid technology. *Curr. Opin. Genet. Dev.* **2018**, *52*, 117–122. [CrossRef]

120. Finnberg, N.K.; Gokare, P.; Lev, A.; Grivennikov, S.I.; Macfarlane, A.W.; Campbell, K.S.; Winters, R.M.; Kaputa, K.; Farma, J.M.; Abbas, A.E.-S.; et al. Application of 3D tumoroid systems to define immune and cytotoxic therapeutic responses based on tumoroid and tissue slice culture molecular signatures. *Oncotarget* **2017**, *8*, 66747–66757. [CrossRef]

121. Schettini, F.; Buono, G.; Cardalesi, C.; Desideri, I.; De Placido, S.; Del Mastro, L. Hormone Receptor/Human Epidermal Growth Factor Receptor 2-positive breast cancer: Where we are now and where we are going? *Cancer Treat. Rev.* **2016**, *46*, 20–26. [CrossRef]

122. Salles, G.; Barrett, M.; Foà, R.; Maurer, J.; O'Brien, S.; Valente, N.; Wenger, M.; Maloney, D.G. Rituximab in B-Cell Hematologic Malignancies: A Review of 20 Years of Clinical Experience. *Adv. Ther.* **2017**, *34*, 2232–2273. [CrossRef] [PubMed]

123. Ciombor, K.K.; Goldberg, R. Hypermutated Tumors and Immune Checkpoint Inhibition. *Drugs* **2018**, *78*, 155–162. [CrossRef] [PubMed]

124. Gibney, G.T.; Weiner, L.M.; Atkins, M.B. Predictive biomarkers for checkpoint inhibitor-based immunotherapy. *Lancet Oncol.* **2016**, *17*, e542–e551. [CrossRef]

125. Singh, M.; Jadhav, H.R. Targeting non-small cell lung cancer with small-molecule EGFR tyrosine kinase inhibitors. *Drug Discov. Today* **2018**, *23*, 745–753. [CrossRef]

126. Luke, J.J.; Flaherty, K.T.; Ribas, A.; Long, G.V. Targeted agents and immunotherapies: Optimizing outcomes in melanoma. *Nat. Rev. Clin. Oncol.* **2017**, *14*, 463–482. [CrossRef]

127. Giesen, C.; A O Wang, H.; Schapiro, D.; Zivanovic, N.; Jacobs, A.; Hattendorf, B.; Schüffler, P.; Grolimund, D.; Buhmann, J.M.; Brandt, S.; et al. Highly multiplexed imaging of tumor tissues with subcellular resolution by mass cytometry. *Nat. Methods* **2014**, *11*, 417–422. [CrossRef]

128. O'Connor, M.J. Targeting the DNA Damage Response in Cancer. *Mol. Cell* **2015**, *60*, 547–560. [CrossRef]

129. Coppé, J.-P.; Mori, M.; Pan, B.; Yau, C.; Wolf, D.M.; Ruiz-Saenz, A.; Brunen, D.; Prahallad, A.; Cornelissen-Steijger, P.; Kemper, K.; et al. Mapping phospho-catalytic dependencies of therapy-resistant tumours reveals actionable vulnerabilities. *Nature* **2019**, *21*, 778–790. [CrossRef]

130. Kopetz, S.; Desai, J.; Chan, E.; Hecht, J.R.; O'Dwyer, P.J.; Maru, D.; Morris, V.; Janku, F.; Dasari, A.; Chung, W.; et al. Phase II Pilot Study of Vemurafenib in Patients with Metastatic BRAF-Mutated Colorectal Cancer. *J. Clin. Oncol.* **2015**, *33*, 4032–4038. [CrossRef]

131. Xu, T.; Jin, J.; Gregory, C.; Hickman, J.J.; Boland, T. Inkjet printing of viable mammalian cells. *Biomaterials* **2005**, *26*, 93–99. [CrossRef]

132. Murphy, S.V.; Atala, A. 3D bioprinting of tissues and organs. *Nat. Biotechnol.* **2014**, *32*, 773–785. [CrossRef] [PubMed]

133. Moroni, L.; Burdick, J.A.; Highley, C.; Lee, S.J.; Morimoto, Y.; Takeuchi, S.; Yoo, J.J. Biofabrication strategies for 3D in vitro models and regenerative medicine. *Nat. Rev. Mater.* **2018**, *3*, 21–37. [CrossRef] [PubMed]

134. Hospodiuk, M.; Dey, M.; Sosnoski, D.; Ozbolat, I.T. The bioink: A comprehensive review on bioprintable materials. *Biotechnol. Adv.* **2017**, *35*, 217–239. [CrossRef] [PubMed]

135. Satpathy, A.; Datta, P.; Wu, Y.; Ayan, B.; Bayram, E.; Ozbolat, I.T. Developments with 3D bioprinting for novel drug discovery. *Expert Opin. Drug Discov.* **2018**, *13*, 1115–1129. [CrossRef] [PubMed]

136. Madden, L.R.; Nguyen, T.V.; Garcia-Mojica, S.; Shah, V.; Le, A.V.; Peier, A.; Visconti, R.; Parker, E.M.; Presnell, S.C.; Nguyen, D.G.; et al. Bioprinted 3D Primary Human Intestinal Tissues Model Aspects of Native Physiology and ADME/Tox Functions. *iScience* **2018**, *2*, 156–167. [CrossRef] [PubMed]

137. Sambuy, Y.; De Angelis, I.; Ranaldi, G.; Scarino, M.L.; Stammati, A.; Zucco, F. The Caco-2 cell line as a model of the intestinal barrier: Influence of cell and culture-related factors on Caco-2 cell functional characteristics. *Cell Boil. Toxicol.* **2005**, *21*, 1–26. [CrossRef]

138. McDonnell, A.M.; Dang, C. Basic Review of the Cytochrome P450 System. *J. Adv. Pract. Oncol.* **2013**, *4*, 263–268. [CrossRef]

139. Langer, E.M.; Allen-Petersen, B.; King, S.M.; Kendsersky, N.D.; Turnidge, M.A.; Kuziel, G.M.; Riggers, R.; Samatham, R.; Amery, T.S.; Jacques, S.L.; et al. Modeling Tumor Phenotypes In Vitro with Three-Dimensional Bioprinting. *Cell Rep.* **2019**, *26*, 608–623.e6. [CrossRef]

140. Colosi, C.; Shin, S.R.; Manoharan, V.; Massa, S.; Costantini, M.; Barbetta, A.; Dokmeci, M.R.; Dentini, M.; Khademhosseini, A. Microfluidic Bioprinting of Heterogeneous 3D Tissue Constructs Using Low-Viscosity Bioink. *Adv. Mater.* **2015**, *28*, 677–684. [CrossRef]

141. Grigoryan, B.; Paulsen, S.J.; Corbett, D.C.; Sazer, D.W.; Fortin, C.L.; Zaita, A.J.; Greenfield, P.T.; Calafat, N.J.; Gounley, J.; Ta, A.H.; et al. Multivascular networks and functional intravascular topologies within biocompatible hydrogels. *Science* **2019**, *364*, 458–464. [CrossRef]

142. Wilhelm, S.M.; Dumas, J.; Adnane, L.; Lynch, M.; Carter, C.A.; Schütz, G.; Thierauch, K.-H.; Zopf, D. Regorafenib (BAY 73-4506): A new oral multikinase inhibitor of angiogenic, stromal and oncogenic receptor tyrosine kinases with potent preclinical antitumor activity. *Int. J. Cancer* **2011**, *129*, 245–255. [CrossRef] [PubMed]

143. Fang, Y.; Eglen, R.M. Three-Dimensional Cell Cultures in Drug Discovery and Development. *SLAS Discov. Adv. Life Sci. R&D* **2017**, *22*, 456–472. [CrossRef]

144. Horvath, P.; Aulner, N.; Bickle, M.; Davies, A.M.; Del Nery, E.; Ebner, D.; Montoya, M.; Östling, P.; Pietiäinen, V.; Price, L.; et al. Screening out irrelevant cell-based models of disease. *Nat. Rev. Drug Discov.* **2016**, *15*, 751–769. [CrossRef] [PubMed]

145. Prahallad, A.; Sun, C.; Huang, S.; Di Nicolantonio, F.; Salazar, R.; Zecchin, D.; Beijersbergen, R.L.; Bardelli, A.; Bernards, R. Unresponsiveness of colon cancer to BRAF(V600E) inhibition through feedback activation of EGFR. *Nature* **2012**, *483*, 100–103. [CrossRef] [PubMed]

146. Van Cutsem, E.; Huijberts, S.; Grothey, A.; Yaeger, R.; Cuyle, P.-J.; Elez, E.; Fakih, M.; Montagut, C.; Peeters, M.; Yoshino, T.; et al. Binimetinib, Encorafenib, and Cetuximab Triplet Therapy for Patients With BRAF V600E–Mutant Metastatic Colorectal Cancer: Safety Lead-In Results From the Phase III BEACON Colorectal Cancer Study. *J. Clin. Oncol.* **2019**, *37*, 1460–1469. [CrossRef] [PubMed]

147. Rolfo, C.; Mack, P.C.; Scagliotti, G.V.; Baas, P.; Barlesi, F.; Bivona, T.G.; Herbst, R.S.; Mok, T.S.; Peled, N.; Pirker, R.; et al. Liquid Biopsy for Advanced Non-Small Cell Lung Cancer (NSCLC): A Statement Paper from the IASLC. *J. Thorac. Oncol.* **2018**, *13*, 1248–1268. [CrossRef] [PubMed]

148. Bychkov, D.; Linder, N.; Turkki, R.; Nordling, S.; Kovanen, P.E.; Verrill, C.; Walliander, M.; Lundin, M.; Haglund, C.; Lundin, J. Deep learning based tissue analysis predicts outcome in colorectal cancer. *Sci. Rep.* **2018**, *8*, 3395. [CrossRef]

149. FOxTROT Collaborative Group. Feasibility of preoperative chemotherapy for locally advanced, operable colon cancer: The pilot phase of a randomised controlled trial. *Lancet Oncol.* **2012**, *13*, 1152–1160. [CrossRef]

150. Kamps, R.; Brandão, R.; Bosch, B.J.V.D.; Paulussen, A.D.C.; Xanthoulea, S.; Blok, M.J.; Romano, A. Next-Generation Sequencing in Oncology: Genetic Diagnosis, Risk Prediction and Cancer Classification. *Int. J. Mol. Sci.* **2017**, *18*, 308. [CrossRef]

151. Berger, M.F.; Mardis, E.R. The emerging clinical relevance of genomics in cancer medicine. *Nat. Rev. Clin. Oncol.* **2018**, *15*, 353–365. [CrossRef]

152. Figueiras, R.G.; Baleato-González, S.; Padhani, A.R.; Luna-Alcalá, A.; Marhuenda, A.; Vilanova, J.C.; Osorio-Vázquez, I.; Martinez-De-Alegria, A.; Gomez-Caamaño, A. Advanced Imaging Techniques in Evaluation of Colorectal Cancer. *Radiographics* **2018**, *38*, 740–765. [CrossRef]

153. Nowak-Sliwinska, P.; Weiss, A.; Ding, X.; Dyson, P.J.; Bergh, H.V.D.; Griffioen, A.W.; Ho, C.-M. Optimization of drug combinations using Feedback System Control. *Nat. Protoc.* **2016**, *11*, 302–315. [CrossRef] [PubMed]

154. Ding, X.; Liu, W.; Weiss, A.; Li, Y.; Wong, I.; Griffioen, A.W.; Bergh, H.V.D.; Xu, H.; Nowak-Sliwinska, P.; Ho, C.-M. Discovery of a low order drug-cell response surface for applications in personalized medicine. *Phys. Boil.* **2014**, *11*, 065003. [CrossRef] [PubMed]

155. Weiss, A.; Nowak-Sliwinska, P. Current Trends in Multidrug Optimization: An Alley of Future Successful Treatment of Complex Disorders. *SLAS Technol. Transl. Life Sci. Innov.* **2016**, *22*, 254–275. [CrossRef] [PubMed]

156. Weiss, A.; Le Roux-Bourdieu, M.; Zoetemelk, M.; Ramzy, G.; Rausch, M.; Harry, D.; Miljkovic-Licina, M.; Falamaki, K.; Wehrle-Haller, B.; Meraldi, P.; et al. Identification of a Synergistic Multi-Drug Combination Active in Cancer Cells via the Prevention of Spindle Pole Clustering. *Cancers* **2019**, *11*, 1612. [CrossRef] [PubMed]

157. Prasetyanti, P.R.; Van Hooff, S.R.; Van Herwaarden, T.; De Vries, N.; Kalloe, K.; Rodermond, H.; Van Leersum, R.; De Jong, J.H.; Franitza, M.; Nürnberg, P.; et al. Capturing colorectal cancer inter-tumor heterogeneity in patient-derived xenograft (PDX) models. *Int. J. Cancer* **2018**, *144*, 366–371. [CrossRef]

 *cancers*

Article

Sequential Isolation and Characterization of Single CTCs and Large CTC Clusters in Metastatic Colorectal Cancer Patients

Federica Francescangeli [1], Valentina Magri [2], Maria Laura De Angelis [1], Gianluigi De Renzi [3], Orietta Gandini [3], Ann Zeuner [1], Paola Gazzaniga [3,†] and Chiara Nicolazzo [3,*,†]

1 Department of Oncology and Molecular Medicine, Istituto Superiore di Sanità, Viale Regina Elena 299, 00161 Rome, Italy; federica.francescangeli@iss.it (F.F.); marialaura.deangelis@iss.it (M.L.D.A.); a.zeuner@iss.it (A.Z.)
2 Department of Radiology, Oncology and Pathology, Sapienza University of Rome, Viale Regina Elena 324, 00161 Rome, Italy; valentina.magri@uniroma1.it
3 Cancer Liquid Biopsy Unit, Department of Molecular Medicine, Sapienza University of Rome, 00161 Rome, Italy; gianluigi.derenzi@uniroma1.it (G.D.R.); orietta.gandini@uniroma1.it (O.G.); paola.gazzaniga@uniroma1.it (P.G.)
* Correspondence: chiara.nicolazzo@uniroma1.it; Tel.: +39-06-499-733-434
† These authors share senior authorship.

Simple Summary: The presence of cancer cells clusters is a frequent event capable of increasing their aptitude to survive in the bloodstream. Consistently, clusters ranging from 2–50 cancer cells are detected in about 50% of patients with metastatic cancers, including colorectal carcinoma. Although a deepened analysis of clusters might certainly offer new insights into the complexity of metastatic cascade, research in this field has come to a halt, since most circulating tumor cells isolation techniques are not compatible with large-sized clusters isolation. In the present study, we describe a sequential method to simultaneously isolate single and clustered circulating tumor cells from a single blood draw, opening new scenarios for an ever more precise characterization of colorectal cancer metastatic cascade.

Citation: Francescangeli, F.; Magri, V.; De Angelis, M.L.; De Renzi, G.; Gandini, O.; Zeuner, A.; Gazzaniga, P.; Nicolazzo, C. Sequential Isolation and Characterization of Single CTCs and Large CTC Clusters in Metastatic Colorectal Cancer Patients. *Cancers* **2021**, *13*, 6362. https://doi.org/10.3390/cancers13246362

Academic Editor: Galatea Kallergi

Received: 11 November 2021
Accepted: 15 December 2021
Published: 18 December 2021

Publisher's Note: MDPI stays neutral with regard to jurisdictional claims in published maps and institutional affiliations.

Abstract: Circulating tumor cells (CTCs) detach from a primary tumor or its metastases and circulate in the bloodstream. The vast majority of CTCs are deemed to die into the bloodstream, with only few cells representing viable metastatic precursors. Particularly, single epithelial CTCs do not survive long in the circulation due to the loss of adhesion-dependent survival signals. In metastatic colorectal cancer, the generation of large CTC clusters is a very frequent occurrence, able to increase the aptitude of CTCs to survive in the bloodstream. Although a deepened analysis of large-sized CTC clusters might certainly offer new insights into the complexity of the metastatic cascade, most CTC isolation techniques are unfortunately not compatible with large-sized CTC clusters isolation. The inappropriateness of standard CTC isolation devices for large clusters isolation and the scarce availability of detection methods able to specifically isolate and characterize both single CTCs and CTC clusters finally prevented in-depth studies on the prognostic and predictive value of clusters in clinical practice, unlike that which has been described for single CTCs. In the present study, we validated a new sequential filtration method for the simultaneous isolation of large CTC clusters and single CTCs in patients with metastatic colorectal cancer at failure of first-line treatments. The new method might allow differential downstream analyses for single and clustered CTCs starting from a single blood draw, opening new scenarios for an ever more precise characterization of colorectal cancer metastatic cascade.

Keywords: circulating tumor cells; CTC cluster; colorectal cancer; size-based method; ScreenCell®; epithelial mesenchymal transition; hypoxia; HIF-1α; immunofluorescence analysis; sequential filtration

1. Introduction

Circulating tumor cells (CTCs) detach from a primary tumor or its metastases and circulate in the bloodstream [1]. Beyond the unquestionable prognostic value of the number of CTCs in patients with metastatic solid tumors, a detailed molecular characterization of CTCs is critical to improve our understanding of key pathways that mediate the dissemination of cancer cells [2,3]. The vast majority of CTCs are deemed to die into the bloodstream, with only few cells representing viable metastatic precursors [4]. Particularly, single epithelial CTCs do not survive long in the circulation due to the loss of adhesion-dependent survival signals [5]. Therefore, the interaction with other CTCs generating CTC clusters has been described as a frequent event able to increase their aptitude to survive in the bloodstream [6]. Consistently, clusters of CTCs ranging from 2–50 cancer cells are detected in about 50% of patients with metastatic cancers, and are associated with worse prognosis [7]. In colorectal cancer, the presence of CTC clusters has been widely described and correlated with elevated circulating levels of transforming growth factor-β (TGF-β) [8]. Recent studies suggest that CTC clusters and single CTCs display distinct gene expression profiles and molecular features, which might account for their different metastatic propensity [2]. Transcriptome analyses have shown that CTC clusters often display mixed epithelial and mesenchymal features and that proteins involved in desmosome junctions, such as plakoglobin, are preferentially expressed in clusters compared to single cells [9]. Moreover, some evidence has been provided that large CTC clusters are protected from reactive oxygen species by activating the metabolic switch to glycolysis through hypoxia-inducible factor-1α (HIF-1α) [10]. Although a deepened analysis of CTC clusters might certainly offer new insights into the complexity of the metastatic cascade, research in this field has come to a halt, since most CTC isolation techniques are not compatible with large-sized CTC clusters isolation. In this regard, despite the large body of evidence that has been provided showing that CTC clusters are usually enriched in mesenchymal markers, the clinical significance of CTC clusters has been mostly demonstrated using the CellSearch® (Menarini Silicon Biosystems, Castel Maggiore, BO, Italy), an antigen-dependent method able to isolate only clusters with epithelial features, missing CTC clusters undergoing epithelial–mesenchymal transition (EMT) [11]. Apart from EMT, other explanations, such as clusters' disruption in devices with turbulent flow, might account for their underestimation when using antigen-dependent assays. The failure of antigen-dependent methods to capture CTC clusters paved the way for alternative antigen-independent methods for CTC isolation [12]. Among them, ScreenCell® (ScreenCell, Sarcelles, France) is a filtration method allowing the isolation of CTCs by size using a filter with 6.5 to 8 μm pores. The rationale is that CTCs are generally larger in size than hematopoietic cells, so most of these cells pass through the filter whereas CTCs and clusters are retained [13]. Different downstream analyses such as immunocytochemistry, immunofluorescence, DNA or RNA extraction for genomic study can be directly performed on the filter in order to characterize CTCs [14]. However, in order to perform differential downstream analyses for single CTCs and CTC clusters, a single filter is not sufficient, making it necessary for this purpose to increase the starting blood volume to obtain more filters. In the present study we described a new method for the simultaneous isolation of CTC clusters and single CTCs from a single blood draw through a sequential filtration, using adapted ScreenCell® filters with increased pore size. We validated the assay in a small population of patients with metastatic colorectal cancer at failure of first line treatments.

2. Materials and Methods

2.1. Blood Samples Collection

Ten patients with metastatic colorectal cancer at failure of first-line treatments have been enrolled. For each patient, peripheral blood was collected into a K_2EDTA tube, stored at +4 °C and processed within 3 h after drawing. Informed consent was obtained from all participants included in the study. The protocol was approved by Ethical Committee of Policlinico Umberto I (protocol n. 668/09, 9 July 2009; amended protocol 179/16, 1 March 2016). Characteristics of CTC-positive patient population are shown in Table 1.

Table 1. Patient characteristics.

Characteristics	No. (%)
Age (in years)	
Mean	67.7
Range	55–84
PS	
0	4 (57)
1	3 (43)
Sex	
Male	3 (43)
Female	4 (57)
Colorectal cancer stage	
IV	7 (100)
Right-sided	4 (57)
Left-sided	3 (43)
Mutations	
RAS	4 (57)
BRAF	1 (14)

PS: performance status.

2.2. Establishment of a Customized Filtration Method for the Isolation of CTCs Clusters

ScreenCell® (ScreenCell, Sarcelles, France) is a simple, non-invasive technology for isolating CTCs from whole blood. The ScreenCell® filtration devices were developed in order to isolate CTCs by size on a microporous membrane filter. These devices are designed for isolation of: (a) fixed cells for cytological studies (ScreenCell® Cyto); (b) live cells for culture (ScreenCell® CC) and (c) molecular biology (ScreenCell® MB) [14]. The filter allows fast and regular filtration, preserving the CTCs morphology and structures. At the end of filtration, the ScreenCell® Cyto filter is released onto a standard microscopy glass slide. Cytological studies including staining, cell enumeration, immunocytochemistry and FISH assays, can then be conducted directly on the filter. The circular filter of the ScreenCell® device is composed of polycarbonate material, with a smooth flat and hydrophilic surface. Circular pores are calibrated for isolation of fixed or live cells and randomly distributed throughout the filter (1×10^5 pores/cm^2). In order to enable the selective filtration of large CTC clusters we aimed to modify the size of the pores, increasing it to 15 μm size. These adapted devices were referred as ScreenCell Cyto-Cl.

2.3. Sequential Isolation of Single CTCs and of CTC Clusters

In order to simultaneously isolate single CTCs and CTC clusters, sequential filtration was performed first using the new ScreenCell Cyto-Cl kit specifically designed and adapted to isolate cell clusters and then the ScreenCell® Cyto kit to isolate single cells. Blood samples were collected using tubes containing K_2EDTA, stored at +4 °C and processed within 3 h. Briefly, 3 mL of blood was diluted in 4 mL of fixed cells (FC2) dilution buffer allowing lysis of red blood cells (RBC) while preserving other cells. After 8 min of incubation at room temperature, 7 mL of diluted sample was put into device tank of ScreenCell Cyto-Cl device and filtered under a pressure gradient created by a vacutainer tube. After washing with phosphate-buffered saline (PBS) to remove RBC debris, the filter was left on absorbing paper to dry at room temperature. Thereafter, all clusters-depleted blood samples were filtrated using ScreenCell® Cyto device to isolate residual single CTCs. After washing with PBS, the filter was left on absorbing paper to dry at room temperature. Filters were stored at −20 °C until downstream analysis. Each filtration was usually completed within 3 min.

2.4. Immunofluorescence Staining

For immunofluorescence, filters were hydrated with Tris-Buffered Saline (TBS) for 10 min and directly stained with antihuman biotinylated CD45 (#130-098-551, Miltenyi Biotec, Bologna, Italy) in order to eliminate hematopoietic cells as follows: filters were washed twice in TBS 0.002% Tween20, endogenous peroxidase activities were blocked using 0.03% hydrogen peroxide for 15 min in the dark, then the sections were incubated at room temperature for 1 h 30 min with CD45 biotinylated antibody. Sections were then processed using streptavidin conjugated to horseradish peroxidase and substrate–chromogen solution both contained in UltraTek HRP Anti-Polyvalent DAB kit (#AMF080, Scytek Laboratories, Logan, UT, USA), following the manufacturer's instructions. Samples were then incubated in a humid chamber overnight at +4 °C with the following primary antibodies goat antihuman CK20 (#SC-17113, Santa Cruz Biotechnology, Dallas, TX, USA), rabbit antihuman HIF-1α (#36169, Cell Signaling Technology, Danvers, MA, USA) and mouse antihuman Vimentin (#SC-373717, Santa Cruz Biotechnology) in a humid chamber overnight at +4 °C. The filters were then washed twice in PBS and incubated with a mixture of appropriate secondary antibodies: donkey anti-mouse IgG Alexa Fluor®488-conjugated (#A21202), donkey anti-goat IgG Alexa Fluor®647-conjugated (#A21447) and donkey anti-rabbit IgG Alexa Fluor®555-conjugated (#A31572) for 45 min at room temperature in the dark. Nuclei were stained with DAPI for 15 min at room temperature. All antibodies were dissolved in PBS containing 3% bovine serum albumin (BSA), 3% fetal bovine serum (FBS), 0.001% NaN$_3$ and 0.1% Triton X-100. Finally, the filters were mounted with Prolong-Gold Antifade (Thermo Fisher Scientific, Waltham, MA, USA) on slides and analyzed using a Zeiss LSM900 confocal microscope or an Olympus FV1000 confocal microscope equipped with 60× oil immersion objectives.

3. Results

We sought to investigate the efficacy of the double filtration system in isolating all circulating tumor cells, including large clusters, regardless of surface markers. All blood samples were successfully filtered. In seven blood samples we were able to detect 42 single CTCs, with a range in number from 3 to 9 per 3 mL of blood, and 31 CTC clusters with large dimension, with a range 3 to 6 per 3 mL of blood, as shown in Table 2. Interestingly, both CTC and CTC clusters were detected in these patients. Conversely, in three patients we were not able to detect CTCs or CTCs clusters.

Table 2. CTCs and CTC clusters detection through the sequential filtration.

Patient	CTC				CTC Cluster			
	N_T	CK20 (*N*)	HIF-1α (*N*)	VIM (*N*)	N_T	CK20 (*N*)	HIF-1α (*N*)	VIM (*N*)
14AA6844	7	6	2	2	5	1	4	4
14AA6865	9	7	3	4	3	1	3	3
14AA6922	6	5	2	3	4	1	4	3
15AA0421	3	3	1	1	6	2	5	4
15AA0433	8	7	3	3	6	2	6	6
15AA0814	5	4	1	2	3	1	2	2
15AA0924	4	3	1	2	4	1	2	2

CTC: circulating tumor cell; CK: cytokeratin; HIF: hypoxia-inducible factor; VIM: vimentin; *N*: number; T: total.

Hypoxia and EMT-like features were investigated in order to assess whether EMT was associated with HIF-1α in both single and clustered CTCs. For this purpose, a triple immunofluorescence staining for CK20, vimentin and HIF-1α was carried out (Figure 1). Hematopoietic cells were preventively excluded by staining each filter for CD45, as shown in Figure S1. According to our hypothesis, CK20 was predominantly expressed in single CTCs. In fact, the antigen was found expressed in 83.3% out of the 42 single CTCs analyzed, whilst it was found expressed in 29% out of the 31 CTC clusters (Table 2). Conversely, vimentin and HIF-1α were mostly detected in CTC clusters. Indeed, HIF-1α and vimentin were found expressed in 84 and 77% out of the 31 CTC clusters, respectively; whilst HIF-1α was detected in 31% out of the 42 single CTCs and vimentin in 40.5% (Table 2).

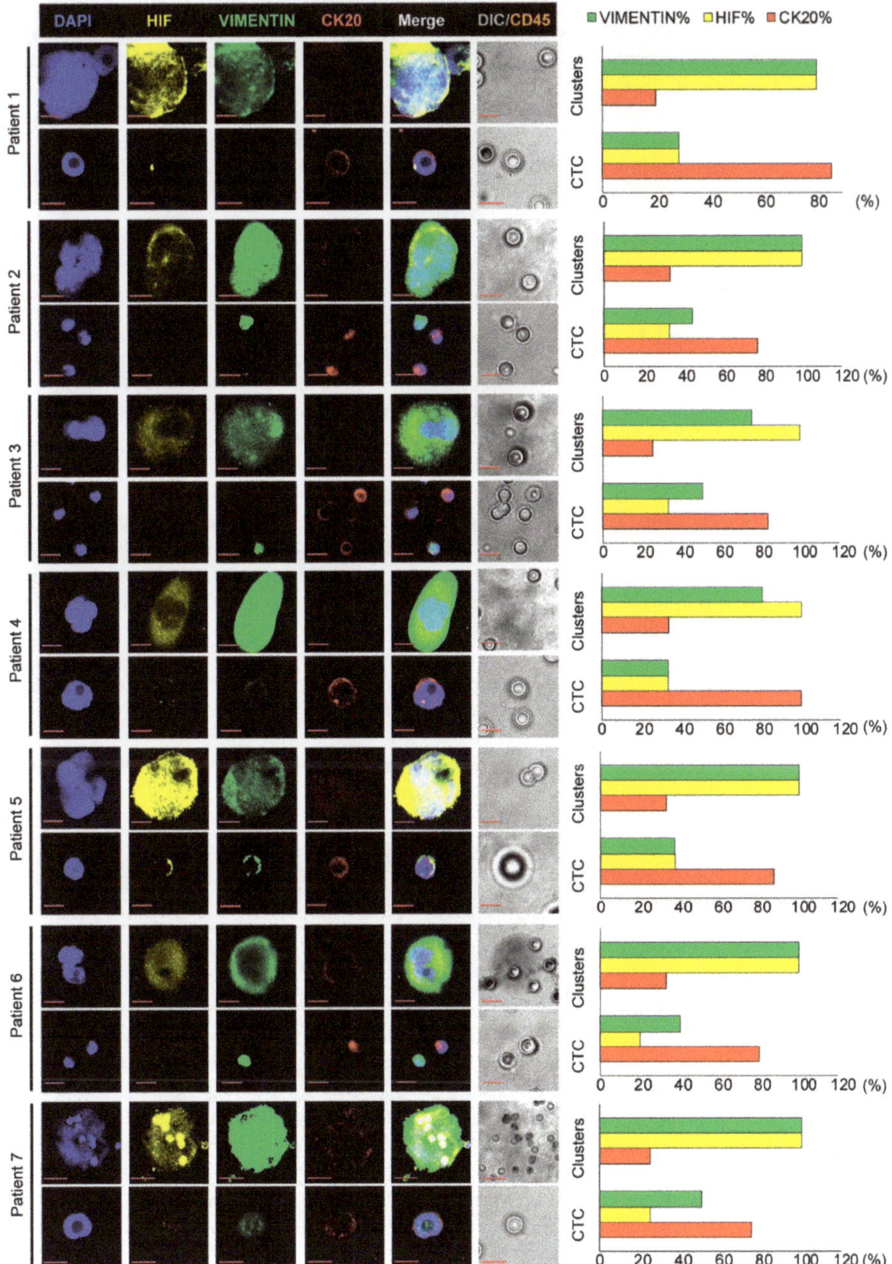

Figure 1. Illustrative images of triple immunofluorescence assay on clusters and circulating colon cancer cells. (**Left**) Representative confocal images of CTC clusters and CTCs stained with anti-CK20 (**red**), anti-vimentin (**green**) and anti-HIF-1α (**yellow**) antibodies. (**Right**) Graphical representation of percent of single or clustered CTCs expressing CK20 (**red bars**), vimentin (**green bars**), HIF-1α (**yellow bars**). Magnification 60×, 5× zoom bar 10 μm. CTC: circulating tumor cell; CK: cytokeratin; HIF: hypoxia-inducible factor.

Clusters showed a different phenotype compared to CTCs, by reflecting hybrid-EMT features, with a poor or barely detectable CK20 expression (Figures 1left and S2) in a percentage ranging from 20 to 33 as shown by graphs on the right (red bar); while Vimentin and HIF increased their expression ranging from 75 to 100 (%) and from 80 to 100 (%), respectively, as shown in images on the left of Figure 1. These data suggest a role for EMT and HIF-1α in large cluster organization. Altogether, these observations indicate that this filtration system is valid and effective, allowing collection and analysis even of large clusters that would have been excluded by epithelial antibodies-based methods for CTC detection.

4. Discussion

Circulating tumor cells clusters represent a peculiar class of CTCs, with specific properties including reduced apoptosis, enhanced survival and high metastatic potential. Unlike single CTCs, CTC clusters have not been deeply investigated, mainly due to the paucity of reliable detection methods [15]. In fact, most assays for CTC clusters enrichment, which depend on epithelial specific markers such as cytokeratins (CKs), and the epithelial cell adhesion molecule (EpCAM), usually underestimate CTC clusters due to their hybrid epithelial–mesenchymal features. A further limitation of antibody-based methods, such as the FDA-approved CellSearch®, is that larger CTC clusters have a small area-to-volume ratio, thus preventing correct binding to the antibodies used in the enrichment step [16]. The inappropriateness of standard CTCs isolation devices for clusters isolation and the scarce availability of detection methods able to specifically isolate and characterize both single CTCs and CTC clusters finally banned in-depth studies on the prognostic and predictive value of clusters in clinical practice, unlike that which has been described for single CTCs. Although evidence has been provided that size-exclusion assays, such as blood filtration, would represent an affordable approach for CTC clusters isolation, only few filtration devices can simultaneously detect single and clustered CTCs starting from a single blood sample [17]. Here we used a sequential filtration-based approach to investigate the simultaneous presence of single CTCs and CTC clusters in a small group of patients with metastatic colorectal cancer. The ScreenCell® technology, which we had previously used for single CTCs isolation and characterization in non-small cell lung cancer (NSCLC) and colorectal cancer [18,19], was adapted for CTC clusters isolation by increasing the filter pore size. This innovative double-step filtration allows a rapid, easy and simultaneous enrichment of both single and clustered CTCs from a single blood draw, obtaining two filters that can be easily subjected to specific downstream analyses. In this pilot study we sought to investigate the efficacy of the double filtration method to simultaneously isolate single and clustered CTCs, and to clarify whether they might display distinct molecular features, mainly in terms of hybrid EMT-related characteristics. The double-filtration method described herein allowed us to isolate in all patients both single CTCs and large clusters, confirming what we and others have previously demonstrated, namely that in colon carcinoma the presence of large clusters is a very frequent phenomenon, usually associated with higher levels of TGF-β in circulation [20,21]. Consistently with literature studies, CTC clusters display manifest hybrid-EMT features compared to single CTCs, as demonstrated by the constant expression in CTC clusters of vimentin and CK20, with vimentin always expressed to a much higher extent. The choice to include HIF-1α in the triple immunofluorescence experiments was dictated by the unequivocal role that hypoxia plays in the generation of CTC clusters [22]. In fact, recent studies have demonstrated that hypoxic cancer cells are characterized by upregulated cell–cell junction components, a property that seems to be associated with their propensity to frequently intravasate as clusters rather than as individual CTCs [23]. Our results confirm that, in all the patients analyzed, single CTCs significantly differ from clusters in terms of HIF-1α expression, HIF-1α being constantly expressed in large clusters, but not in single CTCs.

5. Conclusions

This liquid biopsy test seems promising for the future isolation and characterization of different CTCs subtypes, including clusters. The advantages of this test compared to others currently in use are the possibility of using a single blood sample, in addition to the speed of execution and low costs. Although in this pilot study we aimed to check the validity of the test using immunofluorescence as a downstream analysis, we stress that a further advantage is represented by the possibility of carrying out different downstream analyses on the two filters obtained from the same patient, without having to repeat the blood sampling. Further studies including a larger patient cohort and different cancer types are currently ongoing in order to validate these results.

Supplementary Materials: The following are available online at https://www.mdpi.com/article/10.3390/cancers13246362/s1, Figure S1: Hematopoietic cells identification. Representative confocal images showing CD45 positive cells observed at differential interference contrast (DIC) stained with diaminobenzidine. Magnification $60\times$, $5\times$ zoom bar 10 µm. Figure S2: Illustrative images of triple immunofluorescence assay on circulating colon cancer cells clusters. Representative confocal images of CTC clusters stained with anti-CK20 (red), anti-vimentin (green) and anti-HIF-1α (yellow) antibodies. Magnification $60\times$, $5\times$ zoom bar 10 µm.

Author Contributions: Conceptualization, P.G. and C.N.; methodology, F.F., M.L.D.A. and C.N.; validation, F.F., A.Z. and C.N.; formal analysis, V.M., G.D.R. and O.G.; investigation, F.F., M.L.D.A., V.M. and C.N.; resources, V.M., A.Z. and P.G.; data curation, C.N. and F.F.; writing—original draft preparation, P.G.; writing—review and editing, A.Z. and P.G.; supervision, P.G. and C.N.; project administration, P.G.; funding acquisition, A.Z. and P.G. All authors have read and agreed to the published version of the manuscript.

Funding: This research was partially supported by an Italian Association for Cancer Research (AIRC) Investigator Grant to A.Z. (AIRC IG 2017 Ref: 20744) and by Sapienza University of Rome to P.G. (Grant 2015 n. C26A15HKTF).

Institutional Review Board Statement: The study was conducted according to the guidelines of the Declaration of Helsinki, and approved by the Ethics Committee of Sapienza University of Rome (protocol n. 668/09, 9 July 2009; amended protocol 179/16, 1 March 2016).

Informed Consent Statement: Informed consent was obtained from all subjects involved in the study.

Data Availability Statement: The data presented in this study are available on request from the corresponding author.

Acknowledgments: The authors would like to thank all the participants of this study for their valuable contribution.

Conflicts of Interest: The authors declare no conflict of interest. The funders had no role in the design of the study; in the collection, analyses or interpretation of data; in the writing of the manuscript or in the decision to publish the results.

References

1. Paoletti, C.; Hayes, D.F. Circulating Tumor Cells. In *Novel Biomarkers in the Continuum of Breast Cancer*; Advances in Experimental Medicine and Biology; Springer: Cham, Switzerland, 2015; Volume 882, pp. 235–258. [CrossRef]
2. Castro-Giner, F.; Aceto, N. Tracking cancer progression: From circulating tumor cells to metastasis. *Genome Med.* **2020**, *12*, 31. [CrossRef]
3. Massagué, J.; Obenauf, A.C. Metastatic colonization by circulating tumour cells. *Nature* **2016**, *529*, 298–306. [CrossRef] [PubMed]
4. Wang, W.-C.; Zhang, X.-F.; Peng, J.; Li, X.-F.; Wang, A.-L.; Bie, Y.-Q.; Shi, L.-H.; Lin, M.-B. Survival Mechanisms and Influence Factors of Circulating Tumor Cells. *BioMed Res. Int.* **2018**, *2018*, 6304701. [CrossRef] [PubMed]
5. Strilic, B.; Offermanns, S. Intravascular Survival and Extravasation of Tumor Cells. *Cancer Cell* **2017**, *32*, 282–293. [CrossRef] [PubMed]
6. Aceto, N.; Bardia, A.; Miyamoto, D.T.; Donaldson, M.C.; Wittner, B.S.; Spencer, J.A.; Yu, M.; Pely, A.; Engstrom, A.; Zhu, H.; et al. Circulating Tumor Cell Clusters Are Oligoclonal Precursors of Breast Cancer Metastasis. *Cell* **2014**, *158*, 1110–1122. [CrossRef]
7. Amintas, S.; Bedel, A.; Moreau-Gaudry, F.; Boutin, J.; Buscail, L.; Merlio, J.-P.; Vendrely, V.; Dabernat, S.; Buscail, E. Circulating Tumor Cell Clusters: United We Stand Divided We Fall. *Int. J. Mol. Sci.* **2020**, *21*, 2653. [CrossRef]
8. Kapeleris, J.; Zou, H.; Qi, Y.; Gu, Y.; Li, J.; Schoning, J.; Monteiro, M.J.; Gu, W. Cancer stemness contributes to cluster formation of colon cancer cells and high metastatic potentials. *Clin. Exp. Pharmacol. Physiol.* **2019**, *47*, 838–847. [CrossRef] [PubMed]

9. Lu, L.; Zeng, H.; Gu, X.; Ma, W. Circulating tumor cell clusters-associated gene plakoglobin and breast cancer survival. *Breast Cancer Res. Treat.* **2015**, *151*, 491–500. [CrossRef]

10. Chang, P.-H.; Chen, M.-C.; Tsai, Y.-P.; Tan, G.Y.T.; Hsu, P.-H.; Jeng, Y.-M.; Tsai, Y.-F.; Yang, M.-H.; Hwang-Verslues, W.W. Interplay between desmoglein2 and hypoxia controls metastasis in breast cancer. *Proc. Natl. Acad. Sci. USA* **2021**, *118*, e2014408118. [CrossRef]

11. Dementeva, N.; Kokova, D.; Mayboroda, O. Current Methods of the Circulating Tumor Cells (CTC) Analysis: A Brief Overview. *Curr. Pharm. Des.* **2017**, *23*, 4726–4728. [CrossRef]

12. Ferreira, M.M.; Ramani, V.C.; Jeffrey, S.S. Circulating tumor cell technologies. *Mol. Oncol.* **2016**, *10*, 374–394. [CrossRef]

13. Hendricks, A.; Brandt, B.; Geisen, R.; Dall, K.; Röder, C.; Schafmayer, C.; Becker, T.; Hinz, S.; Sebens, S. Isolation and Enumeration of CTC in Colorectal Cancer Patients: Introduction of a Novel Cell Imaging Approach and Comparison to Cellular and Molecular Detection Techniques. *Cancers* **2020**, *12*, 2643. [CrossRef]

14. DeSitter, I.; Guerrouahen, B.S.; Benali-Furet, N.; Wechsler, J.; Jänne, P.A.; Kuang, Y.; Yanagita, M.; Wang, L.; Berkowitz, J.A.; Distel, R.J.; et al. A new device for rapid isolation by size and characterization of rare circulating tumor cells. *Anticancer. Res.* **2011**, *31*, 427–441.

15. Hong, Y.; Fang, F.; Zhang, Q. Circulating tumor cell clusters: What we know and what we expect (Review). *Int. J. Oncol.* **2016**, *49*, 2206–2216. [CrossRef] [PubMed]

16. Bankó, P.; Lee, S.Y.; Nagygyörgy, V.; Zrínyi, M.; Chae, C.H.; Cho, D.H.; Telekes, A. Technologies for circulating tumor cell separation from whole blood. *J. Hematol. Oncol.* **2019**, *12*, 48. [CrossRef] [PubMed]

17. Rushton, A.; Nteliopoulos, G.; Shaw, J.; Coombes, R. A Review of Circulating Tumour Cell Enrichment Technologies. *Cancers* **2021**, *13*, 970. [CrossRef]

18. Raimondi, C.; Carpino, G.; Nicolazzo, C.; Gradilone, A.; Gianni, W.; Gelibter, A.; Gaudio, E.; Cortesi, E.; Gazzaniga, P. PD-L1 and epithelial-mesenchymal transition in circulating tumor cells from non-small cell lung cancer patients: A molecular shield to evade immune system? *Oncoimmunology* **2017**, *6*, e1315488. [CrossRef] [PubMed]

19. Nicolazzo, C.; Raimondi, C.; Gradilone, A.; Emiliani, A.; Zeuner, A.; Francescangeli, F.; Belardinilli, F.; Seminara, P.; Loreni, F.; Magri, V.; et al. Circulating Tumor Cells in Right- and Left-Sided Colorectal Cancer. *Cancers* **2019**, *11*, 1042. [CrossRef]

20. Gazzaniga, P.; Raimondi, C.; Nicolazzo, C.; Carletti, R.; Di Gioia, C.; Gradilone, A.; Cortesi, E. The rationale for liquid biopsy in colorectal cancer: A focus on circulating tumor cells. *Expert Rev. Mol. Diagn.* **2015**, *15*, 925–932. [CrossRef]

21. Divella, R.; Daniele, A.; Abbate, I.; Bellizzi, A.; Savino, E.; Simone, G.; Giannone, G.; Giuliani, F.; Fazio, V.; Gadaleta-Caldarola, G.; et al. The presence of clustered circulating tumor cells (CTCs) and circulating cytokines define an aggressive phenotype in metastatic colorectal cancer. *Cancer Causes Control* **2014**, *25*, 1531–1541. [CrossRef]

22. Donato, C.; Kunz, L.; Castro-Giner, F.; Paasinen-Sohns, A.; Strittmatter, K.; Szczerba, B.M.; Scherrer, R.; Di Maggio, N.; Heusermann, W.; Biehlmaier, O.; et al. Hypoxia Triggers the Intravasation of Clustered Circulating Tumor Cells. *Cell Rep.* **2020**, *32*, 108105. [CrossRef] [PubMed]

23. Petrova, V.; Annicchiarico-Petruzzelli, M.; Melino, G.; Amelio, I. The hypoxic tumour microenvironment. *Oncogenesis* **2018**, *7*, 10. [CrossRef] [PubMed]

Article

A Patient-Derived Organoid-Based Radiosensitivity Model for the Prediction of Radiation Responses in Patients with Rectal Cancer

Misun Park [1], Junhye Kwon [1], Joonseog Kong [2], Sun Mi Moon [3], Sangsik Cho [3], Ki Young Yang [4], Won Il Jang [5], Mi Sook Kim [5], Younjoo Kim [1,4,*,†] and Ui Sup Shin [1,3,*,†]

[1] Department of Radiological & Clinical Research, Korea Cancer Center Hospital, Korea Institute of Radiological and Medical Sciences, Seoul 01812, Korea; usre@kirams.re.kr (M.P.); jhkwon@kirams.re.kr (J.K.)
[2] Department of Pathology, Korea Cancer Center Hospital, Korea Institute of Radiological and Medical Sciences, Seoul 01812, Korea; balltta9@kirams.re.kr
[3] Department of Surgery, Korea Cancer Center Hospital, Korea Institute of Radiological and Medical Sciences, Seoul 01812, Korea; msm@kirams.re.kr (S.M.M.); whtkdtlr@kirams.re.kr (S.C.)
[4] Department of Internal Medicine, Korea Cancer Center Hospital, Korea Institute of Radiological and Medical Sciences, Seoul 01812, Korea; gooddryang@naver.com
[5] Department of Radiation Oncology, Korea Cancer Center Hospital, Korea Institute of Radiological and Medical Sciences, Seoul 01812, Korea; zzang11@kirams.re.kr (W.I.J.); mskim@kirams.re.kr (M.S.K.)
* Correspondence: younjoo282@kirams.re.kr (Y.K.); uisup.shin@kirams.re.kr (U.S.S.); Tel.: +82-2-970-1208 (Y.K.); +82-2-970-1216 (U.S.S.)
† These authors contributed equally to this work.

Citation: Park, M.; Kwon, J.; Kong, J.; Moon, S.M.; Cho, S.; Yang, K.Y.; Jang, W.I.; Kim, M.S.; Kim, Y.; Shin, U.S. A Patient-Derived Organoid-Based Radiosensitivity Model for the Prediction of Radiation Responses in Patients with Rectal Cancer. *Cancers* **2021**, *13*, 3760. https://doi.org/10.3390/cancers13153760

Academic Editor: Samuel C. Mok

Received: 21 June 2021
Accepted: 19 July 2021
Published: 27 July 2021

Publisher's Note: MDPI stays neutral with regard to jurisdictional claims in published maps and institutional affiliations.

Simple Summary: Predicting the tumor regression grade of locally advanced rectal cancer after neoadjuvant chemoradiation is important for customized treatment strategies; however, there are no reliable prediction tools. A novel preclinical model based on patient-derived tumor organoids has shown promising features of the recapitulation of real tumors and their treatment response. We conducted a small co-clinical trial to determine the correlation between the irradiation response of individual patient-derived rectal cancer organoids and the results of actual radiotherapy. Among the quantitative experimental data, the survival fraction was best matched and correlated with the patients' real treatment outcome. In the machine learning-based prediction model for radiotherapy results using the survival fraction data, the prediction accuracy was excellent at more than 89%. Enhanced machine learning with the accumulation of further new experimental data would help in creating a more reliable prediction model, and this new preclinical model can lead to more advanced precision medicine.

Abstract: Patient-derived tumor organoids closely resemble original patient tumors. We conducted this co-clinical trial with treatment-naive rectal cancer patients and matched patient-derived tumor organoids to determine whether a correlation exists between experimental results obtained after irradiation in patients and organoids. Between November 2017 and March 2020, we prospectively enrolled 33 patients who were diagnosed with mid-to-lower rectal adenocarcinoma based on endoscopic biopsy findings. We constructed a prediction model through a machine learning algorithm using clinical and experimental radioresponse data. Our data confirmed that patient-derived tumor organoids closely recapitulated original tumors, both pathophysiologically and genetically. Radiation responses in patients were positively correlated with those in patient-derived tumor organoids. Our machine learning-based prediction model showed excellent performance. In the prediction model for good responders trained using the random forest algorithm, the area under the curve, accuracy, and kappa value were 0.918, 81.5%, and 0.51, respectively. In the prediction model for poor responders, the area under the curve, accuracy, and kappa value were 0.971, 92.1%, and 0.75, respectively. Our patient-derived tumor organoid-based radiosensitivity model could lead to more advanced precision medicine for treating patients with rectal cancer.

Keywords: machine learning; patient-derived tumor organoid; precision medicine; radiation response; rectal cancer

1. Introduction

Since the German trial of 2004, neoadjuvant chemoradiation therapy (NCRT), followed by radical surgery with total mesorectal excision, has been a standard treatment for locally advanced rectal cancer without metastasis [1,2]. With NCRT, the rate of local recurrence is significantly reduced, and the survival rate of cancer patients is significantly increased among good radiation responders [3–5]. Tumor response is evaluated based on pathologic findings of tumor regression, or the amount of TNM downstaging in postoperative surgical specimens compared with the clinical TNM staging [6]. The downstaging rate is 60–80%, of which 15–20% show a pathological complete response. However, approximately 20–40% of patients do not benefit from NCRT.

Currently, even if a complete response is clinically observed after NCRT, radical resection is recommended, which can be accompanied by serious surgical morbidity or impaired quality of life. However, it has been suggested that radical surgery is unnecessary if NCRT eradicates all tumor cells. Beets et al. suggested the 'wait and see' approach for rectal cancer patients [7]. According to these authors, if rectal cancer patients have a clinical complete response, as determined based on strict preoperative endoscopic criteria, after NCRT, undertaking nonoperative management or delayed surgery does not compromise long-term oncologic results [8]. In contrast, to improve the radiation response rate, many studies have been conducted by adding more intensive drug therapies during the peri-radiation period. The single-agent 5-fluorouracil (5-FU) or its derivatives have been used as a radiosensitizer. However, more intensive chemotherapeutic drugs (oxaliplatin or irinotecan) or biologics (cetuximab, bevacizumab, or panitumumab) have been added to enhance the radiation response [9–16]. However, administering these intensive treatments to all patients with rectal cancer is not cost-effective and is associated with increased toxicity. Moreover, the issue of overtreatment cannot be avoided.

In terms of precision medicine, rectal cancer is an ideal candidate, as treatment strategies can be tailored according to the expected radioresponsiveness. If a pathological complete response is expected, patients could avoid radical surgery, or if the expected radioresponsiveness is poor, more intensive preoperative chemotherapy could be administered. Therefore, the development of reliable prediction tools for radioresponsiveness is important.

As a preclinical model for precision medicine, patient-derived tumor organoids (PDTOs) have shown advantages over patient-derived tumor xenograft models, but have many limitations in clinical usage owing to their high cost and time taken to establish individual patient-derived models [17,18]. For pancreatic cancer and metastatic gastrointestinal cancer, the PDTO models showed a high correlation with clinical outcomes in terms of drug response [19,20]. Regarding radiation response, Ganesh et al. [21] and Yao et al. [22] recently generated PDTOs from patients with rectal cancer, and reported that PDTOs mirrored individual radiotherapy outcomes. Their results suggest that PDTOs can be used to predict individual responses to chemoradiation. However, prior studies have not identified the method that can best determine the correlation between PDTO response and patient outcome.

In this co-clinical trial, we attempted to reproduce previous study results to determine whether there is a correlation between experimental results obtained after irradiation in PDTOs and actual individual NCRT results of patients. In addition, we constructed a simple machine learning model that predicts patients' actual NCRT results based on the experimental data.

2. Materials and Methods

2.1. Patient Enrolment and Treatment

Between November 2017 and March 2020, we prospectively enrolled 33 patients diagnosed with mid-to-lower rectal adenocarcinoma pathologically confirmed by endoscopic biopsy. All patients underwent a staging workup using pelvic MRI; chest, abdominal, and pelvic computed tomography (CT); and 18-fluoro-2-deoxy-glucose positron emission tomography/CT. For patients diagnosed with locally advanced rectal cancer, NCRT was performed over a long course with a dose of 50.4 Gy in 28 fractions administered during weekdays. Chemotherapy was administered with a single-agent infusional 5-FU (425 mg per body square meter) for 5 days every 4 weeks before surgery. Radical surgeries were performed 6–8 weeks after completing radiotherapy with the aim of total mesorectal excision. Adjuvant chemotherapy was recommended for all medically fit patients after radical resection. For one patient who was diagnosed with a small resectable liver metastasis during staging workup, short-course radiotherapy with 25 Gy was administered in 5 Gy fractions over 5 days, followed by three cycles of neoadjuvant therapy: FOLFOX (5-FU, leucovorin, and oxaliplatin) with bevacizumab (the first cycle of FOLFOX only) every 2 weeks. Radical surgery, including liver metastasectomy, was performed 8 weeks after completing radiotherapy.

2.2. Pathologic Examination of Surgical Specimens

Standard pathologic tumor staging of the surgical specimen was performed and recorded according to the 8th edition of the TNM classification of the American Joint Committee on Cancer by dedicated gastrointestinal pathologists [23]. Pathologic response after NCRT was evaluated using the tumor regression grade (TRG) system suggested by the Gastrointestinal Pathology Study Group of the Korean Society of Pathologists [24]. The definitions of the TRG system are as follows: (A) TRG 0, complete response (no residual tumor cells were identified); (B) TRG 1, near complete response (only a few scattered tumor cells were present); (C) TRG 2, partial response (residual tumor glands with predominant fibrosis were easily identified); and (D) TRG 3, poor or no response (tumor cells did not demonstrate any response to chemoradiotherapy).

2.3. Tissue Acquisition

Pre-NCRT rectal cancer samples were obtained from enrolled patients at the endoscopic evaluation stage. Four or five rectal cancer biopsy samples were collected. A pathologist verified that the collected samples were histologically adenocarcinoma or normal crypts using hematoxylin and eosin (H&E) staining. The biopsy samples were pooled and immediately placed in cold phosphate-buffered saline with 50 μg/mL gentamicin (Gibco, Grand Island, NY, USA).

2.4. Organoid Cultures

Tumor organoids were isolated and cultured as previously described [25]. The composition of the PDTO culture medium is described in Supplementary Table S1. To prevent anoikis, 10 μM of Y-27632 was added to the culture medium for the first 2–3 days. When organoids were >200 μm, they were passaged by pipetting using Gentle Cell Dissociation Reagent (STEMCELL Technologies, Vancouver, BC, Canada) according to the manufacturer's instructions. Most of PDTO used in experiments were cultured more than 14 days.

2.5. Immunocytochemistry and Immunohistochemistry

For immunocytochemistry, PDTOs were fixed in 4% paraformaldehyde at 25 °C for 24 h, embedded in paraffin, and then dissected into 3-μm-thick sections. After treatment with Smartblock solution (CANDOR Bioscience GmbH, Wangen im Allgäu, Germany) for 30 min at 25 °C, the slides were incubated with primary antibodies at 4 °C overnight and then incubated with secondary antibodies for 1 h at 25 °C. Images were acquired using the EVOS FL Cell Imaging System (Thermo Fisher Scientific, Carlsbad, CA, USA).

Immunohistochemistry was performed to characterize organoids and their tissues of origin with the colorectal markers caudal type homeobox 2 transcription factor, cytokeratin 7, and cytokeratin 20 on 3-μm-thick formalin-fixed paraffin-embedded tissue and organoid sections. Sections were incubated for 1 h at 37 °C with primary antibodies. Detection was performed using an Envision/Horseradish Peroxidase system (Dako; Agilent Technologies, Inc., Santa Clara, CA, USA) and counterstained with hematoxylin for 10 min at 25 °C. Finally, the sections were dehydrated through a graded series of alcohol, cleared in xylene, and mounted. Images were acquired using an IX73 inverted microscope (Olympus Corporation, Tokyo, Japan). The antibodies and dilutions used are described in Supplementary Table S2.

2.6. Survival Fraction Analysis

For survival fraction analysis, organoids were resuspended in TrypLE Express (Thermo Fisher Scientific, Carlsbad, CA, USA) via pipetting with a p200 pipette and incubated at 37 °C for 10 min. Cells were centrifuged at $600\times g$ for 5 min, and the supernatant was discarded. The pellet was resuspended in Matrigel and distributed into a 48-well plate (500–1000 cells/20 μL of Matrigel per well). After the Matrigel had polymerized, 100 μL of culture medium was added. Over the following days, organoids were treated with 0 Gy, 2 Gy, 4 Gy, and 6 Gy using a ^{137}Cs γ-ray source (Atomic Energy of Canada, Ltd., Renfrew County, ON, Canada) at a dose rate of 3.81 Gy/min. After 14 days, viable organoids were counted using Cell3 iMager Neo cc-3000 (Screen Holdings Co., Ltd., Kyoto, Japan). Analysis recipes were as follows: organoid diameter min 93, max 2907; organoid area min 6833, max 6,640,106; and circularity min 0.24, max 1. The plating efficiency was defined as the number of formed organoids/seeded cells × 100%. Survival fraction was calculated as follows: number of formed organoids/number of seeded cells in plate × (plating efficiency/100)]. A single-hit multitarget model was used to fit the survival curves, and D_0, called the 'mean lethal dose', was the dose required to reduce the fraction of surviving organoids to 37% [26], calculated using GraphPad Prism software (version 8.0; GraphPad Software, Inc., La Jolla, CA, USA). For each PTDO, experimental replication of 4 wells was used. We obtained a total of 76 sets of survival fraction data.

2.7. Viability Assay

For the viability assay, organoids were resuspended in TrypLE. Cells were resuspended in Matrigel and distributed into a 96-well plate (5000 cells/10 μL of Matrigel per well). After the Matrigel had polymerized, 100 μL of culture medium was added. Over the following days, organoids were treated with 0 Gy, 2 Gy, 4 Gy, and 6 Gy. After 7 days, organoid viability was evaluated using CellTiter 96 AQUEOUS One Solution contains a tetrazolium compound [3-(4,5-dimethylthiazol-2-yl)-5-(3-carboxymethoxyphenyl)-2-(4-sulfophenyl)-2H-tetrazolium, inner salt; MTS] (Promega, Madison, WI, USA) according to the manufacturer's instructions. Optical density was measured using a BioTek Eon microplate absorbance reader (BioTek Instruments Inc., Winooski, VT, USA) at 490 nm. Matrigel without organoids (10 μL) was used as a control.

2.8. Second Passage

For analysis at the second passage, organoids were treated with 5 Gy. After 72 h, organoids were passaged by pipetting using Gentle Cell Dissociation Reagent with a 1:2–1:4 split ratio. After 72 h, viable organoids were counted using the EVOS FL Cell Imaging System (Thermo Fisher Scientific, Carlsbad, CA, USA).

2.9. EdU Staining

Organoids were incubated with 10 μM EdU for 2 h and evaluated using Click-iT Plus EdU Imaging Kits (Thermo Fisher Scientific, Carlsbad, CA, USA) according to the manufacturer's instructions. Images were acquired using the EVOS FL Cell Imaging System (Thermo Fisher Scientific, Carlsbad, CA, USA).

2.10. Western Blot Analysis

For Western blot analysis, organoids were washed with cold phosphate-buffered saline and lysed in radioimmunoprecipitation assay buffer (Thermo Fisher Scientific, Carlsbad, CA, USA). Proteins were quantified using the Bradford method, and 20–40 µg of protein was resolved using SDS-PAGE. The membranes were incubated with primary antibodies overnight at 4 °C, followed by incubation with a secondary antibody (Santa Cruz Biotechnology, Santa Cruz, CA, USA) for 1 h at 25 °C. Proteins were visualized using enhanced chemiluminescence (Thermo Fisher Scientific, Grand Island, NE, USA). Western blot images were analyzed using the Bio-Rad ChemiDoc (Bio-Rad, Richmond, CA, USA).

2.11. Targeted Next-Generation Sequencing Analysis

To analyze the mutational status of tissues and organoids, they were harvested using a cell recovery solution (Corning, Inc., Corning, NY, USA). DNA extraction and library construction were performed using the Gentra Puregene kit (Qiagen, Hilden, Germany) and SureSelect XT library prep kit (Agilent Technologies, Santa Clara, CA, USA). Deep targeted sequencing using Axen Cancer Panel 2 (170 cancer-related genes; Macrogen, Seoul, Korea) and the NextSeq 500 mid-output system platform (Illumina, San Diego, CA, USA) was conducted on 19 PDTOs. Libraries comprising 150-bp end reads were sequenced via high-throughput sequencing using synthesis technology to a depth coverage of approximately $2000\times$.

2.12. Statistical Analysis

Data obtained from a minimum of three independent experiments are expressed as mean ± standard deviation. Unpaired two-tailed Student's *t*-tests were used to determine significant differences between the two groups. One-way analysis of variance followed by Tukey's and Bonferroni tests was performed to compare the means between multiple groups, and *p* values < 0.05 were considered significant. Statistical and receiver operating characteristic (ROC) curve analyses were performed using R 4.0.2 (https://www.r-project.org; accessed on 15 May 2020). Analysis of the mutation-annotated files was conducted using the R package 'maftools' (version 3.12), which included the generation of figure oncoplots [27]. Comparison of linear-quadratic (LQ) cell survival curves was performed using analysis of variance calculated with the R package 'CFAssay' (version 1.22.0) [28].

2.13. Development of Predictive Models Using Machine Learning

To build the prediction model for TRG, we used survival fraction data. A total of 76 experimental data points were randomly split in a 1:1 ratio into training and testing datasets. The machine learning model was built using the training data and subsequently tested on the remaining 50% of the data comprising the testing set. The supervised machine learning classification algorithm performed binary logistic regression and random forest classification with the R package, 'randomForest' version 4.6-14. For model training with a random forest method, we used 200 trees and two variables as training hyperparameters. We calculated the area under the ROC curve (AUC), accuracy, and kappa value of the testing dataset to evaluate the model performance.

3. Results

3.1. Patient Characteristics and Treatment Outcomes

Tumor tissues were collected by endoscopic biopsy from 33 patients with rectal cancer. Among 33 tumor tissues, 10 PDTOs could not be established due to bacterial contamination in one case and no expansion in the culture medium in nine cases (70% success rate). In addition, two patients were excluded as they were diagnosed with unresectable metastatic rectal cancer; radical surgeries were not planned for these patients, and it would not have been possible to evaluate their TRG. Two patients refused radical surgeries and were also excluded. Finally, 19 patients and their PDTOs were analyzed in this study (Figure 1A). Representative images of the 19 PDTOs are displayed in Supplementary

Figure S1. Individual patient characteristics and clinical treatment results are summarized in Table 1. The median age of patients was 59 (interquartile range, 53.0–70.5) years. The male-to-female ratio was 14:5. Eighteen patients had stage III disease, and one patient had stage IV disease with resectable liver metastasis. After R0 surgery following NCRT, TRGs were as follows: five patients achieved TRG 0 (26.3%), and one patient had TRG 1. Three patients had TRG 3, and the other 11 patients had TRG 2 (Figure 1B). During a median of 19.0 (interquartile range, 12.5–26.5) months of follow-up, six patients developed tumor recurrence (five distant, one local), and one patient died due to recurrence (Table 1).

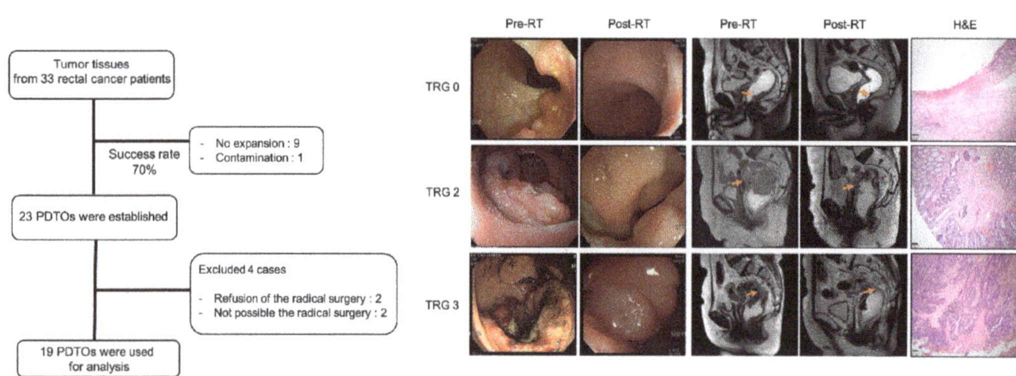

Figure 1. Patient characteristics and treatment outcomes. (**A**) Flow chart indicating the number of patients with rectal cancer, including reasons for non-evaluability, and the success rate of establishing cultures from patients. (**B**) Pre- and post-RT endoscopic clinical responses, magnetic resonance images. and H&E staining images are shown for TRG 0, TRG 2, and TRG 3 patients. Magnification, ×4. Scale bars, 200 µm Abbreviations: H&E, hematoxylin and eosin; PDTO, patient-derived tumor organoid; RT, radiotherapy; and TRG, tumor regression grade.

Table 1. Individual patient characteristics and treatment results.

Sample No. (PDTO)	5	7	8	9	10	11	12	13	14	15	18	19	21	22	23	28	29	30	33
Sex/Age	F/84	F/64	M/47	M/70	M/49	M/54	M/77	M/57	M/49	M/64	M/71	F/62	M/82	F/76	M/52	M/54	F/59	M/58	M/52
BMI (kg/m^2)	24.8	24.4	16.2	18.9	25.5	22.2	16.5	25.7	17.7	24.6	22.7	17.3	20.9	24.3	19.7	26.1	25.3	20.8	21.2
Diabetes	No	No	No	No	No	No	No	No	Yes	No	No	No	No	No	No	No	Yes	No	No
Clinical Stage	T3N1	T3N2	T3N2	T3N1	T3N2	T3N1	T3N1	T3N1	T3N2	T2N1	T3N1	T3N1	T4N2	T3N2	T2N1	T3N1	T3N1	T3N1	T2N0M1
Pre-RT-Tumor Size(cm) MRI	6.5	4.5	4	3.8	8	4	4.2	5	6.5	3.8	5.4	3.5	7.5	6.5	6	3.5	5.3	5.6	2.6
Post-RT-Tumor Size(cm) MRI	3.5	3	2.8	3	7	2	2.3	2.6	4	1.6	5.3	2.5	5	2.5	2.5	2.5	2.8	3.5	1.1
Post-RT-Tumor Size (cm) Surgical Specimen	1.2	3.7	2.4	2.3	3.4	2.5	2.5	0.8	3.8	1	7.5	1.2	3.5	3.5	2.6	0.9	2.5	0.5	2.2
TRG	2	2	2	2	2	2	3(2)	1	0	0	3	0	2	3	0	0	2	2	2
yp Stage	T3N0	T3N2	T2N2	T2N0	T2N2	T3N1	T3N1	T2N1	T0N0	T0N0	T3N1	T0N0	T3N0	T3N0	T0N0	T0N0	T3N0	T2N0	T2N0M1
Site of Recurrence		Distant lymph node			Liver	Lung, distant lymph node	Lung				Local		Lung						
Recurrence-free Survival (Months)	26	25	28	28	15	13	13	26	24	23	10	17	10	12	14	9	6	5	5
Dead	No	No	No	No	No	No	Yes	No	No	No	No	No	No	No	No	No	No	No	No
Overall Survival (Months)	26	28	28	28	27	28	19	26	24	23	14	17	13	12	14	9	6	5	5

BMI, body mass index; RT, radiotherapy; MRI, magnetic resonance imaging; TRG0, (complete response) TRG1, (Nearly complete) TRG2, (Moderate) TRG3, (Minimal).

3.2. Histological and Genomic Characterization of PDTOs

To verify PDTOs, immunostaining was performed using paraffin-embedded organoid sections and tissues. Our PDTOs differentiated into enterocytes (villin 1), goblet cells (mucin 2), and enterochromaffin cells (chromogranin A) and contained amplifying cells (Ki-67; Figure 2A). To analyze the mutational status of the 19 PDTOs, we performed targeted next-generation sequencing analysis using Axen Cancer Panel 2. Variants were filtered based on a multivariate alteration detection of <2%, type of alteration (multi-hit, missense, nonsense, splicing site, in-frame del, and frame-shift), fusion gene, copy number alterations, and functional consequence (pathogenic, likely pathogenic, benign, and likely benign). Genes of the WNT signaling pathway (*APC* and *FBXW7*) were mutated in 68.4% (13/19) of all PDTOs. *APC* and *FBXW7* mutations were identified in 13 of 19 PDTOs (68.4%) and 6 of 19 PDTOs (31.5%), respectively (Figure 2B). All mutation alterations are displayed in Supplementary Figure S2. We performed H&E staining and immunostaining of the proteins cytokeratin 7, cytokeratin 20, and caudal type homeobox 2 transcription factor to confirm that our PDTOs originated from rectal cancer tissue and not from normal rectal mucosa. Our PDTOs showed similar histological morphologies and CK protein expression patterns to those of original tumor tissues (Figure 2C). Overall, these data demonstrated that PDTOs recapitulated the histological morphologies and marker expression of the paired patient tissues, as previously reported [21,29]. To define the capacity of colorectal cancer organoids to mirror the genome heterogeneity of the corresponding patient tumor, we compared the mutational status of three genes (*KRAS, NRAS,* and *BRAF*) in 19 PDTOs and corresponding tumor tissues. *KRAS, NRAS,* and *BRAF* mutations in PDTOs were matched to 86.6%, 100%, and 100% of those in corresponding tumor tissues, respectively (Figure 2D).

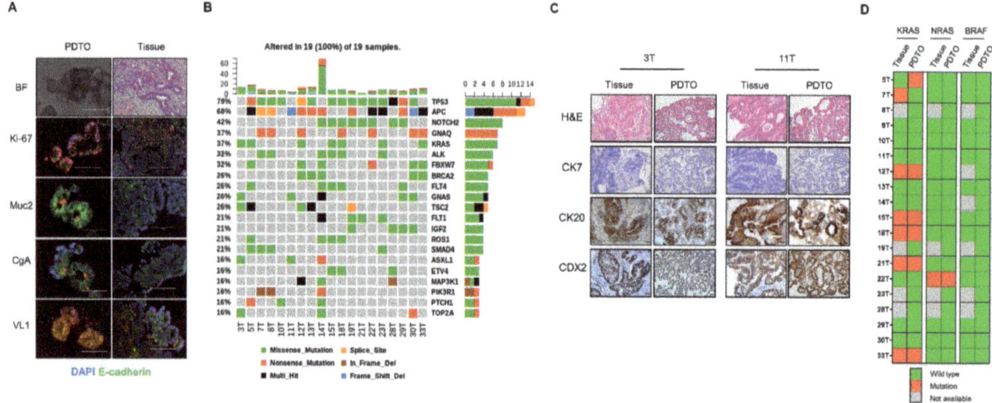

Figure 2. Histological and genomic characterization of PDTOs. (**A**) Fluorescence microscopy images of FFPE sections of organoids and corresponding tissues for goblet cells (mucin 2 [Muc2]+, red), entero-endocrine cells (CgA+, red), enterocytes (villin 1 [VL1]+, red), and proliferating cells (Ki-67, red). Counterstain, DAPI (blue) and epithelial, E-cadherin (green). Scale bars, 100 μm. (**B**) The mutation landscape of 19 PDTOs. The frequency of alterations in PDTO is noted with the type of genetic alteration (indicated by color code). The top 21 mutated genes observed in PDTOs, including the most known significant cancer driver genes, are shown. (**C**) Immunohistochemical profile of FFPE sections of organoids and corresponding tissues for cytokeratin 7, cytokeratin 20, and caudal type homeobox 2 transcription factor along with corresponding H&E staining. Magnification, ×40. Scale bars, 50 μm. (**D**) *KRAS, NRAS,* and *BRAF* mutation status of PDTOs and paired tumor tissues. Abbreviations: CDX, caudal type homeobox; CK, cytokeratin; DAPI, 4′,6-diamidino-2-phenylindole; FFPE, formalin fixed paraffin-embedded; H&E, hematoxylin and eosin; and PDTO, patient-derived tumor organoid.

3.3. PDTOs Response to Irradiation

To validate the response of PDTOs to irradiation in vitro, we performed a radiation dose-dependent (0 Gy, 2 Gy, 4 Gy, and 6 Gy) survival analysis of 19 PDTOs. Supplementary Figure S3 displays representative images of irradiated organoids, and we counted the number of viable organoids after irradiation to measure the survival fraction (Figure 3A and Supplementary Figure S4). We analyzed the D_0 value (the dose required to reduce the fraction of surviving organoids to 37%); a higher D_0 value indicates greater radioresistance [26]. Therefore, we defined radioresistant PDTOs and radiosensitive PDTOs according to the D_0 value (Figure 3B). These survival fraction data were validated by direct comparison using the MTS cell viability assay (Figure 3C and Supplementary Figure S5). The results demonstrated the heterogeneity of the radioresponse in 19 PDTOs. According to our data, PDTO-22 and PTDO-19 showed radioresistant and radiosensitive characteristics, respectively (Figure 3D,E). To confirm these different radioresponses, we tested this result using several in vitro analyses. The organoid viability of PDTO-19 cells was significantly reduced compared with that of PDTO-22 at 2 Gy, 4 Gy, and 6 Gy ($p < 0.0001$; Figure 3F). To directly assess the regenerative ability of organoids, we counted organoids at the second passage after splitting the irradiated organoids. Seventy-two hours after splitting, the relative number of PDTO-19 organoids was significantly lower than that of PDTO-22 after irradiation ($p = 0.034$; Figure 3G). To determine whether the change in cell viability was accompanied by cell proliferation, we performed EdU staining in PTDOs after irradiation and showed that 13% of the cells in the S phase decreased after irradiation in PDTO-22. In contrast, 30% of S phase cells were reduced after irradiation in PDTO-19 ($p = 0.029$; Figure 3H). To evaluate the apoptotic cellular response to radiation, apoptosis-related protein levels were analyzed. Cleaved-PARP and -caspase-3 levels, which are considered hallmarks of apoptosis, were increased in PDTO-19 after irradiation compared to those in PDTO-22 (Figure 3I).

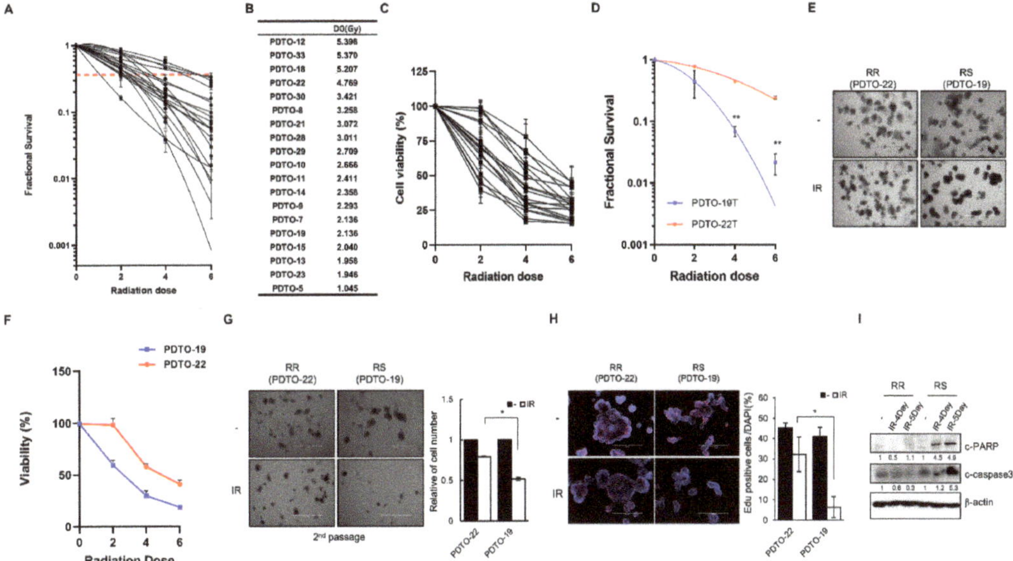

Figure 3. The response of PDTOs to radiation. (**A**) Dose–response of survival fraction in 19 PDTOs ($n = 4$, independent experiments for each PDTO) is shown. Data are presented as mean ± standard deviation. The red line represents a survival fraction of 0.37. (**B**) D_0 values were calculated according to the multitarget single-hit model. (**C**) MTS cell viability assay of 19 PDTOs after 0 Gy, 2 Gy, 4 Gy, and 6 Gy irradiation ($n = 6$, independent experiments for each PDTO). Data are normalized to those of control cells. Data are presented as mean ± standard deviation. (**D**) Dose–response of survival fraction in PDTO-19

and PDTO-22 (n = 4 independent experiments for each PDTO) is shown. ** $p < 0.01$. (**E**) Morphology of PDTO-19 and PDTO-22 after irradiation with 5 Gy after 5 days. Scale bars, 1000 µm. (**F**) MTS cell viability assay of PDTO-19 and PDTO-22 after treatment with 0 Gy, 2 Gy, 4 Gy, and 6 Gy. Data are normalized to those of the control cells and presented as mean ± standard deviation. ** $p < 0.01$. (**G**) (**left**) Image of organoids after the second passage. Scale bars, 1000 µm. (**right**) The relative organoid number after the second passage. Data are presented as mean ± standard deviation. * $p < 0.05$. (**H**) (**left**) Fluorescence microscopy images of EdU incorporation in PDTO-19 and PDTO-22 after irradiation. Scale bars, 400 µm. Blue, DAPI; red, EdU. (**right**) Statistical analysis representing EdU-positive cells per DAPI-stained cell (n = 3). * $p < 0.05$. (**I**) Expression levels of c-PARP and c-caspase-3 in PDTOs. β-actin was the loading control. Abbreviations: c-PARP, cleaved poly-ADP-ribose polymerase; DAPI, 4′,6-diamidino-2-phenylindole; IR, ionizing radiation; PDTO, patient-derived tumor organoid; RR, radioresistant; and RS, radiosensitive.

3.4. Correlation of Experimental Data with Actual TRG Outcomes

To compare the experimental results of the survival fraction, D_0 value, and cell viability, we regrouped TRGs into three categories: TRG 0, TRG 1/2, and TRG 3 (Figure 1B). The results of comparisons according to the three TRG groups and according to whether TRGs were at their two extreme categories, good responders (TRG 0 or not) and poor responders (TRG 3 or not), are shown in Figure 4A. Generally, p values obtained by comparing the mean (SD) values among the three TRG groups were more significant in the survival fraction and D_0 data than in cell viability. Furthermore, comparing after actual TRGs were regrouped according to whether TRGs were in the two extreme categories or not, the p values were more significant for comparisons of survival fraction and D_0 data (Figure 4A). Next, we performed ROC analyses to determine whether our experimental data could classify TRGs and which experimental data would be more appropriate to use for classifying TRGs. While D_0 data had a single value, D_0 only, the survival fraction data and cell viability data had multiple values at each radiation dose (2 Gy, 4 Gy, and 6 Gy). Therefore, we used a multiple logistic regression model to analyze survival fraction and cell viability data in this ROC analysis. In the ROC analysis of good responders (TRG 0), AUCs matched to D_0, survival fraction, and cell viability tests were 0.753 (95% confidence interval (CI), 0.644–0.863), 0.897 (95% CI, 0.83–0.965), and 0.631 (95% CI, 0.525–0.737), respectively (Figure 4B). When analyzing poor responders (TRG 3), the AUCs of the respective experimental data were as follows: D_0, 0.966 (95% CI, 0.926–1); the survival fraction model, 0.974 (95% CI, 0.941–1); and the cell viability model, 0.898 (95% CI, 0.827–0.968; Figure 4C). Sensitivity, specificity, positive predictive value, and negative predictive value were highest in the survival fraction model (Supplementary Table S3). We reconstructed the LQ curve according to the regrouped TRG using the 19 individual PDTO survival fraction curves (Figure 3A). When comparing the curves of the three groups, the LQ curves were clearly divided with statistical significance ($p < 0.0001$). In addition, TRG 0 or not ($p < 0.0001$) and TRG 3 or not ($p < 0.0001$) of the LQ curve were still significantly divided with respect to the TRG groups (Figure 4D).

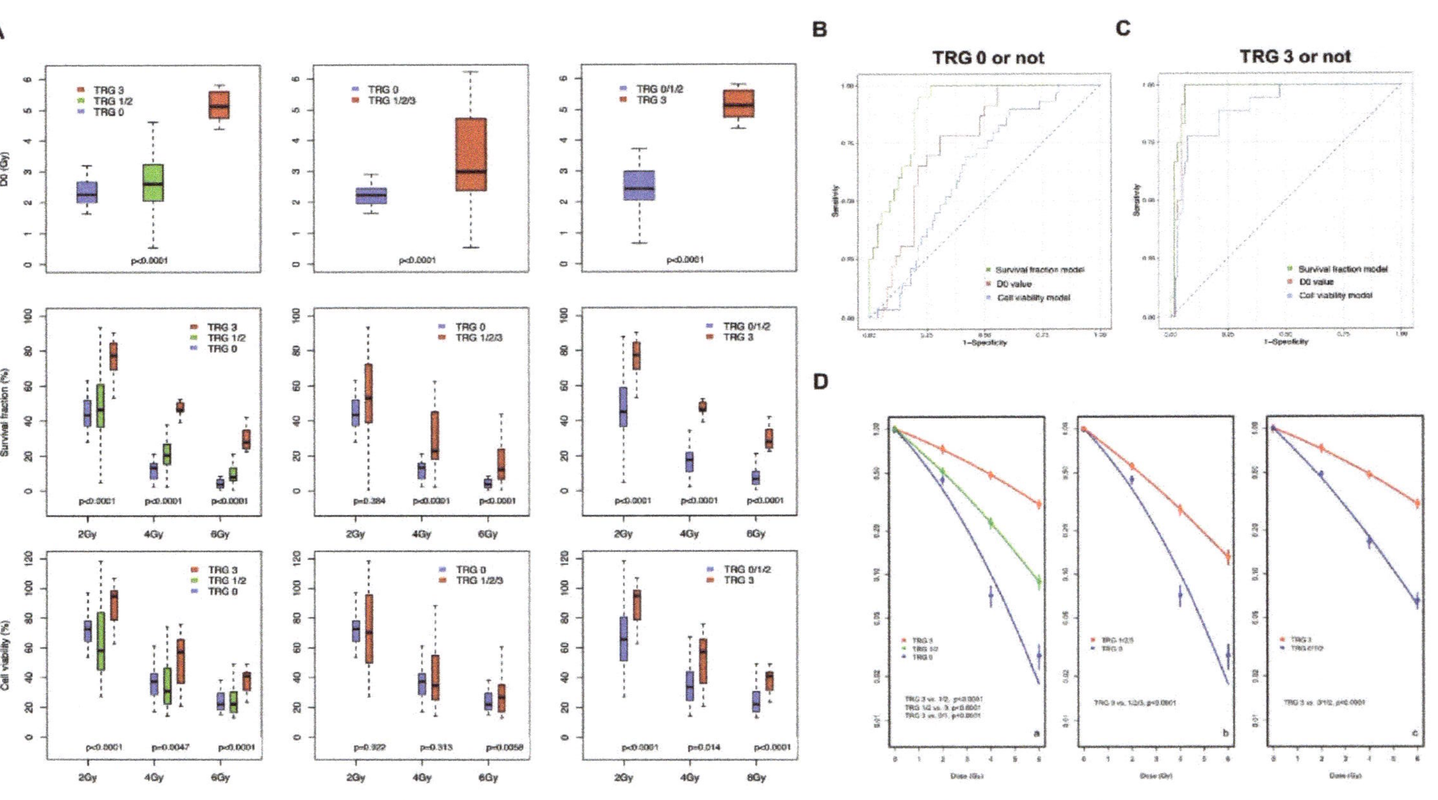

Figure 4. Correlation of experimental data with actual TRG outcomes following patients' treatments. (**A**) Comparison of the mean and standard deviation of D0, survival fraction, and cell viability data according to the TRG groups. (**B**) Receiver operating curve analysis for predicting TRG 0 or not by D0, survival fraction, and cell viability data. (**C**) Receiver op-erating curve analysis for predicting TRG 3 vs. other using D0, survival fraction, and cell viability data. (**D**) Compari-son of reconstructed linear-quadratic cell survival curves according to the TRG groups (**a**), TRG 0 or not (**b**) and TRG 3 or not (**c**). Abbreviation: TRG, tumor regression grade.

3.5. Machine Learning-Assisted Prediction Model

As shown in Figure 4, the AUC, sensitivity, specificity, positive predictive value, and negative predictive value of the survival fraction model were highest among the values from the three experimental datasets. Therefore, we developed machine learning-based classification models using the survival fraction data. After building a prediction model using a training dataset, we evaluated the model performance using the testing dataset. In the prediction model for good responders (TRG 0) trained using logistic regression, the AUC was 0.916 (Figure 5A), the accuracy was 78.9%, and the kappa value was 0.38. The AUC, accuracy, and kappa value of the model trained using the random forest were 0.918, 81.5%, and 0.51, respectively. In the prediction model for poor responders (TRG 3) trained using logistic regression, the AUC, accuracy, and kappa value were 0.927, 89.5%, and 0.65, respectively (Figure 5B); those of the model trained using the random forest were 0.971, 92.1%, and 0.75, respectively.

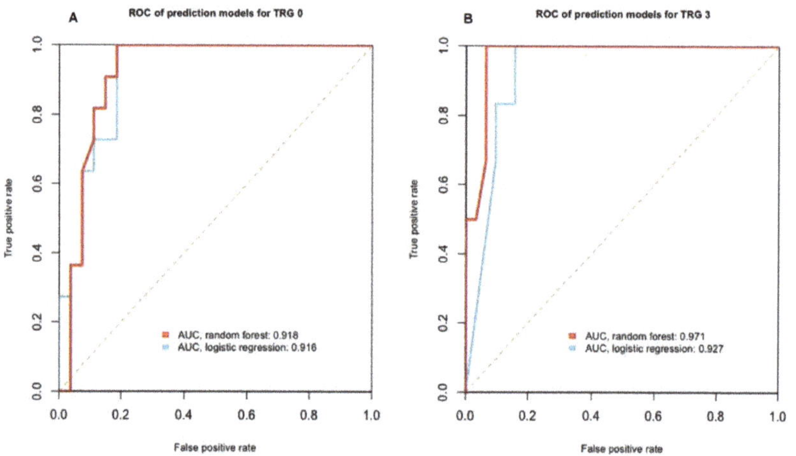

Figure 5. Machine learning-assisted prediction model. (**A,B**) Performance of machine learning-based prediction models for TRG 0 (**A**) and TRG 3 (**B**) on the testing dataset presented using receiver operating characteristic curves and their area under the curves. Red line: prediction model trained using the random forest algorithm, blue line: prediction model trained using binary logistic regression. Abbreviations: AUC, area under the ROC curve; ROC, receiver operating characteristic; and TRG, tumor regression grade.

4. Discussion

In this co-clinical trial, we reproduced the results of previous studies [21,22]. The histology, genetic features, and irradiation response of PDTOs mirrored real treatment outcomes of original tumors and patients. Furthermore, our quantitative experimental data correlated well with actual TRG results. With these results, we built a machine learning-based prediction model by inputting the survival fraction values of PTDOs. At the beginning of this study, we did not know which experimental indicator would best match the patient's actual TRG results; thus, we conducted various experiments regarding organoid irradiation responses. Among them, we selected D_0, survival fraction, and cell viability data, which were easily measurable and reproducible by repeated tests. We found that survival fraction data were the best-matched experimental results to the patient's TRG results in statistical analyses. The machine learning-based prediction model using the survival fraction data showed an excellent performance.

Organoid technology has been a highlight for cancer research due to the close resemblance of organoids to original tumors [29–35]. Due to its rapid establishment with a high success rate, the organoid model is noted as a pre- or co-clinical model for precision medicine. Although not commented on in this study, the growth rate of PDTOs was heterogeneous, but could acquire enough volume to assess the irradiation response within 1–2 weeks in most cases. It added a testing time of approximately 4 weeks to obtain irradiation response data that can predict the TRG results of real patients. It is a clinically significant period to produce treatment recommendations, as stated by Yao et al. [22].

This study has some limitations. First, the study sample size was small. The goal of a machine learning model is to generalize patterns using training data to correctly predict new data that have never been presented to the model. Overfitting occurs when a model adjusts excessively to the training data, sees patterns that do not exist, and consequently performs poorly in predicting new data. The fewer samples for training, the more models that can fit. Our treatment-naive sample number was not smaller than that of previous studies [21,22]; however, it was not sufficient to obtain a reproducible prediction model, although we used the random forest method for model training and obtained acceptable model performance results. Random forest is an ensemble machine learning model that increases the model performance, but is not a solution for small sample size issues. To develop a reliable predictive model using organoids, a reliable volume of training samples is required [36]. Given that it is difficult to generate sufficient data in a single laboratory, it is necessary to collect and share data produced under consented standard experimental conditions among clinical organoid researchers.

Second, in the current organoid model itself, one can only observe the irradiation response of cancer cells themselves. For the actual therapeutic response of tumor cells to radiation, the role of the microenvironment is very important. Although organoid cultures provide more favorable conditions than traditional cell line models for tissue physiology and structure, which are close to in vivo situations, the model does not robustly retain the complexity and diversity of the tumor microenvironment (T-ME). The T-ME has been gradually recognized as a key contributor to cancer progression and a determinant of treatment outcomes [37,38]. In radiotherapy, vascular, stromal, and immunological changes in T-ME induced by radiation promote radioresistance and tumor recurrence; furthermore, radiotherapy has recently been proposed to target the T-ME to overcome radioresistance [39]. However, organoid cultures typically contain epithelium. Thus, to overcome these limitations, models for co-culture of tumor organoids and T-ME have recently been introduced. Öhlund et al. developed a co-culture system of pancreatic cancer organoids and cancer-associated fibroblasts that can recapitulate some of the features observed in patients [40]. Dijkstra et al. described a patient-personalized in vitro model that enabled the induction and analysis of tumor-specific T-cell responses using colorectal cancer organoids and T lymphocytes isolated from patients' peripheral blood [41]. In addition, a unique co-culture method based on an air-liquid interface system permitted the propagation of PDO and tumor-infiltrating lymphocytes [42]. Organoid culture methods that partially retain the patient's T-ME might overcome the hurdles of organoid culture and offer reliable results.

Finally, in this study, we only evaluated the response against irradiation. In a real situation, various chemotherapeutic agents are combined to obtain improved NCRT results [9–14]. However, we did not perform a drug sensitivity test as our study population comprised a homogenous patient group that used a single agent, 5-FU, as a concurrent treatment for all patients except one, and the difference in radioresponse affected by the combination of various drugs could not be observed. Based on this study result of radiosensitivity, and through further validations, we believe that we will be able to identify which element or combinations of current multimodal treatments would be most helpful and identify a more advanced tailored treatment via current ex vivo tests with PDTOs.

5. Conclusions

As revealed by previous studies, individual PDTOs recapitulated responses of original tumors to irradiation. The radiation response of PDTOs could predict the patient's TRG with statistical significance. The PDTO-based radiosensitivity model could be a reliable diagnostic tool for the tailored treatment of rectal cancer.

Supplementary Materials: The following are available online at https://www.mdpi.com/article/10.3390/cancers13153760/s1, Figure S1: Observed morphologies of 19 PDTOs. Figure S2: The mutation of 19 PDTOs for all alterations is displayed. Figure S3: Morphologies of PDTOs after irradiation at 2 Gy, 4 Gy, and 6 Gy. Figure S4: Dose–response of survival fraction in 19 PDTOs. Figure S5: Dose–response of cell viability in 19 PDTOs. Figure S6: Whole blot showing all the bands with molecular weight marker. Table S1: List of chemical and reagents used for studies. Table S2. List of antibodies used for studies. Table S3: Results of ROC about two extreme categories.

Author Contributions: Conceptualization: M.P., Y.K. and U.S.S.; Methodology: M.P., J.K. (Junhye Kwon), J.K. (Joonseog Kong), Y.K. and U.S.S.; Software: W.I.J., M.S.K. and U.S.S.; Validation: M.P., J.K. (Junhye Kwon), J.K. (Joonseog Kong), Y.K. and U.S.S.; Formal analysis: M.P., J.K. (Junhye Kwon), J.K. (Joonseog Kong), Y.K. and U.S.S.; Resources: J.K. (Joonseog Kong), S.M.M., S.C., K.Y.Y., W.I.J., M.S.K., Y.K. and U.S.S.; Writing—Original Draft: M.P., J.K. (Junhye Kwon), J.K. (Joonseog Kong), Y.K. and U.S.S.; Visualization: M.P., J.K. (Junhye Kwon), J.K. (Joonseog Kong), Y.K. and U.S.S.; and Supervision: Y.K. and U.S.S. All authors have read and agreed to the published version of the manuscript.

Funding: This study was supported by a grant from the Korea Institute of Radiological and Medical Sciences, funded by the Ministry of Science and ICT, Republic of Korea (grant no. 50542-2020).

Institutional Review Board Statement: The study was conducted according to the guidelines of the Declaration of Helsinki. The Ethics Committee of Korea Cancer Center Hospital approved this study (approval no. KIRAMS-2017-07-001). All research was performed according to the approved guidelines and regulations of the institution.

Informed Consent Statement: All samples were obtained from patients who provided written informed consent.

Data Availability Statement: All data in this study are included as Supplementary Tables and Figures. Any additional information is available upon request.

Conflicts of Interest: The authors declare no conflict of interest.

References

1. Sauer, R.; Becker, H.; Hohenberger, W.; Rodel, C.; Wittekind, C.; Fietkau, R.; Martus, P.; Tschmelitsch, J.; Hager, E.; Hess, C.F.; et al. Preoperative versus postoperative chemoradiotherapy for rectal cancer. *N. Engl. J. Med.* **2004**, *351*, 1731–1740. [CrossRef]
2. Sauer, R.; Liersch, T.; Merkel, S.; Fietkau, R.; Hohenberger, W.; Hess, C.; Becker, H.; Raab, H.R.; Villanueva, M.T.; Witzigmann, H.; et al. Preoperative versus postoperative chemoradiotherapy for locally advanced rectal cancer: Results of the German CAO/ARO/AIO-94 randomized phase III trial after a median follow-up of 11 years. *J. Clin. Oncol.* **2012**, *30*, 1926–1933. [CrossRef]
3. Kapiteijn, E.; Marijnen, C.A.; Nagtegaal, I.D.; Putter, H.; Steup, W.H.; Wiggers, T.; Rutten, H.J.; Pahlman, L.; Glimelius, B.; van Krieken, J.H.; et al. Preoperative radiotherapy combined with total mesorectal excision for resectable rectal cancer. *N. Engl. J. Med.* **2001**, *345*, 638–646. [CrossRef] [PubMed]
4. van Gijn, W.; Marijnen, C.A.; Nagtegaal, I.D.; Kranenbarg, E.M.; Putter, H.; Wiggers, T.; Rutten, H.J.; Pahlman, L.; Glimelius, B.; van de Velde, C.J.; et al. Preoperative radiotherapy combined with total mesorectal excision for resectable rectal cancer: 12-year follow-up of the multicentre, randomised controlled TME trial. *Lancet Oncol.* **2011**, *12*, 575–582. [CrossRef]
5. Yeo, S.G.; Kim, D.Y.; Park, J.W.; Choi, H.S.; Oh, J.H.; Kim, S.Y.; Chang, H.J.; Kim, T.H.; Sohn, D.K. Stage-to-stage comparison of preoperative and postoperative chemoradiotherapy for T3 mid or distal rectal cancer. *Int. J. Radiat. Oncol. Biol. Phys.* **2012**, *82*, 856–862. [CrossRef]
6. Dworak, O.; Keilholz, L.; Hoffmann, A. Pathological features of rectal cancer after preoperative radiochemotherapy. *Int. J. Colorectal. Dis.* **1997**, *12*, 19–23. [CrossRef] [PubMed]
7. Beets, G.L.; Figueiredo, N.L.; Habr-Gama, A.; van de Velde, C.J. A new paradigm for rectal cancer: Organ preservation: Introducing the International Watch & Wait Database (IWWD). *Eur. J. Surg. Oncol.* **2015**, *41*, 1562–1564. [CrossRef]
8. Habr-Gama, A.; Perez, R.O.; Nadalin, W.; Sabbaga, J.; Ribeiro, U., Jr.; Silva e Sousa, A.H., Jr.; Campos, F.G.; Kiss, D.R.; Gama-Rodrigues, J. Operative versus nonoperative treatment for stage 0 distal rectal cancer following chemoradiation therapy: Long-term results. *Ann. Surg.* **2004**, *240*, 711–718. [CrossRef]
9. Gerard, J.P.; Azria, D.; Gourgou-Bourgade, S.; Martel-Lafay, I.; Hennequin, C.; Etienne, P.L.; Vendrely, V.; Francois, E.; de La Roche, G.; Bouche, O.; et al. Clinical outcome of the ACCORD 12/0405 PRODIGE 2 randomized trial in rectal cancer. *J. Clin. Oncol.* **2012**, *30*, 4558–4565. [CrossRef]

10. Zhang, J.; Huang, M.; Cai, Y.; Wang, L.; Xiao, J.; Lan, P.; Hu, H.; Wu, X.; Ling, J.; Peng, J.; et al. Neoadjuvant Chemotherapy with mFOLFOXIRI without Routine Use of Radiotherapy for Locally Advanced Rectal Cancer. *Clin. Colorectal. Cancer* **2019**, *18*, 238–244. [CrossRef] [PubMed]

11. Fokas, E.; Allgauer, M.; Polat, B.; Klautke, G.; Grabenbauer, G.G.; Fietkau, R.; Kuhnt, T.; Staib, L.; Brunner, T.; Grosu, A.L.; et al. Randomized Phase II Trial of Chemoradiotherapy Plus Induction or Consolidation Chemotherapy as Total Neoadjuvant Therapy for Locally Advanced Rectal Cancer: CAO/ARO/AIO-12. *J. Clin. Oncol.* **2019**, *37*, 3212–3222. [CrossRef]

12. Landry, J.C.; Feng, Y.; Prabhu, R.S.; Cohen, S.J.; Staley, C.A.; Whittington, R.; Sigurdson, E.R.; Nimeiri, H.; Verma, U.; Benson, A.B. Phase II Trial of Preoperative Radiation with Concurrent Capecitabine, Oxaliplatin, and Bevacizumab followed by Surgery and Postoperative 5-Fluorouracil, Leucovorin, Oxaliplatin (FOLFOX), and Bevacizumab in Patients with Locally Advanced Rectal Cancer: 5-Year Clinical Outcomes ECOG-ACRIN Cancer Research Group E3204. *Oncologist* **2015**, *20*, 615–616. [CrossRef]

13. Dewdney, A.; Cunningham, D.; Tabernero, J.; Capdevila, J.; Glimelius, B.; Cervantes, A.; Tait, D.; Brown, G.; Wotherspoon, A.; Gonzalez de Castro, D.; et al. Multicenter randomized phase II clinical trial comparing neoadjuvant oxaliplatin, capecitabine, and preoperative radiotherapy with or without cetuximab followed by total mesorectal excision in patients with high-risk rectal cancer (EXPERT-C). *J. Clin. Oncol.* **2012**, *30*, 1620–1627. [CrossRef] [PubMed]

14. Helbling, D.; Bodoky, G.; Gautschi, O.; Sun, H.; Bosman, F.; Gloor, B.; Burkhard, R.; Winterhalder, R.; Madlung, A.; Rauch, D.; et al. Neoadjuvant chemoradiotherapy with or without panitumumab in patients with wild-type KRAS, locally advanced rectal cancer (LARC): A randomized, multicenter, phase II trial SAKK 41/07. *Ann. Oncol.* **2013**, *24*, 718–725. [CrossRef] [PubMed]

15. Rosello, S.; Papaccio, F.; Roda, D.; Tarazona, N.; Cervantes, A. The role of chemotherapy in localized and locally advanced rectal cancer: A systematic revision. *Cancer Treat. Rev.* **2018**, *63*, 156–171. [CrossRef]

16. Papaccio, F.; Rosello, S.; Huerta, M.; Gambardella, V.; Tarazona, N.; Fleitas, T.; Roda, D.; Cervantes, A. Neoadjuvant Chemotherapy in Locally Advanced Rectal Cancer. *Cancers* **2020**, *12*, 3611. [CrossRef]

17. Moro, M.; Casanova, M.; Roz, L. Patient-derived xenografts, a multi-faceted in vivo model enlightening research on rare liver cancer biology. *Hepatobiliary Surg. Nutr.* **2017**, *6*, 344–346. [CrossRef]

18. Marshall, L.J.; Triunfol, M.; Seidle, T. Patient-Derived Xenograft vs. Organoids: A Preliminary Analysis of Cancer Research Output, Funding and Human Health Impact in 2014–2019. *Animals* **2020**, *10*, 1923. [CrossRef]

19. Tiriac, H.; Belleau, P.; Engle, D.D.; Plenker, D.; Deschenes, A.; Somerville, T.D.D.; Froeling, F.E.M.; Burkhart, R.A.; Denroche, R.E.; Jang, G.H.; et al. Organoid Profiling Identifies Common Responders to Chemotherapy in Pancreatic Cancer. *Cancer Discov.* **2018**, *8*, 1112–1129. [CrossRef] [PubMed]

20. Vlachogiannis, G.; Hedayat, S.; Vatsiou, A.; Jamin, Y.; Fernandez-Mateos, J.; Khan, K.; Lampis, A.; Eason, K.; Huntingford, I.; Burke, R.; et al. Patient-derived organoids model treatment response of metastatic gastrointestinal cancers. *Science* **2018**, *359*, 920–926. [CrossRef] [PubMed]

21. Ganesh, K.; Wu, C.; O'Rourke, K.P.; Szeglin, B.C.; Zheng, Y.; Sauve, C.G.; Adileh, M.; Wasserman, I.; Marco, M.R.; Kim, A.S.; et al. A rectal cancer organoid platform to study individual responses to chemoradiation. *Nat. Med.* **2019**, *25*, 1607–1614. [CrossRef]

22. Yao, Y.; Xu, X.; Yang, L.; Zhu, J.; Wan, J.; Shen, L.; Xia, F.; Fu, G.; Deng, Y.; Pan, M.; et al. Patient-Derived Organoids Predict Chemoradiation Responses of Locally Advanced Rectal Cancer. *Cell Stem Cell* **2020**, *26*, 17–26.e16. [CrossRef] [PubMed]

23. Amin, M.B.; Greene, F.L.; Edge, S.B.; Compton, C.C.; Gershenwald, J.E.; Brookland, R.K.; Meyer, L.; Gress, D.M.; Byrd, D.R.; Winchester, D.P. The Eighth Edition AJCC Cancer Staging Manual: Continuing to build a bridge from a population-based to a more "personalized" approach to cancer staging. *CA Cancer J. Clin.* **2017**, *67*, 93–99. [CrossRef]

24. Kim, B.H.; Kim, J.M.; Kang, G.H.; Chang, H.J.; Kang, D.W.; Kim, J.H.; Bae, J.M.; Seo, A.N.; Park, H.S.; Kang, Y.K.; et al. Standardized Pathology Report for Colorectal Cancer, 2nd Edition. *J. Pathol. Transl. Med.* **2020**, *54*, 1–19. [CrossRef]

25. Park, M.; Kwon, J.; Shin, H.J.; Moon, S.M.; Kim, S.B.; Shin, U.S.; Han, Y.H.; Kim, Y. Butyrate enhances the efficacy of radiotherapy via FOXO3A in colorectal cancer patientderived organoids. *Int. J. Oncol.* **2020**, *57*, 1307–1318. [CrossRef] [PubMed]

26. Martin, M.L.; Adileh, M.; Hsu, K.S.; Hua, G.; Lee, S.G.; Li, C.; Fuller, J.D.; Rotolo, J.A.; Bodo, S.; Klingler, S.; et al. Organoids Reveal That Inherent Radiosensitivity of Small and Large Intestinal Stem Cells Determines Organ Sensitivity. *Cancer Res.* **2020**, *80*, 1219–1227. [CrossRef]

27. Mayakonda, A.; Lin, D.C.; Assenov, Y.; Plass, C.; Koeffler, H.P. Maftools: Efficient and comprehensive analysis of somatic variants in cancer. *Genome Res.* **2018**, *28*, 1747–1756. [CrossRef]

28. Braselmann, H.; Michna, A.; Hess, J.; Unger, K. CFAssay: Statistical analysis of the colony formation assay. *Radiat. Oncol.* **2015**, *10*, 223. [CrossRef]

29. Sachs, N.; de Ligt, J.; Kopper, O.; Gogola, E.; Bounova, G.; Weeber, F.; Balgobind, A.V.; Wind, K.; Gracanin, A.; Begthel, H.; et al. A Living Biobank of Breast Cancer Organoids Captures Disease Heterogeneity. *Cell* **2018**, *172*, 373–386. [CrossRef] [PubMed]

30. Kuo, C.J.; Curtis, C. Organoids reveal cancer dynamics. *Nature* **2018**, *556*, 441–442. [CrossRef]

31. Muthuswamy, S.K. Organoid Models of Cancer Explode with Possibilities. *Cell Stem Cell* **2018**, *22*, 290–291. [CrossRef]

32. Crespo, M.; Vilar, E.; Tsai, S.Y.; Chang, K.; Amin, S.; Srinivasan, T.; Zhang, T.; Pipalia, N.H.; Chen, H.J.; Witherspoon, M.; et al. Colonic organoids derived from human induced pluripotent stem cells for modeling colorectal cancer and drug testing. *Nat. Med.* **2017**, *23*, 878–884. [CrossRef]

33. Seidlitz, T.; Merker, S.R.; Rothe, A.; Zakrzewski, F.; von Neubeck, C.; Grutzmann, K.; Sommer, U.; Schweitzer, C.; Scholch, S.; Uhlemann, H.; et al. Human gastric cancer modelling using organoids. *Gut* **2019**, *68*, 207–217. [CrossRef] [PubMed]

34. Broutier, L.; Mastrogiovanni, G.; Verstegen, M.M.; Francies, H.E.; Gavarro, L.M.; Bradshaw, C.R.; Allen, G.E.; Arnes-Benito, R.; Sidorova, O.; Gaspersz, M.P.; et al. Human primary liver cancer-derived organoid cultures for disease modeling and drug screening. *Nat. Med.* **2017**, *23*, 1424–1435. [CrossRef]
35. Zhang, H.C.; Kuo, C.J. Personalizing pancreatic cancer organoids with hPSCs. *Nat. Med.* **2015**, *21*, 1249–1251. [CrossRef]
36. Figueroa, R.L.; Zeng-Treitler, Q.; Kandula, S.; Ngo, L.H. Predicting sample size required for classification performance. *BMC Med. Inform. Decis. Mak.* **2012**, *12*, 8. [CrossRef] [PubMed]
37. Hirata, E.; Sahai, E. Tumor Microenvironment and Differential Responses to Therapy. *Cold Spring Harb. Perspect. Med.* **2017**, *7*. [CrossRef] [PubMed]
38. Jarosz-Biej, M.; Smolarczyk, R.; Cichon, T.; Kulach, N. Tumor Microenvironment as A "Game Changer" in Cancer Radiotherapy. *Int. J. Mol. Sci.* **2019**, *20*, 3212. [CrossRef] [PubMed]
39. Barker, H.E.; Paget, J.T.; Khan, A.A.; Harrington, K.J. The tumour microenvironment after radiotherapy: Mechanisms of resistance and recurrence. *Nat. Rev. Cancer* **2015**, *15*, 409–425. [CrossRef]
40. Ohlund, D.; Handly-Santana, A.; Biffi, G.; Elyada, E.; Almeida, A.S.; Ponz-Sarvise, M.; Corbo, V.; Oni, T.E.; Hearn, S.A.; Lee, E.J.; et al. Distinct populations of inflammatory fibroblasts and myofibroblasts in pancreatic cancer. *J. Exp. Med.* **2017**, *214*, 579–596. [CrossRef]
41. Dijkstra, K.K.; Cattaneo, C.M.; Weeber, F.; Chalabi, M.; van de Haar, J.; Fanchi, L.F.; Slagter, M.; van der Velden, D.L.; Kaing, S.; Kelderman, S.; et al. Generation of Tumor-Reactive T Cells by Co-culture of Peripheral Blood Lymphocytes and Tumor Organoids. *Cell* **2018**, *174*, 1586–1598. [CrossRef] [PubMed]
42. Neal, J.T.; Li, X.; Zhu, J.; Giangarra, V.; Grzeskowiak, C.L.; Ju, J.; Liu, I.H.; Chiou, S.H.; Salahudeen, A.A.; Smith, A.R.; et al. Organoid Modeling of the Tumor Immune Microenvironment. *Cell* **2018**, *175*, 1972–1988.e16. [CrossRef] [PubMed]

Review

The Role of Cancer Stem Cells in Colorectal Cancer: From the Basics to Novel Clinical Trials

Céline Hervieu [1], Niki Christou [1,2], Serge Battu [1] and Muriel Mathonnet [1,2,*]

1   EA 3842 CAPTuR "Control of Cell Activation in Tumor Progression and Therapeutic Resistance", Faculty of Medicine, Genomics, Environment, Immunity, Health and Therapeutics (GEIST) Institute, University of Limoges, 87025 Limoges CEDEX, France; celine.hervieu@unilim.fr (C.H.); christou.niki19@gmail.com (N.C.); serge.battu@unilim.fr (S.B.)
2   Department of General, Endocrine and Digestive Surgery, University Hospital of Limoges, 87025 Limoges CEDEX, France
*   Correspondence: muriel.mathonnet@unilim.fr

Simple Summary: Cancer stem cells (CSCs) fuel tumor growth, metastasis and resistance to therapy in colorectal cancer (CRC). These cells therefore represent a promising target for the treatment of CRC but are difficult to study because of the complexity of their isolation. This review presents the methods currently used to isolate colorectal CSCs as well as the techniques for characterizing these cells with their advantages and limitations. The aim of this review is to provide a state-of-the-art on the clinical relevance of CSCs in CRC by outlining current treatments for CRC, the resistance mechanisms developed by CSCs to overcome them, and ongoing clinical trials of drugs targeting CSCs in CRC. Overall, this review addresses the complexity of studying CSCs in CRC research and developing clinically effective treatments to enable CRC patients to achieve a short and long-term therapeutic response.

Citation: Hervieu, C.; Christou, N.; Battu, S.; Mathonnet, M. The Role of Cancer Stem Cells in Colorectal Cancer: From the Basics to Novel Clinical Trials. *Cancers* **2021**, *13*, 1092. https://doi.org/10.3390/cancers13051092

Academic Editors: Marta Baiocchi and Ann Zeuner

Received: 28 January 2021
Accepted: 27 February 2021
Published: 4 March 2021

Publisher's Note: MDPI stays neutral with regard to jurisdictional claims in published maps and institutional affiliations.

Abstract: The treatment options available for colorectal cancer (CRC) have increased over the years and have significantly improved the overall survival of CRC patients. However, the response rate for CRC patients with metastatic disease remains low and decreases with subsequent lines of therapy. The clinical management of patients with metastatic CRC (mCRC) presents a unique challenge in balancing the benefits and harms while considering disease progression, treatment-related toxicities, drug resistance and the patient's overall quality of life. Despite the initial success of therapy, the development of drug resistance can lead to therapy failure and relapse in cancer patients, which can be attributed to the cancer stem cells (CSCs). Thus, colorectal CSCs (CCSCs) contribute to therapy resistance but also to tumor initiation and metastasis development, making them attractive potential targets for the treatment of CRC. This review presents the available CCSC isolation methods, the clinical relevance of these CCSCs, the mechanisms of drug resistance associated with CCSCs and the ongoing clinical trials targeting these CCSCs. Novel therapeutic strategies are needed to effectively eradicate both tumor growth and metastasis, while taking into account the tumor microenvironment (TME) which plays a key role in tumor cell plasticity.

Keywords: colorectal cancer; cancer stem cells; drug resistance; clinical trials

1. Introduction

Colorectal cancer (CRC) is the fourth leading cause of cancer-related death worldwide [1]. While the occurrence and mortality rates of CRC is declining in the European countries, these rates are increasing in rapidly transitioning countries, such as many African and South Asian countries [2]. The tumor–node–metastases (TNM) classification allows the stratification of patient groups according to the stage of the disease, based on anatomical information [3,4]. The location and stage of the tumor enable both the assessment of the patient's prognosis and the determination of the therapeutic approach, depending on the patient's overall health as well as the status of the tumor in terms of mutation and mismatch repair (MMR) [1,5]. Therapeutic options for the treatment of CRC are surgical resection, systemic therapy including chemotherapy, targeted therapy and immunotherapy, local therapy for metastases and palliative therapy [1,6]. Importantly, surgical resection is the only curative treatment, if all macroscopic and microscopic tumor foci can be removed [1,6]. Unfortunately, even after well directed curative treatment, some patients experience treatment failure that may be associated with the development of multidrug resistance (MDR) during or after treatment. In addition, despite initially successful therapy, the development of drug resistance often leads to relapse in cancer patients, known as minimal residual disease (MRD) [7]. Both MDR and MRD can be attributed to a subpopulation of tumor cells with self-renewal and multi-lineage differentiation capabilities, the cancer stem cells (CSCs), known as colorectal cancer stem cells (CCSCs) for CRC [8]. CSCs contribute to tumor initiation and dissemination, treatment resistance and metastasis development. Tumor microenvironment (TME) and metabolic plasticity may also be involved in therapeutic failure by imposing selective pressures on cancer cells that lead to chemoresistance and cancer progression [9,10]. Therefore, the development of new therapies targeting CSCs, taking into account the TME and tumor metabolism, represents an interesting approach to overcome resistance to therapies [11]. In this review, we will present the origin of CCSCs and provide an overview of the techniques currently used to isolate them. Then, we will review current knowledge on the clinical relevance of CCSCs, through the clinical management of CRC and the mechanisms of resistance to therapies associated with CCSCs. Finally, we will introduce some clinical trials based on drugs targeting CCSCs.

2. Colorectal Cancer Stem Cells

The CSC theory suggests that tumor growth is driven by a small number of dedicated stem cells (SCs), the CSCs [8]. By definition, a CSC has the ability to self-renew in order to expand its pool and to generate all the differentiated cells that comprise the tumor (multipotency). The transformation of a colorectal stem cell into CCSC requires the acquisition of tumor-related features.

2.1. Colorectal Cancer Stem Cell Origin

The history of CSCs began two decades ago with the discovery of CSCs in human acute myeloid leukemia (AML) by Dick and colleagues [12]. For the first time, a cell capable of initiating human AML in immunodeficient mice and possessing differentiation, proliferation and self-renewal capabilities was described. A few years later, using similar experimental approaches, the presence of CSC was demonstrated in solid cancers such as colorectal cancer. The origin of CSCs in CRC is controversial, and several hypotheses have been proposed. CCSCs are associated with the acquisition of malignant molecular and cellular changes either due to the accumulation of genetic and epigenetic alterations in restricted stem/progenitor cells and normal tumor cells, or to the dedifferentiation of somatic cells caused by various genetic and environmental factors [13–15]. CSCCs exhibit tumor-related characteristics such as uncontrolled growth, tumorigenicity and therapy resistance, and may constitute the small reservoir of drug-resistant cells that are responsible for relapses after chemotherapy-induced remission, known as MRD, and distant metastasis [7,11]. Thus, CCSCs play a key role in the initiation, invasion and progression of CRC as well as resistance to therapy. These CCSCs give rise to heterogeneous tumors that can be serially transplanted into immunodeficient mice that resemble the original tumor [16]. In addition, CCSCs have the ability to form disseminated metastatic tumors due to their extensive proliferative potential [15]. One of the main challenges in the study of CCSCs is their isolation, due to their low percentage within the tumor [16]. However, the CCSC population appears to be phenotypically and functionally heterogeneous and dynamic, which is another barrier to their isolation [17]. Therefore, the development of therapies that selectively eradicate CCSCs offers promising opportunities for a sustainable clinical response but requires effective technologies to detect and isolate them [11].

2.2. Colorectal Cancer Stem Cell Isolation Methods

Different methods are used to isolate CCSCs, based either on the expression pattern of CCSC markers, the functional aspect of CCSCs, or their biophysical features [18]. The objective of this chapter is to present the techniques currently in use with the advantages and disadvantages of each approach.

2.2.1. CCSC Isolation Based on Phenotypic Features

Many stem cells markers were found to be associated with CCSC features. However, the heterogeneous and dynamic nature of CCSCs challenges their isolation and enrichment. The first publications from the literature identifying subpopulations of CSCs in CRC are summarized in Table 1. Experimental models, CCSC isolation methods and characterization techniques used by the authors are detailed in this table. Studies conducted by O'Brien et al. and Ricci-Vitiani et al. identified the first CCSC marker: the five-transmembrane glycoprotein CD133 [19,20]. However, its use has become controversial as the tumorigenic and clonogenic potential of CD133$^+$-CSCs depends on the positivity for a specific glycosylated epitope of the CD133 protein [21].

Table 1. Experimental models, markers and CCSC isolation and characterization methods used in the first publications identifying CSCs in CRC.

References	Experimental Models	Identified CCSC Subpopulations	CCSC Isolation Methods	CCSC Characterization Assays
O'Brien et al. [20]	CRC patient tissues CRC cells from patient tumors Animal model (mice)	CD133$^+$	MACS and FACS	Flow cytometry Immunohistochemistry Tumorigenicity assay
Ricci-Vitiani et al. [19]	CRC patient tissues CRC cells from patient tumors Primary tumor cell cultures Animal model (mice)	CD133$^+$	MACS and FACS	Sphere formation assay Flow cytometry Immunohistochemistry Tumorigenicity assay
Dalerba et al. [22]	CRC patient tissues CRC xenograft lines Single-cell suspensions	EpCAM^{high}/CD44$^+$ EpCAM^{high}/CD44$^+$/CD166$^+$	FACS	ALDH assay Flow cytometry Tumorigenicity assay
Barker et al. [23]	Animal model (Ah-cre/Apc$^{flox/flox}$ and Lgr5-EGFP-IRES-creER^{T2}/APC$^{flox/flox}$ mice)	Lgr5$^+$	/	LacZ analysis Immunohistochemistry
Sangiorgi and Capecchi [24]	Animal model (Bmi1-IRES-Cre-ER mice)	Bmi1$^+$	/	LacZ analysis Immunohistochemistry
Vermeulen et al. [25]	CRC patient tissues CRC cells and single-cell-derived cultures from patient tumors Animal model (mice)	CD133$^+$/CD24$^+$ CD44$^+$/CD166$^+$ CD24$^+$/CD29$^+$	MACS and FACS	Sphere formation assay In vitro differentiation assay Immunohistochemistry Flow cytometry Tumorigenicity assay
Pang et al. [26]	CRC patient tissues CRC cells from patient tumors Animal model (mice)	CD133$^+$/CD26$^+$ CD133$^+$/CD26$^+$/CD44$^+$	MACS and FACS	Sphere formation assay In vitro invasion assays Chemotherapeutic treatments Tumorigenicity assay
Todaro et al. [27]	CRC patient tissues Sphere-derived adherent cultures CRC cells from patient tumors or spheres Animal model (mice)	CD44v6$^+$	MACS and FACS	Immunofluorescence Immunohistochemistry Invasion assay Sphere formation assay Tumorigenicity assay

CRC: colorectal cancer; CCSC: colorectal cancer stem cells; CD: cluster of differentiation; MACS: magnetic-activated cell sorting; FACS: fluorescence-activated cell sorting; ALDH: aldehyde dehydrogenase.

Then, Clarke's group showed that EpCAM^{high}/CD44$^+$ cells isolated from human CRC could establish a tumor in mice with morphological and phenotypic heterogeneity of the original tumor and concluded that CD44 and EPCAM markers could be considered robust CCSC markers [22]. In addition, the study by Dalerba et al. highlights an additional differentially expressed marker, CD166, which could be used to further enrich CCSCs in the EpCA^{high}/CD44$^+$ population [22]. Using lineage-tracing experiments in mice, Clevers and coworkers identified stem cells in the small intestine and colon using the marker gene *Lgr5* [28] and proposed them as the cells-of-origin of intestinal cancer [23]. At the same time, Sangiorgi and Capecchi's study found another intestinal stem cell marker in vivo,

Bmi1 [24]. Importantly, Bmi-1 and Lgr5 markers define two types of SCs, quiescent and rapidly cycling SCs, respectively [23,24], and may identify CCSCs. Vermeulen et al. showed that spheroid cultures from primary CRC have a tumor-initiating capacity and that a cell subpopulation expresses CD24, CD29, CD44 and CD166 markers, suggested as CCSC markers [25]. The study by Pang et al. identifies a subpopulation of CD26$^+$ cells capable of developing distant metastases when injected into the mouse cecal wall and associated with increased invasiveness and chemoresistance, whereas CD26$^-$ cells cannot [26]. Interestingly, the presence of CD26$^+$ cells in the primary tumor of patients without distant metastases at that time may predict future distant metastases, highlighting a critical role of CSCs in the progression of metastatic cancer and important clinical implications [26]. The transmembrane glycoprotein CD44 has several splicing variants, including CD44v6, which appears to negatively impact the prognosis of CRC patients [29,30]. Todaro et al. demonstrated that all identified CCSCs express the CD44v6 marker, which supports their migration and promotes metastasis [27]. Each of these markers has its own function and role in the prognosis of CRC, as shown in Table 2.

Table 2. Functions and roles in CRC prognosis of CCSC markers.

CCSC Markers	Functions	Roles in Prognosis of CRC	References
Bmi-1	Polycomb-repressor protein Involved in self-renewal	High expression of Bmi-1 is associated with poor survival	[23,24,31,32]
CD24 (Heat stable antigen 24)	Cell adhesion molecule Alternative ligand of P-selectin	Strong cytoplasmic expression of CD24 is correlated with shortened patient survival	[25,33]
CD26	Cell adhesion glycoprotein Promote invasion and metastases	Elevated-CD26 expression is associated with advanced tumor staging and worse overall survival	[26,34]
CD29 (Integrin-β1)	Transmembrane proteinInvolved in cell adhesion	Overexpression of CD29 is correlated with poor prognosis and aggressive clinicopathological features	[25,35]
CD44	Transmembrane glycoprotein Regulate cell interactions, adhesion and migration	CD44 overexpression is associated with lymph node metastasis, distant metastases and poor prognosis	[36–38]
CD44v6	Bind hepatocyte growth factor Promote migration and metastases	High level of CD44v6 has an unfavorable impact on overall survival	[27,29,38]
CD133 (Prominin-1)	Cell surface glycoprotein Regulate self-renewal and tumor angiogenesis	CD133 expression is correlated with low survival in CRC patients	[21,39,40]
CD166 (Activated leukocyte adhesion molecule)	Cell adhesion molecule Mediate homophilic interactions	Overexpression of CD166 is correlated with shortened patient survival	[22,25,41]
EpCAM (Epithelial cell adhesion molecule)	Transmembrane glycoprotein Regulate cell adhesion, proliferation and migration	Loss of EpCAM expression is associated with tumor stage, lymph node and distant metastases and poor prognosis	[22,37,42]
Lgr5 (Leucine-rich repeat-containing G-protein coupled receptor 5)	Seven-transmembrane protein Target of Wnt pathway involved in self-renewal	Lgr5 expression is associated with lymph node and distant metastases, and overexpression with reduced overall survival	[23,28,37,43]

CCSC: colorectal cancer stem cells; CD: cluster of differentiation; ECM: extracellular matrix; CRC: colorectal cancer.

All these markers can be expressed by CCSCs, but they do not all have the same capacity. Some, such as CD133, Lgr5, Bmi-1, CD26 and CD44v6 alone identify CCSCs, while the other presented markers allow the identification of CCSCs only in combination with one or more of the aforementioned markers. In conclusion, these markers play a key role in the identification of CCSCs and can be used alone or in combination to sort CCSCs by magnetic-activated cell sorting (MACS) or fluorescence-activated cell sorting (FACS) techniques.

MACS is a magnetic-based cell isolation technique, using a positive selection strategy, presented in Figure 1 panel 1 [44]. Magnetic beads are conjugated to highly specific monoclonal antibodies that recognize CCSC marker on the surface of cells of interest. Then, the heterogeneous suspension of cells is passed through a separation column, in a magnetic field, to retain the cells labeled with magnetic beads and antibodies [45]. By switching off the magnetic field, target cells will be eluted. MACS is a fast and easy method of cell separation, especially for the isolation of CCSCs that represent a small cell population in the tumor mass. However, MACS is only a mono-parameter separation method that requires cell labelling and is unable to separate cells based on the variable expression of markers [44,45].

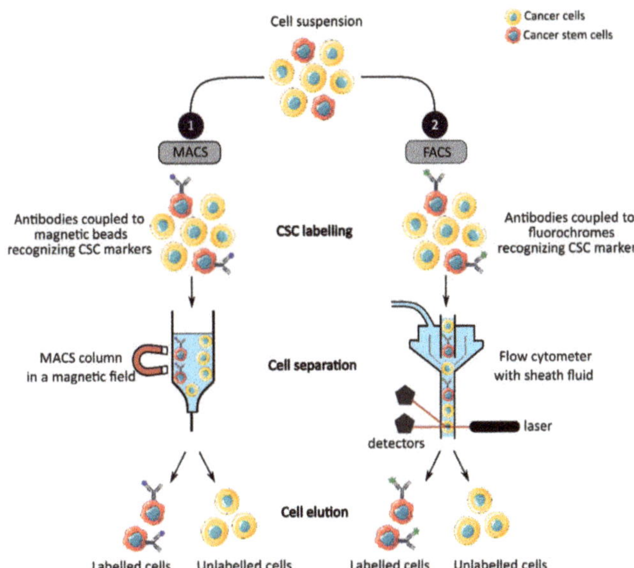

Figure 1. Phenotypic sorting of CSCs through the expression of CSC markers recognized by antibodies coupled to either magnetic beads, MACS (1), or fluorochromes, FACS (2). Once the antibodies are added, the cell suspension is passed through either a MACS column in a magnetic field that retains the antibody-labeled cells (1) or through a flow cytometer that distinguishes and isolates labeled cells from unlabeled cells (2). CSC: cancer stem cell; MACS: magnetic-activated cell sorting; FACS: fluorescence-activated cell sorting.

FACS uses fluorescently labeled antibodies that target the cell surface or intracellular markers to isolate CCSCs [44]. Antibodies are conjugated to fluorochromes and recognize the marker of interest within a cell suspension, as shown in Figure 1 panel 2 [44]. The cell suspension is then hydrodynamically focused into a stream of individual cells by the flow cytometer and passed through a laser which provides information on the size, granularity and fluorescent properties of single cells [18]. Fluorochromes with different emission wavelengths can be used simultaneously to allow multiparameter separations [44]. Both technologies allow the sorting of CCSCs with high purity but require the availability of antibodies and cell labeling, which can modify their properties and induce cell differentiation [16,44,46]. In addition, phenotypic characterization is insufficient to define a CCSC because these markers are also expressed by normal SCs.

Therefore, in order to confirm the detection and isolation of CCSCs, their functional capabilities need to be evaluated by in vitro and in vivo assays [18].

2.2.2. CCSC Isolation Based on Functional Features

CCSCs have many intrinsic properties that can be used to identify them, such as their capacity for self-renewal, multi-lineage differentiation, detoxification due to aldehyde dehydrogenase 1 (ALDH1) activity and dye exclusion ability, colony/sphere formation and tumorigenicity, which are illustrated in Figure 2. These functional characteristics have been used to develop effective methods for isolating CCSCs. The ALDH activity assay is based on the use of a fluorescent and non-toxic ALDH substrate that freely diffuses into intact and viable cells [47]. Then, in the presence of the detoxifying enzyme ALDH, the substrate is converted into a negatively charged fluorescent product that is retained inside the cells. Thus, cells with high ALDH activity become brightly fluorescent and can be measured by flow cytometry as presented in Figure 2 panel 1a [47,48]. CCSCs increase their ALDH1 activity to resist to chemotherapeutic agents and prevent apoptosis by maintaining low levels of reactive oxygen species [47]. The advantage of the ALDH assay is high stability compared to the use of surface markers, but its specificity is low due to its expression in both normal SCs and CSCs [48].

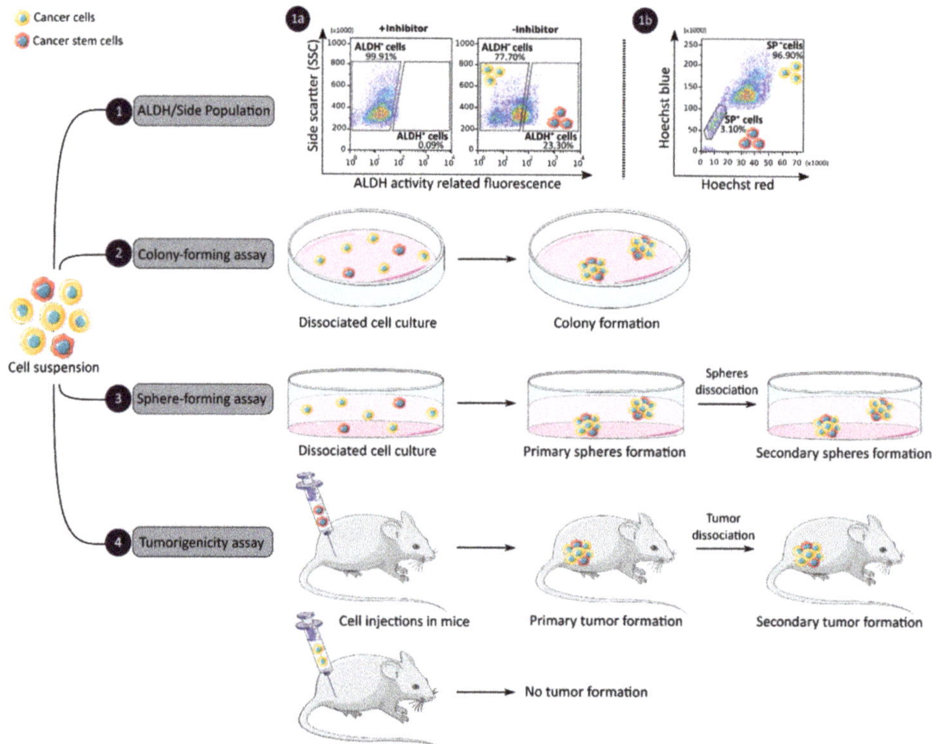

Figure 2. Functional sorting of CSCs due to their specific properties such as enhanced detoxification (1), ALDH (1a) and SP (1b), in vitro self-renewal and differentiation capacity, colony- (2) and sphere-forming (3) assays, and the ability to form tumors in vivo, tumorigenicity assay (4). CSC: cancer stem cell; ALDH: aldehyde dehydrogenase; SP: side population.

The side population (SP) assay relies on the differential ability of the cells to efflux dye via ATP-binding cassette (ABC) transporters [49]. Hoechst33342 is a fluorescent dye that binds all nucleic acids and has the particularity of passing through the plasma membrane of living cells. When excited by UV lights, Hoechst dye emits a fluorescence that can be detected by a flow cytometer [49]. SP cells are capable of actively removing the dye from the cell and have a unique low Hoechst fluorescence emission, as shown in Figure 2 panel 1b. CCSCs highly express efflux transporters, such as multidrug resistance protein 1 (ABCB1), multidrug resistance-associated proteins (ABCC1) and breast cancer resistance protein (ABCG2), to protect themselves against cytotoxic substances and therefore look like SP cells [18]. The SP assay is an easy and reliable method that does not require cell labeling, but due to its low purity and specificity, the SP assay is often combined with cell labeling to significantly increase the purity of sorted CSCCs [18,49].

Colony and sphere formation assays evaluate in vitro the self-renewal and differentiation capacities of individual cells in two (2D) and three (3D) dimensions, respectively, which are shown in Figure 2, panels 2 and 3 [50,51]. Both assays are based on non-adherent cultures using either a soft agar layer (2D) or low adherent plates (3D) [52,53]. In the soft agar method illustrated in Figure 2 panel 2, the suspension of individual cells is mixed with the soft agar which may, after several weeks of incubation, give colonies that can be stained with crystal violet to determine their number and size [50,52]. In comparison, in the 3D culture shown in Figure 2 panel 3, the individual cells in suspension are grown at very low cell density and in serum-free medium (DMEM/F12 medium) supplemented with growth factors (human recombinant basic fibroblast growth factor and human recombinant epidermal growth factor), N2 supplement, glucose, insulin and optionally antibiotics such as penicillin/streptomycin for several weeks to obtain spheroids [51,54]. The produced spheroids mimic various characteristics of solid tumors, such as growth kinetics, gene expression pattern and cellular organization with the outer layer containing highly proliferative cells, the middle layer with senescent or quiescent cells and the inner layer comprising necrotic cells due to a lack of oxygen and nutrients [53]. CCSCs can be identified in both techniques as they have the ability to form larger and more numerous colonies and are capable of giving rise to a tumor sphere (colonosphere) resembling the primary sphere when passed in series, due to their ability to grow and divide independently of their environment which normal cells are unable to do because of anoikis [18,52,55]. Thus, in vitro, 3D models appear to be a relevant preclinical model for testing new drugs, evaluating potential combinations and understanding drug resistance, by mimicking CSC-containing tumors in vitro, before testing them in vivo [18,53,55]. However, these models require well-established protocols and appropriate cell dilution to certify that each colony/sphere is derived from a single cell [18].

The tumorigenicity assay is considered the gold standard method for studying the CSC properties of human tumors in vivo [18,56]. This approach allows to determine the tumor-initiating ability of cancer cells in immunodeficient mice and their capacity for self-renewal in vivo after the dissociation of primary tumors and transplantation in secondary recipient mice, as illustrated in Figure 2 panel 4 [57]. In vivo limiting dilution is the best method for identifying the lowest concentration of cells capable of forming a tumor and determining the frequency of CSCs [18,58]. Importantly, only CSCs have the ability to generate a xenograft that is histologically similar to the parental tumor from which it originated, to be serially transplanted in a xenograft assay due to their long-term self-renewal capacity, and to generate daughter cells [56,58]. However, the use of mouse models requires ethical consideration and complicated laboratory equipment. In addition, the results of xenograft experiments are highly dependent on the number of cells, the implantation site and the incubation period, which leads to certain limitations [18]. Nevertheless, mouse models remain unique models for studying the biology of CSCs in vivo [57,58].

2.2.3. CCSC Isolation Based on Biophysical Features

The development of enrichment and isolation methods for CCSCs without cell labeling offers new perspectives, such as sorting techniques based on biophysical characteristics. The sedimentation field-flow fractionation (SdFFF) is a gentle, non-invasive and label-free method that prevents interference for further cell use and the allows separation of cells according to their size, density, shape and rigidity [16,59]. Cell separation by SdFFF depends on the differential elution of cell subpopulations submitted both to the action of a parabolic profile generated by the mobile phase in the channel and to a multigravitational external field generated by the rotation of the channel, as presented in Figure 3 [16,59]. In the past decade, SdFFF cell sorting has been adapted and applied in many fields such as neurology, oncology and stem cells [16,60–62]. The study by Mélin et al. describes a strategy, based on SdFFF elution, to obtain activated and quiescent CSC subpopulations from eight different human CRC cell lines [16]. The combination of cell sorting by SdFFF with the grafting of these CSC-enriched fractions into chick embryo chorioallantoic membrane (CAM) model demonstrates the potential of SdFFF to produce innovative matrices for the study of carcinogenesis and the analysis of treatment sensitivity [16,63]. The advantages of this isolation method are the use of biophysical characteristics for cell sorting without cell labeling; however, this technique requires a large number of cells and is time consuming [46].

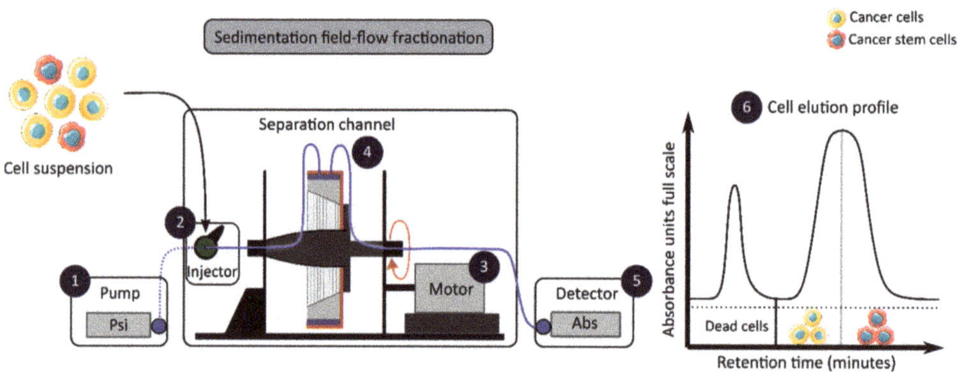

Figure 3. Biophysical sorting of CSCs according to their size, density, shape and rigidity using the SdFFF technique, which does not require cell labelling or fixation. The SdFFF is composed of a pump (1) to transport the mobile phase (PBS) and the cells, an injector (2) to introduce the cell suspension, a motor (3) to rotate the separation channel (4) and a detector (5) coupled to a computer to obtain the elution profile of the cell suspension (6). Psi is a common unit of pressure. CSC: cancer stem cell; SdFFF: sedimentation field-flow fractionation; PBS: phosphate-buffered saline; Abs: absorbance.

2.2.4. CCSC Isolation Methods: Discussion

Taken together, this chapter provides an overview of the techniques commonly used to identify and sort CCSCs, which are summarized in Table 3. The use of cell surface markers remains the most widely used in cancer research, however, it remains controversial due to the lack of a universal marker for CCSCs. Moreover, nowadays, none of the CSC isolation techniques are capable of 100% enrichment of CCSCs due to the shared properties between normal SCs, non-CCSCs and CCSCs [14,17]. As an example, Shmelkov and colleagues have shown that CD133 expression in the colon is not limited to SCs but is also expressed on differentiated tumors cells [64]. In addition, the authors found that both CD133$^+$ and CD133$^-$ isolated from metastatic colon tumors are capable of initiating tumors in a serial xenotransplantation model [64]. A few years later, the study by Kemper et al. demonstrated that CD133 is expressed on the cell surface of CSCs and differentiated tumor cells but is differentially glycosylated [21]. Similarly, using the ALDH activity assay, Huang et al. found that ALDH1 is a marker of both normal and malignant human colonic SCs [48]. Consequently, cell surface markers and ALDH activity cannot be used alone to sort and define CSCs. Thus, the SdFFF technique offers new perspectives for CSC sorting that does not require cell labeling or fixation and thereby allows the combination of this technique with other CSC characterization methods. Therefore, the combined use of CCSC isolation methods can provide a more powerful and efficient tool for identifying and sorting CCSCs. The advantages and weaknesses of each method must be known in order to select the best method based on the experimental question, as shown in Table 3.

Table 3. Advantages and disadvantages of CCSC isolation methods.

Features	Isolation Methods	Advantages	Disadvantages	References
Phenotypic	MACS	High specificity Fast and easy method	No universal CCSC marker Monoparameter separation	[18,31,32]
	FACS	High specificity Multiparameter separation	No universal CCSC marker Require large number of cells	[18,31]
Functional	ALDH activity assay	High stability	Low specificity	[47,48]
	Side population assay	No cell labelling required	Low purity and specificity	[49]
	Colony and sphere formation assay	No need for complicated laboratory equipment	Absence of standardized protocol Require proper cell dilution	[50,52,53]
	Tumorigenicity assay	Gold standard method	Complicated laboratory equipment Ethical consideration	[56,58]
Biophysical	SdFFF	No cell labelling required Cell size and density separation	Time consuming	[16,46,59]

CCSC: colorectal cancer stem cell; MACS: magnetic-activated cell sorting; FACS: fluorescence-activated cell sorting; ALDH: aldehyde dehydrogenase; SdFFF: sedimentation field flow fractionation.

3. Clinical Relevance of Colorectal Cancer Stem Cells

Therapeutic advances made in recent decades now enable most cancer patients to achieve major clinical responses [6]. However, although therapeutic approaches are increasing, none of these treatment modalities is curative in most cases of advanced CRC [65]. Furthermore, despite initially successful treatment reflecting the therapeutic effect on the cells that form the tumor bulk, tumor recurrence is almost inevitable due to the development of drug resistance attributed to CCSCs [8].

3.1. Clinical Management of Colorectal Cancer

Treatment options and recommendations depend on several factors, including the patient's overall health, possible side effects, the type and stage of the tumor, and its mutational and MMR status [1,5]. Therapeutic approaches for the treatment of CRC include surgical resection, local therapies for metastatic disease, systemic therapy comprising chemotherapy, targeted therapy and immunotherapy as presented in Table 4, and palliative chemotherapy [6]. To ensure the optimal survival and quality of life for patients, personalized therapy is crucial to enable cancer patients to maximize the benefits while minimizing the harms [5].

Surgical resection is the mainstay of curative intent treatment for localized and advanced CRCs but needs to be complete to be considered curative when there is regional invasion or histological factors with a poor prognosis [66,67]. Surgery can be associated with neoadjuvant therapy in order to shrink tumor mass and facilitate medical operation and/or with adjuvant therapy to limit cancer recurrence [1]. Importantly, neoadjuvant chemotherapy, possibly coupled with radiotherapy, is mainly indicated for rectal cancers [68]. Treatment regimens for patients with localized CRC generally include chemotherapy such as 5-fluorouracil (5-FU) or capecitabine, oxaliplatin and irinotecan, alone or in combination [69–73]. Leucovorin is commonly administered with 5-FU to enhance its antitumor effect [74]. Despite many advances in CRC treatment, approximately 20% of new CRC cases are already metastatic [75]. The most common sites of metastatic colorectal cancer (mCRC) are the liver, lungs and peritoneum. Unfortunately, up to 50% of patients with early-stage disease at diagnosis will eventually develop metastatic disease, and 80–90%

of them have unresectable metastatic disease because of the size, location, and/or extent of disease [76,77].

Local therapies are approved for mCRC with inoperable lesions. The choice of local therapies depends on the location and the extent of the metastases [78]. For patients with unresectable liver or lung metastases, radiofrequency ablation is recommended for the treatment of small and medium-sized lesions, but for larger lesions and those near vascular structures, microwave ablation or stereotactic body radiation therapy may be good alternatives [1,6]. Liver metastases can also be treated by administering a higher dose of chemotherapy directly into the hepatic artery compared to systemic therapy (hepatic arterial infusion) or by combining drug/radiation administration with blood vessel obstruction (chemo/radio-embolization) [79]. For patients with peritoneal metastases, cytoreductive surgery and hyperthermic intraperitoneal chemotherapy are recommended [6]. Local therapies can be administered with curative or palliative intent and are the most often used in combination with systemic therapy [6,79].

Systemic therapy for CRC aims to downsize the primary tumor or metastases in order to convert them to a resectable status and increase progression-free survival [6]. Patients with advanced CRC usually receive several lines of therapy, most often including a combination of chemotherapy with targeted therapy or immunotherapy, depending on tumor mutational and MMR status [5]. Targeted therapies are recommended for patients with *KRAS/NRAS/BRAF* mutated or wild-type tumors, *HER2*-amplified tumors and *NTRK* gene fusion-positive tumors, while immunotherapy is only offered for tumors with high microsatellite instability (MSI), as shown in Table 4. Thus, both statuses must be determined prior to the start of therapy [80]. Unfortunately, for advanced CRC patients whose overall health has deteriorated despite treatment, palliative treatments and the best supportive care are the only remaining options [5]. Therefore, the clinical management of patients with mCRC represents a unique challenge to balance benefits and harms, including the identification of strategies that improve disease response, limit treatment-associated toxicities, and improve the overall quality of life [81].

Table 4. Systemic therapies for localized and advanced colorectal cancer.

Systemic Therapies	Drug Names	Functions	Recommendations	References
Chemotherapy	5-Fluorouracil	Antimetabolite	Localized and advanced tumors	[82]
	Capecitabine	Antimetabolite		[72]
	Irinotecan	Topoisomerase inhibitor		[83]
	Oxaliplatin	Alkylating agent		[84]
	Trifluridine/ Tipiracil	Nucleoside analog/ TP inhibitor		[85]
Targeted therapy	Bevacizumab	mAb anti-VEGF-A	*KRAS/NRAS/BRAF* Mutated tumors	[86]
	Regorafenib	Multikinase inhibitor targeting e.g., VEGFR and BRAF		[87]
	Aflibercept	Recombinant fusion protein blocking VEGF-A/B		[88]
	Ramucirumab	mAb anti-VEGFR-2		[89]
	Cetuximab	mAb anti-EGFR	*KRAS/NRAS/BRAF* Wild-type tumors	[90]
	Panitumumab			[90]
Immunotherapy	Pembrolizumab	mAb anti-PD-1	MSI-high tumors	[91]
	Nivolumab			[92]
	Ipilimumab	mAb anti-CTLA4		[92]

Table 4. *Cont.*

Systemic Therapies	Drug Names	Functions	Recommendations	References
Newly developed therapy	Vemurafenib	BRAF inhibitors	*BRAF* V600E mutated tumors	[93]
	Dabrafenib			[93]
	Encorafenib			[94]
	Trametinib	MEK inhibitors		[93]
	Binimetinib			[94]
	Trastuzumab	mAb anti-HER2	*HER2* amplified tumors	[95]
	Pertuzumab			[95]
	Lapatinib	Dual HER2/EGFR inhibitor		[96]
	Larotrectinib	TRK inhibitors	*NTRK* gene fusion-positive tumors	[97]
	Entrectinib			[98]

TP: thymidine phosphorylase; mAb: monoclonal antibody; VEGF: vascular endothelial growth factor; VEGFR: vascular endothelial growth factor receptor; EGFR: epidermal growth factor receptor; PD-1: programmed death cell receptor 1; CTLA4: cytotoxic T-lymphocyte-associated antigen 4; MEK: mitogen-activated kinases; TRK: tropomyosin receptor kinases; MSI: microsatellite instability; NTRK: neurotrophic receptor tyrosine kinase gene.

3.2. Mechanisms of Drug Resistance Associated with Colorectal Cancer Stem Cells

The effectiveness of current anticancer therapies is limited by the resistance of tumors to chemotherapy and targeted molecular therapies [99]. Resistance to anticancer drugs may be intrinsic, meaning that it occurs prior to treatment and involves pre-existing resistance factors in the mass of tumor cells, or it may be acquired during the treatment of tumors that were initially sensitive due to the induction of various adaptive responses [99]. Furthermore, due to the high degree of tumor heterogeneity, drug resistance may also result from the therapy-induced selection of a drug-resistant tumor subpopulation, such as CCSCs [99]. A wide range of molecular mechanisms are involved in drug resistance, as illustrated in Figure 4, and will be detailed in this chapter [74].

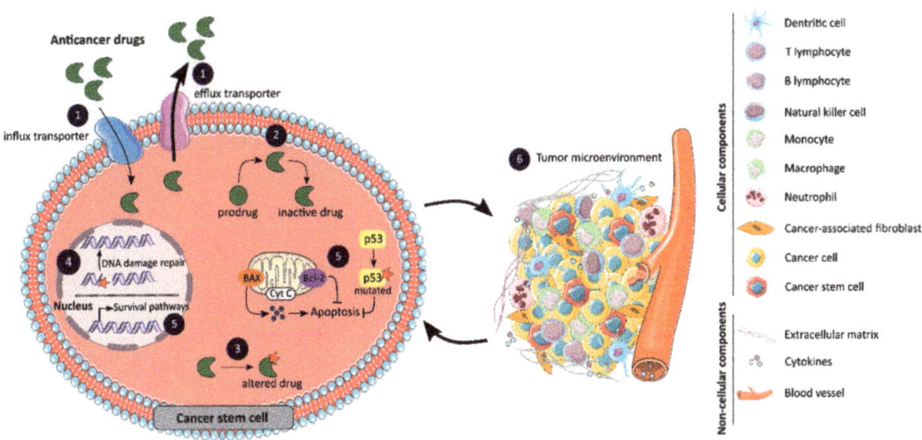

Figure 4. Major mechanisms of anticancer drug resistance attributed to CSCs such as changes in drug transport (1); impaired drug metabolism (2); alterations in drug targets (3); enhanced DNA damage repair (4); impaired balance between apoptosis and survival pathways (5); and the role of the tumor microenvironment comprising cellular and non-cellular components (6). CSC: cancer stem cell; DNA: deoxyribonucleic acid.

3.2.1. Changes in Drug Transport

The anticancer activity of a drug can be limited by poor drug influx or excessive efflux, which alters the amount of drug reaching the tumor, as shown in Figure 4 panel 1 [99]. Several transporter proteins, belonging to the superfamilies ABC and solute carrier (SLC), have been linked to anticancer drug resistance by interfering with drug transport [74]. The ABC transporters ABCB1, ABCC1 and ABCG2 play a pivotal role in the efflux of anticancer drugs [100,101]. In colon cancer, ABCB1 may be overexpressed, leading to reduced cellular accumulation of chemotherapy and therefore therapeutic failure, or may be induced by chemotherapy resulting in the acquired development of multidrug resistance [99]. The impact of SLCs on cancer therapy has been less documented, however, some members of the SLC superfamily are also involved in the transport of anticancer drugs [100]. Changes in the expression of SLC transporters, such as the organic cation transporter OCT2 and the organic zwitterion/cation transporters OCTN1, may affect the ability of tumor cells to uptake anticancer drugs and lead to the development of chemoresistance [100]. The Zhang et al. study shows that the overexpression of human OCT2 transporters increases oxaliplatin accumulation and cytotoxicity in colon cancer cell lines [102]. Taken together, efflux and influx transporters may confer resistance to anticancer agents, and the intrinsic drug resistance of CCSCs may be explained by the higher expression of these transporters [99,100,102].

3.2.2. Impaired Drug Metabolism

The efficacy of anticancer drugs may also be affected by changes in their metabolism, such as the production of an inactive metabolite, as highlighted in Figure 4 panel 2 [99]. The inactivation of anticancer drugs may be associated with the overexpression of drug-metabolizing enzymes, such as cytochrome P450-related enzymes (CYP), UDP-glucuronosyltransferase (UGT) and glutathione S-transferase (GST) [74]. CYP enzymes play a crucial role in the metabolism of many therapeutic drugs, including SN-38, the active metabolite of irinotecan. Indeed, SN-38 can be inactivated by CYP3A4- and CYP3A5-dependent oxidations that form inactive metabolites [103]. The study by Buck et al. shows a significant correlation between CYP3A5 expression and tumor response to irinotecan therapy in CRC [103]. In addition, increased CYP expression in CSCs appears to be associated with chemoresistance [104]. SN-38 is predominantly eliminated by glucuronidation which is mainly mediated by the polypeptide A1 of the UGT1 family, encoded by the *UGT1A1* gene [105]. However, inter-individual variations in UGT1A1 activity exist and are related to the presence of genetic polymorphisms. For example, patients with UGT1A1*28/*28 genotype have a higher risk of developing irinotecan-induced hematological toxicity and require a reduction in irinotecan dose which may impact its anti-cancer effect [105]. The GSTP1 subclass of the GST superfamily is overexpressed in patients with colon cancer and is an important mediator of intrinsic and acquired platinum resistance [106]. Stoehlmacher et al. demonstrated that GSTP1 Ile^{105}Val polymorphism is associated with increased survival in patients with advanced CRC receiving 5-FU/oxaliplatin chemotherapy [106]. Thus, the enhanced ability of tumor cells, particularly CCSCs, to inactivate anti-cancer drugs is mainly due to the overexpression of drug-metabolizing enzymes or polymorphisms [74].

3.2.3. Alterations in Drug Targets

One of the most common mechanisms of resistance to targeted therapy is mediated by alterations in the target protein as suggested in Figure 4 panel 3 [107]. Somatic mutations have been identified in the *KRAS* gene as a biomarker of intrinsic resistance to EGFR-targeting agents in patients with CRC [108]. The Misale et al. study reports for the first time that a substantial fraction of CRC patients who exhibit an initial response to anti-EGFR therapies have, at the time of disease progression, tumors with focal amplification or somatic mutations in *KRAS* which were not detectable prior to therapy initiation [108]. Thus, drug resistance resulting from *KRAS* alterations can be attributed not only to the selection of pre-existing *KRAS* mutant and amplified clones, but also to new mutations resulting from ongoing mutagenesis [108]. The acquisition of mutations in target proteins also contributes to chemotherapy drug resistance. Irinotecan exerts its cytotoxic activity by inhibiting topoisomerase 1 (TOP1). However, increased TOP1 gene copy number at 20q11.2-q13.1 or mutations in the gene that result in reduced affinity for its active metabolite may be involved in increased drug resistance in CCR [74]. Therefore, the alteration of drug targets primarily due to the acquisition of mutations may result in resistance to targeted therapy and chemotherapy.

3.2.4. Enhanced DNA Damage Repair

Drug resistance can also be explained by an enhanced ability of tumor cells, especially CCSCs, to repair drug-induced DNA damage, as presented in Figure 4 panel 4. The repair of DNA adducts induced by platinum-based chemotherapy, such as oxaliplatin, is primarily mediated by the nucleotide excision repair (NER) pathway [74]. The upregulation of excision repair cross-complementing 1 (ERCC1), a key protein of the NER pathway, has been associated with oxaliplatin resistance in CRC [74]. In addition, the level of intra-tumoral ERCC1 mRNA expression is a predictive marker of survival in mCRC patients receiving combination chemotherapy with 5-FU and oxaliplatin [109]. Mismatched or wrongly matched nucleotides are corrected by the MMR system, which plays a crucial role in maintaining genome integrity [74]. DNA repair deficiency can be caused by mutations in *MMR* genes, such as *MLH1* and *MSH2*, and can lead to the MSI phenotype [99]. The study by Valeri et al. shows that the microRNA-21 (miR-21) downregulates hMSH2, and miR-21 overexpression reduces the therapeutic efficacy of 5-FU in a CRC xenograft model, suggesting that the downregulation of MSH2 with miR-21 overexpression may be an important indicator of therapeutic efficacy in CRC [110]. Consequently, the defects or upregulation of DNA repair pathways can serve as biomarkers of therapeutic response, and therapeutic effects can be enhanced by combining the inhibition of a DNA damage response pathway with DNA-damaging agents to eradicate CCSCs [111].

3.2.5. Impaired Balance between Apoptosis and Survival Pathways

Resistance to cell death is one of the hallmarks of human cancers that contribute to tumor progression and drug resistance [101]. Cell death by apoptosis is a physiological program controlled by a tightly regulated balance between pro-apoptotic, anti-apoptotic and pro-survival mechanisms [112]. However, this balance is frequently altered in cancer cells and particularly in CCSCs, as shown in Figure 4 panel 5. The tumor suppressor p53, encoded by the *TP53* gene, is essential for the induction of apoptosis in response to chemotherapy [74]. Nevertheless, p53 is found mutated in approximately 85% of CRC cases, and *TP53*-mutated colon cancer cells tend to be more resistant to many anticancer drugs, including 5-FU and oxaliplatin, compared to *TP53* wild-type cells [74,101]. The BCL-2 family, which contains pro- and anti-apoptotic members, plays a crucial role in the regulation of apoptosis. The loss of expression and/or activity of the pro-apoptotic factor BAX can be explained by frameshift mutations in the *BAX* gene and may result in chemoresistance [74]. The study by Nehls et al. suggests a major prognostic impact of BAX, whose protein expression appears to be important for the clinical outcome of 5-FU-based adjuvant chemotherapy in stage III colon cancer [113]. The balance between apoptosis and survival may also be altered by aberrantly overexpressed or overactivated anti-apoptotic factors, such as Bcl-2, Bcl-XL, the inhibitor of apoptosis proteins and the caspase 8 inhibitor FLIP [74,99]. Importantly, alterations in the genes encoding these anti-apoptotic factors have been linked to resistance to chemotherapy and targeted therapy [99]. Finally, the overactivation of several pro-survival signaling pathways, including Notch, Hedgehog, Wnt, Bone morphogenetic proteins, Janus kinase/signal transducers and activators of transcription (JAK/STAT) and nuclear factor-κB pathways, may also be associated with drug resistance [112]. Taken together, the altered balance between apoptosis and survival in cancer cells, and especially in CCSCs, prevents apoptosis even when DNA repair fails, which is another mechanism of resistance to therapy [112].

3.2.6. Role of the Tumor Microenvironment

In recent years, the TME has emerged as a key driver of tumor progression and drug resistance, challenging the development of new therapies in clinical oncology. TME contains both cellular components with cancerous and non-cancerous cells such as stromal myofibroblasts, endothelial cells, immune cells and cancer-associated fibroblasts (CAFs), and non-cellular components including extracellular matrix (ECM), cytokines, growth factors and extracellular vesicles, as illustrated in Figure 4 panel 6 [114]. In the tumor stroma, CAFs secrete the cytokines CXCL1 and CXCL2 as well as the interleukin-6, which promote angiogenesis and tumor progression [46,114]. Vermeulen et al. showed that myofibroblast-secreted factors, in particular hepatocyte growth factor (HGF), enhance Wnt signaling activity in colon cancer cells and can restore the CSC phenotype in more differentiated tumor cells, both in vitro and in vivo [115]. Furthermore, CSCs reside in anatomically specialized regions of the TME, known as the CSC niche, which retain their properties and protect them from anticancer drugs, contributing to their enhanced resistance to treatment [46,114,116]. Importantly, CSCs can also be maintained in a quiescent state with minimum energy consumption and a low proliferation rate to resist therapies [114]. In response to environmental signals such as hypoxia, the niche adapts to ensure optimal conditions for CSC proliferation and differentiation [46]. CSCs may contribute to vessel recruitment during tumorigenesis by secreting angiogenic factors, such as vascular endothelial growth factor (VEGF) and CXCL12, in order to accelerate angiogenesis in endothelial cells, which in turn secrete factors such as nitric oxide and osteopontin to maintain the stemness of CSCs [15]. Hypoxia is a key factor in cancer progression that regulates cell survival, angiogenesis, invasion and metastasis, via hypoxia-inducible factor (HIF) [14,116]. In addition, hypoxia can induce the epithelial-to-mesenchymal transition (EMT) that leads to the dissemination and invasion of tumor cells due to the loss of cell adhesion properties and the acquisition of a mesenchymal phenotype with motility and invasiveness [8,116]. The expression of SNAI1 protein, the main inducer of EMT, has been detected at the tumor–stromal interface in colon cancer [116] and elevated endogenous levels of SNAI1 in cancer cells have been shown to increase tumor initiation capacity and metastatic potential in mouse and human models [8].

3.2.7. Mechanisms of Drug Resistance Associated with CCSCs: Discussion

Several publications point out that one of the main technical issues in the CSC field is the plasticity of CCSCs and tumor cells, which may be involved in drug resistance [117–120]. Using the CRISPR-Cas9 technology, Shimokawa et al. demonstrated that the selective ablation of LGR5+ CCSCs in human CRC organoids leads to tumor regression in xenografts produced by these organoids [120]. However, after several weeks, tumor regrowth is observed and associated with differentiated tumor cells that dynamically replenish the pool of LGR5+ CCSCs, indicative of cellular plasticity [120]. Another study confirmed these results using CRC organoids that express the diphtheria toxin receptor under the control of the LGR5 locus to selectively ablate LGR5+ CCSCs [117]. Importantly, the removal of CCSCs limits primary tumor growth but does not prevent the regrowth of the tumor at the primary tumor site upon discontinuation of treatment due to proliferative LGR5− cells, whereas it does in metastatic lesions [117]. Thus, the authors demonstrated a protective role of selective CSC depletion in primary tumors on the appearance of distant metastases, suggesting an interesting therapeutic perspective for the management of metastatic diseases [117]. The process of cellular plasticity is crucial for the repopulation of impaired SC niches and tissue homeostasis, but its role in the formation of metastases is poorly studied [118]. Using a CRC mouse model and human tumor xenografts, Fumagalli et al. investigated the role of cellular plasticity in metastasis [118]. Surprisingly, the authors show that the majority of disseminated CRC cells in the circulation are LGR5− cancer cells and are capable of forming distant metastases, in which LGR5+ CSCs subsequently emerge and contribute to long-term metastatic growth [118]. Importantly, microenvironmental factors may enhance cellular plasticity [118]. Thus, cellular plasticity complicates the development of new therapeutic strategies and the eradication of CCSCs does not appear to be sufficient to completely cure cancer due to the impact of the microenvironment [8]. The heterogeneous and dynamic nature of SCCCs constitutes another obstacle to their targeting. Using a marker-free and quantitative analysis of colon cancer growth dynamics, Lenos et al. showed that cells with CSC functionality are not necessarily the same cells as those expressing CSC markers [121]. Interestingly, the authors demonstrated that all tumor cells have the capacity to fuel tumor growth when placed in an appropriate environment, preferentially at the edge of the tumor close to the CAFs [121,122]. Thus, from the authors' point of view, the stem cell function in established cancers is not intrinsically but entirely spatiotemporally orchestrated, suggesting a major role of the microenvironment [121]. Consequently, cellular plasticity and the microenvironment appear to allow tumors to easily adapt to the loss of key compartments, thereby compromising therapeutic efficacy [122]. Therefore, TME plays a crucial role in primary tumor growth and metastasis formation by protecting CSCs from therapeutic agents and appears to be an important target along with the other resistance mechanisms discussed in this chapter for the development of new therapies [116].

4. Clinical Trials on Colorectal Cancer Stem Cells

Clinical trials targeting CCSCs are rare. The complexity relies on the identification of molecular targets required to maintain cancer stemness in CSCs, but not or less by normal tissue SCs, to selectively target CSCs [123]. All clinical trials from the National Institute of Health are listed on the ClinicalTrials.gov website [124]. We narrowed our search by focusing on the terms "colorectal cancer" and "cancer stem cells", resulting in the identification of eight intervention studies as of September 30, 2020. Among them, we excluded all clinical trials whose status was withdrawn (N = 1) and terminated (N = 2) and focused on the remaining clinical trials (N = 5). Subsequently, from these five clinical trials, we selected and reviewed the clinical trials on pharmacological agents under investigation (N = 3), as presented in Table 5.

Table 5. Clinical trials on colorectal cancer and cancer stem cells from ClinicalTrials.gov.

Trial Registration and Status	Study Titles	Interventions	Phases	Investigators
NCT02753127 Active, not recruiting	A Study of Napabucasin (BBI-608) in Combination with FOLFIRI in Adult Patients with Previously Treated Metastatic Colorectal Cancer (CanStem303C)	Drug: Napabucasin	Phase III	Sumitomo Dainippon Pharma Oncology, Inc
NCT01189942 Completed	A Study of FOLFIRI Plus OMP-21M18 as 1st or 2nd-line Treatment in Subjects with Metastatic Colorectal Cancer	Drug: OMP-21M18	Phase I	Mereo BioPharma (OncoMed Pharmaceuticals, Inc.)
NCT02859415 Recruiting	Continuous 24 h Intravenous Infusion of Mithramycin, an Inhibitor of Cancer Stem Cell Signaling, in People with Primary Thoracic Malignancies or Carcinomas, Sarcomas or Germ Cell Neoplasms with Pleuropulmonary Metastases	Drug: Mithramycin	Phase I and II	National Cancer Institute

Napabucasin (BBI608) is the first-in-class cancer stemness inhibitor that targets the STAT3 pathway [123,125]. In a preclinical study, Li et al. showed that BBI608 inhibits the expression of stemness genes and the self-renewal of CSCs and succeeds in depleting CSCs whereas standard chemotherapy leads to the enrichment of these cells [123]. In addition, the authors demonstrated the ability of BBI608 to block both cancer relapse and metastasis in vivo, using a mouse CRC model [123]. These preclinical results provide an interesting approach for the development of new anticancer therapies targeting cancer stemness [123,125]. Several clinical trials were designed prior to the ongoing phase III clinical trial, shown in Table 5. Firstly, a phase I dose-escalation study was conducted in adult patients with advanced malignancies who had failed standard therapies in order to investigate the safety and anti-tumor activity of BBI608 as monotherapy (NCT01775423) [126,127]. BBI608 showed encouraging signs of clinical activity with only mild adverse events observed and an unreached maximum tolerated dose (MTD), suggesting an excellent safety profile of BBI608 at 500 mg twice daily [126,127]. Subsequently, two additional phase Ib/II clinical trials were conducted to determine the safety and anti-tumor activity of BBI608 in combination with panitumumab in *KRAS* wild-type patients with mCRC (NCT01776307) or with FOLFIRI (5-FU/leucovorin/irinotecan) +/− bevacizumab in mCRC (NCT02024607). Both clinical trials showed a high disease control rate (DCR) including patients with partial response, stable disease or tumor regression, which confirms the safety of these combinations with encouraging anti-tumor activity [128–130]. Thereafter, a phase III study was designed to evaluate the efficacy and safety of BBI608 versus placebo with the best supportive care in patients with advanced CRC who had failed all available standard therapy (NCT01830621) [131]. In this trial, BBI608 did not improve overall survival (OS) or progression-free survival (PFS) in unselected patients with advanced CRC but did improve OS in pSTAT3-positive patients compared to the placebo group, suggesting that STAT3 may be an important target for CRC treatment [131,132]. Finally, the ongoing phase III clinical trial aims to assess the efficacy of BBI608 plus FOLFIRI versus FOLFIRI alone in previously treated mCRC patients (N = 1250) (NCT02753127) [133]. Patients are randomized 1:1 in each group and stratified by time to progression to first-line therapy, *RAS* mutation status and primary tumor location [133]. The endpoints of this clinical trial are OS, PFS, DCR and objective response rate in both the general population and p-STAT3-positive subpopulation [133].

Demcizumab (OMP-21M18) is a humanized anti-DLL4 (delta-like ligand 4) antibody that inhibits the Notch pathway and CSC activity through the inhibition of tumor growth and reduction in tumor-initiating cell frequency [134–136]. The study by Ridgway et al. shows that treatment with a DLL4-selective antibody disrupts tumor angiogenesis and inhibits tumor growth in several mouse tumor models [137], these results were confirmed by Hoey et al. using xenograft models of colon tumors [136]. In addition, treatment with anti-human DLL4, alone or in combination with irinotecan, delays tumor recurrence and reduces the frequency of CSCs, as demonstrated by the limiting dilution assay and in vivo tumorigenesis studies [136]. As a result of these preclinical results, several clinical trials were conducted with OMP-21M18. A phase I dose-escalation study was designed to determine the safety, MTD and pharmacokinetics of OMP-21M18 in patients with a previously treated solid tumor for which there is no remaining standard curative therapy (NCT00744562) [138]. In this trial, no more than one dose-limiting toxicity (DLT) was observed at each dose, corresponding to the appearance of side effects severe enough to prevent an increase in the dose of the drug, and the MTD was not reached [138]. OMP-21M18 was generally well tolerated by patients at doses below 5 mg per week and showed anti-tumor activity highlighted by the stabilization of the disease and decrease in tumor size. However, the prolonged administration of this drug was associated with an increased risk of congestive heart failure [138]. Subsequently, a phase Ib study failed to demonstrate enhanced anti-tumor activity of OMP-21M18 in combination with the anti-PD-1 pembrolizumab in patients with advanced or metastatic solid tumors, despite the fact that the combination therapy was well tolerated (NCT02722954) [139]. Finally, a phase I study was conducted to determine the safety and optimal dose of OMP-21M18 in combination with FOLFIRI in patients with mCRC (N = 32) (NCT01189942). Safety was scheduled to be assessed in each patient group after 56 days of treatment and disease status every 8 weeks. The endpoints of this clinical trial were to determine the MTD of OMP-21M18 plus FOLFIRI and to evaluate the safety, pharmacokinetics and preliminary efficacy of this combination. Unfortunately, to date, no results from this clinical trial have been found, although the actual completion date of the study indicated on ClinicalTrials.gov is February 2011.

Mithramycin A (Mit-A) is an antineoplastic antibiotic agent and a potent inhibitor of specificity protein 1 (SP1) [140]. In various human malignancies, SP1 is overexpressed and contributes to the malignant phenotype by regulating genes involved in proliferation, invasion, metastasis, stemness and chemoresistance [141,142]. The study by Zhao et al. demonstrates that the inhibition of SP1 by Mit-A suppresses the growth of colon CSCs and attenuates their characteristics by significantly reducing the percentage of CD44$^+$/CD166$^+$ cells in vitro and in vivo [142]. Another study shows that Mit-A inhibits tumor growth and significantly induces cell death and the PARP cleavage of CSC and non-CSC cells [140]. Thus, these preclinical results highlighted Mit-A as a potentially promising drug candidate for the treatment of CRC [140]. Several clinical trials have been conducted to investigate the safety and efficacy of Mit-A in chest cancers, solid tumors and Ewing sarcoma (NCT01624090 and NCT01610570) [143]. Despite the promising preclinical activity of Mit-A in various advanced malignancies, several patients developed severe hepatotoxicity due to the altered expression of hepatocellular bile transporters resulting in the early termination of the clinical trial [144]. The objective of the ongoing phase I/II clinical trial NCT02859415 was to determine the safest dose of Mit-A in patients with chest cancers, including CRC patients, by specifically selecting patients without these alterations. The endpoints of this clinical trial are to evaluate the DLT, MTD, and pharmacokinetics of Mit-A in patients with primary thoracic malignancies or carcinomas, sarcomas or germ cell neoplasms with pleuropulmonary metastases and to determine their overall response rates.

Clinical Trials on CCSCs: Discussion

Thus, the development of therapeutic agents specifically targeting CCSCs is complex, as outlined in this chapter. Unfortunately, despite encouraging preclinical results, the majority of ongoing clinical trials fail to demonstrate relevant results in phase I/II development, which examines the safety of the drug, and does not allow them to proceed to the next phase. In our search of ClinicalTrials.gov, we found only three clinical trials focusing on CSCs and recruiting patients with CRC, underscoring the rarity and complexity of clinical trial design [124]. Among these trials, no study results were found for one of the drugs tested, demcizumab, although the actual completion date of the study has passed [139]. In addition, of the three drugs in clinical trials, two drugs showed drug-related toxicities in the current or previous study. The prolonged administration of demcizumab was associated with an increased risk of congestive heart failure [138] and some patients treated with Mit-A developed severe hepatotoxicity [144]. However, one of these three drugs, napabucasin, has shown interesting results in previous clinical trials and is currently in a phase III study [128–130]. In conclusion, the development of clinical trials can encounter many problems related either to drugs, to patients with unexpected side effects or toxicities, or to the design of the study.

5. Future Perspectives

The main challenge in preclinical studies is to obtain relevant results that translate into meaningful clinical activity in patients with CRC [134]. Unfortunately, despite interesting preclinical results, many clinical trials fail to demonstrate the benefits of a new pharmacological agent due to the absence of anticancer activity in cancer patients or the presence of toxicities incompatible with the continuation of the trial. The development of new clinical trials must consider the intra- and inter-tumor heterogeneity of CRC patients, which influences their responses to therapies. Nowadays, targeted therapies and immunotherapy have significantly improved the survival of CRC patients, and the newly developed therapies are increasing the therapeutic options for patients with advanced CRC harboring specific genetic abnormalities [5]. However, despite the initial success of commonly used therapies, most drugs fail to target the MRD associated with CSCs which often leads to relapse in cancer patients. Unfortunately, up to 50% of patients with early-stage CRC at diagnosis will eventually develop metastatic disease, and most of them have unresectable metastatic disease because of the size, location, and/or extent of the disease [76,77]. New clinical trials must therefore be designed to test drugs that could become relevant treatment options for patients with early-stage and advanced CRC. However, the lack of accurate preclinical models that take into account intrinsic and extrinsic characteristic of tumors, such as CSC subpopulation, tumor stroma and TME, is a major technical problem [134]. The CCSC isolation and characterization methods presented in this review highlighted the limitations of the methods currently in use, particularly those using CCSC markers. Cell sorting using phenotypic characteristics allows the sorting of only part of the CCSC population because they are heterogeneous, plastic, and subject to many signals from the TME. Thus, the use of new innovative techniques such as SdFFF which sorts cells according to cell characteristics other than marker expression or the combination of several isolation techniques is crucial. In conclusion, more accurate preclinical models are required because current approaches are not precise enough to identify therapies that may be clinically effective, particularly those targeting CCSCs [145].

6. Conclusions

Targeting CCSCs holds promise for preventing disease relapse and metastasis in CRC patients. In addition, as a major driving force of drug resistance, CCSCs are attractive potential targets for the treatment of CRC. However, the development of therapeutic agents specifically targeting CCSCs is complex, as highlighted by the clinical trial results presented in this review. Despite the increasing number of therapies, resistance mechanisms may emerge and thus complicate the therapeutic management of patients with CRC. In order to achieve a short- and long-term therapeutic response, the ideal therapeutic strategy should target both the cancer cells of the tumor mass to obtain tumor regression, CCSCs to prevent relapse and metastasis, and TME to limit cellular plasticity and the reappearance of CCSCs.

Author Contributions: Writing—original draft preparation, C.H.; writing—review and editing, N.C., S.B. and M.M. All authors have read and agreed to the published version of the manuscript.

Funding: This research received no external funding.

Conflicts of Interest: The authors declare no conflict of interest.

References

1. Dekker, E.; Tanis, P.J.; Vleugels, J.L.A.; Kasi, P.M.; Wallace, M.B. Colorectal Cancer. *Lancet* **2019**, *394*, 1467–1480. [CrossRef]
2. World Health Organization: Regional Office for Europe. *World Cancer Report: Cancer Research for Cancer Development*; IARC: Lyon, France, 2020; ISBN 978-92-832-0447-3.
3. Brierley, J.D.; Gospodarowicz, M.K.; Wittekind, C. *TNM Classification of Malignant Tumours*; John Wiley & Sons: Hoboken, NJ, USA, 2017; ISBN 978-1-119-26357-9.
4. Amin, M.B.; Edge, S.; Greene, F.; Byrd, D.R.; Brookland, R.K.; Washington, M.K.; Gershenwald, J.E.; Compton, C.C.; Hess, K.R.; Sullivan, D.C.; et al. *AJCC Cancer Staging Manual*, 8th ed.; Springer International Publishing: Cham, Switzerland, 2017.
5. Benson, A.B.; Venook, A.P.; Al-Hawary, M.M.; Cederquist, L.; Chen, Y.-J.; Ciombor, K.K.; Cohen, S.; Cooper, H.S.; Deming, D.; Engstrom, P.F.; et al. NCCN Guidelines Insights: Colon Cancer, Version 2.2018. *J. Natl. Compr. Canc. Netw.* **2018**, *16*, 359–369. [CrossRef] [PubMed]
6. Van Cutsem, E.; Cervantes, A.; Adam, R.; Sobrero, A.; Van Krieken, J.H.; Aderka, D.; Aranda Aguilar, E.; Bardelli, A.; Benson, A.; Bodoky, G.; et al. ESMO Consensus Guidelines for the Management of Patients with Metastatic Colorectal Cancer. *Ann. Oncol.* **2016**, *27*, 1386–1422. [CrossRef] [PubMed]
7. Van der Heijden, M.; Vermeulen, L. A Cancer Stem Cell Perspective on Minimal Residual Disease in Solid Malignancies. In *Cancer Stem Cell Resistance to Targeted Therapy*; Maccalli, C., Todaro, M., Ferrone, S., Eds.; Resistance to Targeted Anti-Cancer Therapeutics; Springer International Publishing: Cham, Switzerland, 2019; Volume 19, pp. 31–49. ISBN 978-3-030-16623-6.
8. Batlle, E.; Clevers, H. Cancer Stem Cells Revisited. *Nat. Med.* **2017**, *23*, 1124–1134. [CrossRef]
9. Desbats, M.A.; Giacomini, I.; Prayer-Galetti, T.; Montopoli, M. Metabolic Plasticity in Chemotherapy Resistance. *Front. Oncol.* **2020**, *10*, 281. [CrossRef] [PubMed]

10. Serpa, J. *Tumor Microenvironment: The Main Driver of Metabolic Adaptation*; Serpa, J., Ed.; Advances in Experimental Medicine and Biology; Springer International Publishing: Cham, Switzerland, 2020; Volume 1219, ISBN 978-3-030-34024-7.
11. Jordan, C.T. Cancer Stem Cells. *N. Engl. J. Med.* **2006**, *355*, 1253–1261. [CrossRef] [PubMed]
12. Bonnet, D.; Dick, J.E. Human Acute Myeloid Leukemia Is Organized as a Hierarchy That Originates from a Primitive Hematopoietic Cell. *Nat. Med.* **1997**, *3*, 730–737. [CrossRef]
13. Schwitalla, S.; Fingerle, A.A.; Cammareri, P.; Nebelsiek, T.; Göktuna, S.I.; Ziegler, P.K.; Canli, O.; Heijmans, J.; Huels, D.J.; Moreaux, G.; et al. Intestinal Tumorigenesis Initiated by Dedifferentiation and Acquisition of Stem-Cell-like Properties. *Cell* **2013**, *152*, 25–38. [CrossRef]
14. Jahanafrooz, Z.; Mosafer, J.; Akbari, M.; Hashemzaei, M.; Mokhtarzadeh, A.; Baradaran, B. Colon Cancer Therapy by Focusing on Colon Cancer Stem Cells and Their Tumor Microenvironment. *J. Cell. Physiol.* **2020**, *235*, 4153–4166. [CrossRef]
15. Najafi, M.; Farhood, B.; Mortezaee, K. Cancer Stem Cells (CSCs) in Cancer Progression and Therapy. *J. Cell. Physiol.* **2019**, *234*, 8381–8395. [CrossRef] [PubMed]
16. Mélin, C.; Perraud, A.; Akil, H.; Jauberteau, M.-O.; Cardot, P.; Mathonnet, M.; Battu, S. Cancer Stem Cell Sorting from Colorectal Cancer Cell Lines by Sedimentation Field Flow Fractionation. *Anal. Chem.* **2012**, *84*, 1549–1556. [CrossRef] [PubMed]
17. Hirata, A.; Hatano, Y.; Niwa, M.; Hara, A.; Tomita, H. Heterogeneity of Colon Cancer Stem Cells. In *Stem Cells Heterogeneity in Cancer*; Birbrair, A., Ed.; Advances in Experimental Medicine and Biology; Springer International Publishing: Cham, Switzerland, 2019; Volume 1139, pp. 115–126. ISBN 978-3-030-14365-7.
18. Akbarzadeh, M.; Maroufi, N.F.; Tazehkand, A.P.; Akbarzadeh, M.; Bastani, S.; Safdari, R.; Farzane, A.; Fattahi, A.; Nejabati, H.R.; Nouri, M.; et al. Current Approaches in Identification and Isolation of Cancer Stem Cells. *J. Cell. Physiol.* **2019**, *234*, 14759–14772. [CrossRef] [PubMed]
19. Ricci-Vitiani, L.; Lombardi, D.G.; Pilozzi, E.; Biffoni, M.; Todaro, M.; Peschle, C.; De Maria, R. Identification and Expansion of Human Colon-Cancer-Initiating Cells. *Nature* **2007**, *445*, 111–115. [CrossRef]
20. O'Brien, C.A.; Pollett, A.; Gallinger, S.; Dick, J.E. A Human Colon Cancer Cell Capable of Initiating Tumour Growth in Immunodeficient Mice. *Nature* **2007**, *445*, 106–110. [CrossRef]
21. Kemper, K.; Sprick, M.R.; de Bree, M.; Scopelliti, A.; Vermeulen, L.; Hoek, M.; Zeilstra, J.; Pals, S.T.; Mehmet, H.; Stassi, G.; et al. The AC133 Epitope, but Not the CD133 Protein, Is Lost upon Cancer Stem Cell Differentiation. *Cancer Res.* **2010**, *70*, 719–729. [CrossRef]
22. Dalerba, P.; Dylla, S.J.; Park, I.-K.; Liu, R.; Wang, X.; Cho, R.W.; Hoey, T.; Gurney, A.; Huang, E.H.; Simeone, D.M.; et al. Phenotypic Characterization of Human Colorectal Cancer Stem Cells. *Proc. Natl. Acad. Sci. USA* **2007**, *104*, 10158–10163. [CrossRef] [PubMed]
23. Barker, N.; Ridgway, R.A.; van Es, J.H.; van de Wetering, M.; Begthel, H.; van den Born, M.; Danenberg, E.; Clarke, A.R.; Sansom, O.J.; Clevers, H. Crypt Stem Cells as the Cells-of-Origin of Intestinal Cancer. *Nature* **2009**, *457*, 608–611. [CrossRef]
24. Sangiorgi, E.; Capecchi, M.R. Bmi1 Is Expressed in Vivo in Intestinal Stem Cells. *Nat. Genet.* **2008**, *40*, 915–920. [CrossRef]
25. Vermeulen, L.; Todaro, M.; de Sousa Mello, F.; Sprick, M.R.; Kemper, K.; Perez Alea, M.; Richel, D.J.; Stassi, G.; Medema, J.P. Single-Cell Cloning of Colon Cancer Stem Cells Reveals a Multi-Lineage Differentiation Capacity. *Proc. Natl. Acad. Sci. USA* **2008**, *105*, 13427–13432. [CrossRef]
26. Pang, R.; Law, W.L.; Chu, A.C.Y.; Poon, J.T.; Lam, C.S.C.; Chow, A.K.M.; Ng, L.; Cheung, L.W.H.; Lan, X.R.; Lan, H.Y.; et al. A Subpopulation of CD26+ Cancer Stem Cells with Metastatic Capacity in Human Colorectal Cancer. *Cell Stem Cell* **2010**, *6*, 603–615. [CrossRef] [PubMed]
27. Todaro, M.; Gaggianesi, M.; Catalano, V.; Benfante, A.; Iovino, F.; Biffoni, M.; Apuzzo, T.; Sperduti, I.; Volpe, S.; Cocorullo, G.; et al. CD44v6 Is a Marker of Constitutive and Reprogrammed Cancer Stem Cells Driving Colon Cancer Metastasis. *Cell Stem Cell* **2014**, *14*, 342–356. [CrossRef] [PubMed]
28. Barker, N.; van Es, J.H.; Kuipers, J.; Kujala, P.; van den Born, M.; Cozijnsen, M.; Haegebarth, A.; Korving, J.; Begthel, H.; Peters, P.J.; et al. Identification of Stem Cells in Small Intestine and Colon by Marker Gene Lgr5. *Nature* **2007**, *449*, 1003–1007. [CrossRef]
29. Saito, S.; Okabe, H.; Watanabe, M.; Ishimoto, T.; Iwatsuki, M.; Baba, Y.; Tanaka, Y.; Kurashige, J.; Miyamoto, Y.; Baba, H. CD44v6 Expression Is Related to Mesenchymal Phenotype and Poor Prognosis in Patients with Colorectal Cancer. *Oncol. Rep.* **2013**, *29*, 1570–1578. [CrossRef]
30. Ma, L.; Dong, L.; Chang, P. CD44v6 Engages in Colorectal Cancer Progression. *Cell Death Dis.* **2019**, *10*, 30. [CrossRef]
31. Yanai, H.; Atsumi, N.; Tanaka, T.; Nakamura, N.; Komai, Y.; Omachi, T.; Tanaka, K.; Ishigaki, K.; Saiga, K.; Ohsugi, H.; et al. Intestinal Cancer Stem Cells Marked by Bmi1 or Lgr5 Expression Contribute to Tumor Propagation via Clonal Expansion. *Sci. Rep.* **2017**, *7*, 41838. [CrossRef] [PubMed]
32. Du, J.; Li, Y.; Li, J.; Zheng, J. Polycomb Group Protein Bmi1 Expression in Colon Cancers Predicts the Survival. *Med. Oncol.* **2010**, *27*, 1273–1276. [CrossRef]
33. Weichert, W. Cytoplasmic CD24 Expression in Colorectal Cancer Independently Correlates with Shortened Patient Survival. *Clin. Cancer Res.* **2005**, *11*, 6574–6581. [CrossRef] [PubMed]
34. Lam, C.S.-C.; Cheung, A.H.-K.; Wong, S.K.-M.; Wan, T.M.-H.; Ng, L.; Chow, A.K.-M.; Cheng, N.S.-M.; Pak, R.C.-H.; Li, H.-S.; Man, J.H.-W.; et al. Prognostic Significance of CD26 in Patients with Colorectal Cancer. *PLoS ONE* **2014**, *9*, e98582. [CrossRef]
35. Liu, Q.-Z.; Gao, X.-H.; Chang, W.-J.; Gong, H.-F.; Fu, C.-G.; Zhang, W. Expression of ITGB1 Predicts Prognosis in Colorectal Cancer: A Large Prospective Study Based on Tissue Microarray. *Int. J. Clin. Exp. Pathol.* **2015**, *8*, 12802.

36. Du, L.; Wang, H.; He, L.; Zhang, J.; Ni, B.; Wang, X.; Jin, H.; Cahuzac, N.; Mehrpour, M.; Lu, Y.; et al. CD44 Is of Functional Importance for Colorectal Cancer Stem Cells. *Clin. Cancer Res.* **2008**, *14*, 6751–6760. [CrossRef]
37. Leng, Z.; Xia, Q.; Chen, J.; Li, Y.; Xu, J.; Zhao, E.; Zheng, H.; Ai, W.; Dong, J. Lgr5+CD44+EpCAM+ Strictly Defines Cancer Stem Cells in Human Colorectal Cancer. *Cell. Physiol. Biochem.* **2018**, *46*, 860–872. [CrossRef]
38. Wang, Z.; Tang, Y.; Xie, L.; Huang, A.; Xue, C.; Gu, Z.; Wang, K.; Zong, S. The Prognostic and Clinical Value of CD44 in Colorectal Cancer: A Meta-Analysis. *Front. Oncol.* **2019**, *9*, 309. [CrossRef]
39. Zhu, L.; Gibson, P.; Currle, D.S.; Tong, Y.; Richardson, R.J.; Bayazitov, I.T.; Poppleton, H.; Zakharenko, S.; Ellison, D.W.; Gilbertson, R.J. Prominin 1 Marks Intestinal Stem Cells That Are Susceptible to Neoplastic Transformation. *Nature* **2009**, *457*, 603–607. [CrossRef] [PubMed]
40. Glumac, P.M.; LeBeau, A.M. The Role of CD133 in Cancer: A Concise Review. *Clin. Transl. Med.* **2018**, *7*, 18. [CrossRef] [PubMed]
41. Weichert, W. ALCAM/CD166 Is Overexpressed in Colorectal Carcinoma and Correlates with Shortened Patient Survival. *J. Clin. Pathol.* **2004**, *57*, 1160–1164. [CrossRef]
42. Han, S.; Zong, S.; Shi, Q.; Li, H.; Liu, S.; Yang, W.; Li, W.; Hou, F. Is Ep-CAM Expression a Diagnostic and Prognostic Biomarker for Colorectal Cancer? A Systematic Meta-Analysis. *EBioMedicine* **2017**, *20*, 61–69. [CrossRef]
43. Han, Y.; Xue, X.; Jiang, M.; Guo, X.; Li, P.; Liu, F.; Yuan, B.; Shen, Y.; Guo, X.; Zhi, Q.; et al. LGR5, a Relevant Marker of Cancer Stem Cells, Indicates a Poor Prognosis in Colorectal Cancer Patients: A Meta-Analysis. *Clin. Res. Hepatol. Gastroenterol.* **2015**, *39*, 267–273. [CrossRef]
44. Cammareri, P.; Lombardo, Y.; Francipane, M.G.; Bonventre, S.; Todaro, M.; Stassi, G. Isolation and Culture of Colon Cancer Stem Cells. In *Methods in Cell Biology*; Elsevier: Amsterdam, The Netherlands, 2008; Volume 86, pp. 311–324. ISBN 978-0-12-373876-9.
45. Korkusuz, P.; Köse, S.; Yersal, N.; Önen, S. Magnetic-Based Cell Isolation Technique for the Selection of Stem Cells. In *Skin Stem Cells*; Turksen, K., Ed.; Methods in Molecular Biology; Springer: New York, NY, USA, 2018; Volume 1879, pp. 153–163. ISBN 978-1-4939-8869-3.
46. Mathonnet, M. Hallmarks in Colorectal Cancer: Angiogenesis and Cancer Stem-like Cells. *World J. Gastroenterol.* **2014**, *20*, 4189. [CrossRef] [PubMed]
47. Mele, L.; Liccardo, D.; Tirino, V. Evaluation and Isolation of Cancer Stem Cells Using ALDH Activity Assay. In *Cancer Stem Cells*; Papaccio, G., Desiderio, V., Eds.; Methods in Molecular Biology; Springer: New York, NY, USA, 2018; Volume 1692, pp. 43–48. ISBN 978-1-4939-7400-9.
48. Huang, E.H.; Hynes, M.J.; Zhang, T.; Ginestier, C.; Dontu, G.; Appelman, H.; Fields, J.Z.; Wicha, M.S.; Boman, B.M. Aldehyde Dehydrogenase 1 Is a Marker for Normal and Malignant Human Colonic Stem Cells (SC) and Tracks SC Overpopulation during Colon Tumorigenesis. *Cancer Res.* **2009**, *69*, 3382–3389. [CrossRef]
49. Golebiewska, A.; Brons, N.H.C.; Bjerkvig, R.; Niclou, S.P. Critical Appraisal of the Side Population Assay in Stem Cell and Cancer Stem Cell Research. *Cell Stem Cell* **2011**, *8*, 136–147. [CrossRef]
50. Franken, N.A.P.; Rodermond, H.M.; Stap, J.; Haveman, J.; van Bree, C. Clonogenic Assay of Cells in Vitro. *Nat. Protoc.* **2006**, *1*, 2315–2319. [CrossRef] [PubMed]
51. Shaheen, S.; Ahmed, M.; Lorenzi, F.; Nateri, A.S. Spheroid-Formation (Colonosphere) Assay for in Vitro Assessment and Expansion of Stem Cells in Colon Cancer. *Stem Cell Rev. Rep.* **2016**, *12*, 492–499. [CrossRef] [PubMed]
52. Borowicz, S.; Van Scoyk, M.; Avasarala, S.; Karuppusamy Rathinam, M.K.; Tauler, J.; Bikkavilli, R.K.; Winn, R.A. The Soft Agar Colony Formation Assay. *J. Vis. Exp.* **2014**, 51998. [CrossRef] [PubMed]
53. Costa, E.C.; Moreira, A.F.; de Melo-Diogo, D.; Gaspar, V.M.; Carvalho, M.P.; Correia, I.J. 3D Tumor Spheroids: An Overview on the Tools and Techniques Used for Their Analysis. *Biotechnol. Adv.* **2016**, *34*, 1427–1441. [CrossRef]
54. Relier, S.; Yazdani, L.; Ayad, O.; Choquet, A.; Bourgaux, J.-F.; Prudhomme, M.; Pannequin, J.; Macari, F.; David, A. Antibiotics Inhibit Sphere-Forming Ability in Suspension Culture. *Cancer Cell Int.* **2016**, *16*, 6. [CrossRef]
55. Visvader, J.E.; Lindeman, G.J. Cancer Stem Cells in Solid Tumours: Accumulating Evidence and Unresolved Questions. *Nat. Rev. Cancer* **2008**, *8*, 755–768. [CrossRef]
56. Nguyen, L.V.; Vanner, R.; Dirks, P.; Eaves, C.J. Cancer Stem Cells: An Evolving Concept. *Nat. Rev. Cancer* **2012**, *12*, 133–143. [CrossRef]
57. Aiken, C.; Werbowetski-Ogilvie, T. Animal Models of Cancer Stem Cells: What Are They Really Telling Us? *Curr. Pathobiol. Rep.* **2013**, *1*, 91–99. [CrossRef]
58. O'Brien, C.A.; Kreso, A.; Jamieson, C.H.M. Cancer Stem Cells and Self-Renewal. *Clin. Cancer Res.* **2010**, *16*, 3113–3120. [CrossRef]
59. Mélin, C.; Perraud, A.; Bounaix Morand du Puch, C.; Loum, E.; Giraud, S.; Cardot, P.; Jauberteau, M.-O.; Lautrette, C.; Battu, S.; Mathonnet, M. Sedimentation Field Flow Fractionation Monitoring of in Vitro Enrichment in Cancer Stem Cells by Specific Serum-Free Culture Medium. *J. Chromatogr. B* **2014**, *963*, 40–46. [CrossRef] [PubMed]
60. Lacroix, A.; Deluche, E.; Zhang, L.Y.; Dalmay, C.; Mélin, C.; Leroy, J.; Babay, M.; Morand Du Puch, C.; Giraud, S.; Bessette, B.; et al. A New Label-Free Approach to Glioblastoma Cancer Stem Cell Sorting and Detection. *Anal. Chem.* **2019**, *91*, 8948–8957. [CrossRef]
61. Vedrenne, N.; Sarrazy, V.; Battu, S.; Bordeau, N.; Richard, L.; Billet, F.; Coronas, V.; Desmoulière, A. Neural Stem Cell Properties of an Astrocyte Subpopulation Sorted by Sedimentation Field-Flow Fractionation. *Rejuvenation Res.* **2016**, *19*, 362–372. [CrossRef]

62. Faye, P.-A.; Vedrenne, N.; De la Cruz-Morcillo, M.A.; Barrot, C.-C.; Richard, L.; Bourthoumieu, S.; Sturtz, F.; Funalot, B.; Lia, A.-S.; Battu, S. New Method for Sorting Endothelial and Neural Progenitors from Human Induced Pluripotent Stem Cells by Sedimentation Field Flow Fractionation. *Anal. Chem.* **2016**, *88*, 6696–6702. [CrossRef] [PubMed]

63. Mélin, C.; Perraud, A.; Christou, N.; Bibes, R.; Cardot, P.; Jauberteau, M.-O.; Battu, S.; Mathonnet, M. New Ex-Ovo Colorectal-Cancer Models from Different SdFFF-Sorted Tumor-Initiating Cells. *Anal. Bioanal. Chem.* **2015**, *407*, 8433–8443. [CrossRef] [PubMed]

64. Shmelkov, S.V.; Butler, J.M.; Hooper, A.T.; Hormigo, A.; Kushner, J.; Milde, T.; St. Clair, R.; Baljevic, M.; White, I.; Jin, D.K.; et al. CD133 Expression Is Not Restricted to Stem Cells, and Both CD133+ and CD133− Metastatic Colon Cancer Cells Initiate Tumors. *J. Clin. Investig.* **2008**, JCI34401. [CrossRef]

65. Ricci-Vitiani, L.; Fabrizi, E.; Palio, E.; De Maria, R. Colon Cancer Stem Cells. *J. Mol. Med.* **2009**, *87*, 1097–1104. [CrossRef] [PubMed]

66. Labianca, R.; Nordlinger, B.; Beretta, G.D.; Mosconi, S.; Mandalà, M.; Cervantes, A.; Arnold, D. Early Colon Cancer: ESMO Clinical Practice Guidelines for Diagnosis, Treatment and Follow-up. *Ann. Oncol.* **2013**, *24*, vi64–vi72. [CrossRef] [PubMed]

67. Costas-Chavarri, A.; Nandakumar, G.; Temin, S.; Lopes, G.; Cervantes, A.; Cruz Correa, M.; Engineer, R.; Hamashima, C.; Ho, G.F.; Huitzil, F.D.; et al. Treatment of Patients with Early-Stage Colorectal Cancer: ASCO Resource-Stratified Guideline. *J. Glob. Oncol.* **2019**, 1–19. [CrossRef]

68. FOxTROT Collaborative Group Feasibility of Preoperative Chemotherapy for Locally Advanced, Operable Colon Cancer: The Pilot Phase of a Randomised Controlled Trial. *Lancet Oncol.* **2012**, *13*, 1152–1160. [CrossRef]

69. André, T.; de Gramont, A.; Vernerey, D.; Chibaudel, B.; Bonnetain, F.; Tijeras-Raballand, A.; Scriva, A.; Hickish, T.; Tabernero, J.; Van Laethem, J.L.; et al. Adjuvant Fluorouracil, Leucovorin, and Oxaliplatin in Stage II to III Colon Cancer: Updated 10-Year Survival and Outcomes According to *BRAF* Mutation and Mismatch Repair Status of the MOSAIC Study. *J. Clin. Oncol.* **2015**, *33*, 4176–4187. [CrossRef] [PubMed]

70. André, T.; Boni, C.; Navarro, M.; Tabernero, J.; Hickish, T.; Topham, C.; Bonetti, A.; Clingan, P.; Bridgewater, J.; Rivera, F.; et al. Improved Overall Survival with Oxaliplatin, Fluorouracil, and Leucovorin as Adjuvant Treatment in Stage II or III Colon Cancer in the MOSAIC Trial. *J. Clin. Oncol.* **2009**, *27*, 3109–3116. [CrossRef] [PubMed]

71. Schmoll, H.-J.; Cartwright, T.; Tabernero, J.; Nowacki, M.P.; Figer, A.; Maroun, J.; Price, T.; Lim, R.; Van Cutsem, E.; Park, Y.-S.; et al. Phase III Trial of Capecitabine Plus Oxaliplatin as Adjuvant Therapy for Stage III Colon Cancer: A Planned Safety Analysis in 1,864 Patients. *J. Clin. Oncol.* **2007**, *26*, 102–109. [CrossRef]

72. Twelves, C.; Wong, A.; Nowacki, M.; Abt, M.; Burris, H., 3rd; Carrato, A.; Cassidy, J.; Cervantes, A.; Fagerberg, J.; Georgoulias, V.; et al. Capecitabine as Adjuvant Treatment for Stage III Colon Cancer. *N. Engl. J. Med.* **2005**, *352*, 2696–2704. [CrossRef]

73. Haller, D.G.; Tabernero, J.; Maroun, J.; de Braud, F.; Price, T.; Van Cutsem, E.; Hill, M.; Gilberg, F.; Rittweger, K.; Schmoll, H.-J. Capecitabine Plus Oxaliplatin Compared with Fluorouracil and Folinic Acid as Adjuvant Therapy for Stage III Colon Cancer. *J. Clin. Oncol.* **2011**, *29*, 1465–1471. [CrossRef] [PubMed]

74. Marin, J.J.G.; Sanchez de Medina, F.; Castaño, B.; Bujanda, L.; Romero, M.R.; Martinez-Augustin, O.; Moral-Avila, R.D.; Briz, O. Chemoprevention, Chemotherapy, and Chemoresistance in Colorectal Cancer. *Drug Metab. Rev.* **2012**, *44*, 148–172. [CrossRef] [PubMed]

75. Chiorean, E.G.; Nandakumar, G.; Fadelu, T.; Temin, S.; Alarcon-Rozas, A.E.; Bejarano, S.; Croitoru, A.-E.; Grover, S.; Lohar, P.V.; Odhiambo, A.; et al. Treatment of Patients with Late-Stage Colorectal Cancer: ASCO Resource-Stratified Guideline. *JCO Glob. Oncol.* **2020**, 414–438. [CrossRef] [PubMed]

76. Atreya, C.E.; Yaeger, R.; Chu, E. Systemic Therapy for Metastatic Colorectal Cancer: From Current Standards to Future Molecular Targeted Approaches. *Am. Soc. Clin. Oncol. Educ. Book* **2017**, *37*, 246–256. [CrossRef] [PubMed]

77. Van Cutsem, E.; Oliveira, J. Advanced Colorectal Cancer: ESMO Clinical Recommendations for Diagnosis, Treatment and Follow-up. *Ann. Oncol.* **2009**, *20*, iv61–iv63. [CrossRef]

78. Johnston, F.M.; Mavros, M.N.; Herman, J.M.; Pawlik, T.M. Local Therapies for Hepatic Metastases. *J. Natl. Compr. Canc. Netw.* **2013**, *11*, 153–160. [CrossRef]

79. Nosher, J.L.; Ahmed, I.; Patel, A.N.; Gendel, V.; Murillo, P.G.; Moss, R.; Jabbour, S.K. Non-Operative Therapies for Colorectal Liver Metastases. *Surg. Treat. Colorectal Liver Metastases* **2015**, *6*, 17.

80. Sveen, A.; Kopetz, S.; Lothe, R.A. Biomarker-Guided Therapy for Colorectal Cancer: Strength in Complexity. *Nat. Rev. Clin. Oncol.* **2020**, *17*, 11–32. [CrossRef] [PubMed]

81. Advani, S.; Kopetz, S. Ongoing and Future Directions in the Management of Metastatic Colorectal Cancer: Update on Clinical Trials. *J. Surg. Oncol.* **2019**, *119*, 642–652. [CrossRef]

82. IMPACT Investigators. Efficacy of Adjuvant Fluorouracil and Folinic Acid in Colon Cancer. *The Lancet* **1995**, *345*, 939–944. [CrossRef]

83. Saltz, L.B.; Moore, M.J.; Pirotta, N. Irinotecan plus Fluorouracil and Leucovorin for Metastatic Colorectal Cancer. *N. Engl. J. Med.* **2000**, *343*, 905–914. [CrossRef]

84. André, T.; Boni, C.; Mounedji-Boudiaf, L.; Navarro, M.; Tabernero, J.; Hickish, T.; Topham, C.; Zaninelli, M.; Clingan, P.; Bridgewater, J.; et al. Oxaliplatin, Fluorouracil, and Leucovorin as Adjuvant Treatment for Colon Cancer. *N. Engl. J. Med.* **2004**, *350*, 2343–2351. [CrossRef] [PubMed]

85. Mayer, R.J.; Van Cutsem, E.; Falcone, A.; Yoshino, T.; Garcia-Carbonero, R.; Mizunuma, N.; Yamazaki, K.; Shimada, Y.; Tabernero, J.; Komatsu, Y.; et al. Randomized Trial of TAS-102 for Refractory Metastatic Colorectal Cancer. *N. Engl. J. Med.* **2015**, *372*, 1909–1919. [CrossRef]

86. Hurwitz, H.; Fehrenbacher, L.; Novotny, W.; Cartwright, T.; Hainsworth, J.; Heim, W.; Berlin, J.; Baron, A.; Griffing, S.; Holmgren, E.; et al. Bevacizumab plus Irinotecan, Fluorouracil, and Leucovorin for Metastatic Colorectal Cancer. *N. Engl. J. Med.* **2004**, *350*, 2335–2342. [CrossRef] [PubMed]

87. Grothey, A.; Cutsem, E.V.; Sobrero, A.; Siena, S.; Falcone, A.; Ychou, M.; Humblet, Y.; Bouché, O.; Mineur, L.; Barone, C.; et al. Regorafenib Monotherapy for Previously Treated Metastatic Colorectal Cancer (CORRECT): An International, Multicentre, Randomised, Placebo-Controlled, Phase 3 Trial. *Lancet* **2013**, *381*, 303–312. [CrossRef]

88. Van Cutsem, E.; Tabernero, J.; Lakomy, R.; Prenen, H.; Prausová, J.; Macarulla, T.; Ruff, P.; van Hazel, G.A.; Moiseyenko, V.; Ferry, D.; et al. Addition of Aflibercept to Fluorouracil, Leucovorin, and Irinotecan Improves Survival in a Phase III Randomized Trial in Patients with Metastatic Colorectal Cancer Previously Treated with an Oxaliplatin-Based Regimen. *J. Clin. Oncol.* **2012**, *30*, 3499–3506. [CrossRef] [PubMed]

89. Tabernero, J.; Yoshino, T.; Cohn, A.L.; Obermannova, R.; Bodoky, G.; Garcia-Carbonero, R.; Ciuleanu, T.-E.; Portnoy, D.C.; Van Cutsem, E.; Grothey, A.; et al. Ramucirumab versus Placebo in Combination with Second-Line FOLFIRI in Patients with Metastatic Colorectal Carcinoma That Progressed during or after First-Line Therapy with Bevacizumab, Oxaliplatin, and a Fluoropyrimidine (RAISE): A Randomised, Double-Blind, Multicentre, Phase 3 Study. *Lancet Oncol.* **2015**, *16*, 499–508. [CrossRef] [PubMed]

90. Messersmith, W.A.; Ahnen, D.J. Targeting EGFR in Colorectal Cancer. *N. Engl. J. Med.* **2008**, *359*, 1834–1836. [CrossRef]

91. Le, D.T.; Uram, J.N.; Wang, H.; Bartlett, B.R.; Kemberling, H.; Eyring, A.D.; Skora, A.D.; Luber, B.S.; Azad, N.S.; Laheru, D.; et al. PD-1 Blockade in Tumors with Mismatch-Repair Deficiency. *N. Engl. J. Med.* **2015**, *372*, 2509–2520. [CrossRef]

92. Overman, M.J.; Lonardi, S.; Wong, K.Y.M.; Lenz, H.-J.; Gelsomino, F.; Aglietta, M.; Morse, M.A.; Van Cutsem, E.; McDermott, R.; Hill, A.; et al. Durable Clinical Benefit with Nivolumab Plus Ipilimumab in DNA Mismatch Repair–Deficient/Microsatellite Instability–High Metastatic Colorectal Cancer. *J. Clin. Oncol.* **2018**, *36*, 773–779. [CrossRef] [PubMed]

93. Corcoran, R.B.; André, T.; Atreya, C.E.; Schellens, J.H.M.; Yoshino, T.; Bendell, J.C.; Hollebecque, A.; McRee, A.J.; Siena, S.; Middleton, G.; et al. Combined BRAF, EGFR, and MEK Inhibition in Patients with $BRAF^{V600E}$-Mutant Colorectal Cancer. *Cancer Discov.* **2018**, *8*, 428–443. [CrossRef] [PubMed]

94. Kopetz, S.; Grothey, A.; Yaeger, R.; Van Cutsem, E.; Desai, J.; Yoshino, T.; Wasan, H.; Ciardiello, F.; Loupakis, F.; Hong, Y.S.; et al. Encorafenib, Binimetinib, and Cetuximab in BRAF V600E–Mutated Colorectal Cancer. *N. Engl. J. Med.* **2019**, *381*, 1632–1643. [CrossRef]

95. Meric-Bernstam, F.; Hurwitz, H.; Raghav, K.P.S.; McWilliams, R.R.; Fakih, M.; VanderWalde, A.; Swanton, C.; Kurzrock, R.; Burris, H.; Sweeney, C.; et al. Pertuzumab plus Trastuzumab for HER2-Amplified Metastatic Colorectal Cancer (MyPathway): An Updated Report from a Multicentre, Open-Label, Phase 2a, Multiple Basket Study. *Lancet Oncol.* **2019**, *20*, 518–530. [CrossRef]

96. Sartore-Bianchi, A.; Trusolino, L.; Martino, C.; Bencardino, K.; Lonardi, S.; Bergamo, F.; Zagonel, V.; Leone, F.; Depetris, I.; Martinelli, E.; et al. Dual-Targeted Therapy with Trastuzumab and Lapatinib in Treatment-Refractory, KRAS Codon 12/13 Wild-Type, HER2-Positive Metastatic Colorectal Cancer (HERACLES): A Proof-of-Concept, Multicentre, Open-Label, Phase 2 Trial. *Lancet Oncol.* **2016**, *17*, 738–746. [CrossRef]

97. Drilon, A.; Laetsch, T.W.; Kummar, S.; DuBois, S.G.; Lassen, U.N.; Demetri, G.D.; Nathenson, M.; Doebele, R.C.; Farago, A.F.; Pappo, A.S.; et al. Efficacy of Larotrectinib in TRK Fusion–Positive Cancers in Adults and Children. *N. Engl. J. Med.* **2018**, *378*, 731–739. [CrossRef]

98. Demetri, G.D.; Paz-Ares, L.; Farago, A.F.; Liu, S.V.; Chawla, S.P.; Tosi, D.; Kim, E.S.; Blakely, C.M.; Krauss, J.C.; Sigal, D.; et al. Efficacy and Safety of Entrectinib in Patients with NTRK Fusion-Positive Tumours: Pooled Analysis of STARTRK-2, STARTRK-1, and ALKA-372-001. *Ann. Oncol.* **2018**, *29*, ix175. [CrossRef]

99. Holohan, C.; Van Schaeybroeck, S.; Longley, D.B.; Johnston, P.G. Cancer Drug Resistance: An Evolving Paradigm. *Nat. Rev. Cancer* **2013**, *13*, 714–726. [CrossRef]

100. Li, Q.; Shu, Y. Role of Solute Carriers in Response to Anticancer Drugs. *Mol. Cell. Ther.* **2014**, *2*, 15. [CrossRef] [PubMed]

101. Hu, T.; Li, Z.; Gao, C.-Y.; Cho, C.H. Mechanisms of Drug Resistance in Colon Cancer and Its Therapeutic Strategies. *World J. Gastroenterol.* **2016**, *22*, 6876. [CrossRef] [PubMed]

102. Zhang, S.; Lovejoy, K.S.; Shima, J.E.; Lagpacan, L.L.; Shu, Y.; Lapuk, A.; Chen, Y.; Komori, T.; Gray, J.W.; Chen, X.; et al. Organic Cation Transporters Are Determinants of Oxaliplatin Cytotoxicity. *Cancer Res.* **2006**, *66*, 8847–8857. [CrossRef]

103. Buck, E.; Sprick, M.; Gaida, M.; Grüllich, C.; Weber, T.; Herpel, E.; Bruckner, T.; Koschny, R. Tumor Response to Irinotecan Is Associated with CYP3A5 Expression in Colorectal Cancer. *Oncol. Lett.* **2019**. [CrossRef] [PubMed]

104. Thomas, M.L.; Coyle, K.M.; Sultan, M.; Marcato, P. Cancer Stem Cells and Chemoresistance: Strategies to Overcome Therapeutic Resistance. In *Cancer Stem Cells: Emerging Concepts and Future Perspectives in Translational Oncology*; Babashah, S., Ed.; Springer International Publishing: Cham, Switzerland, 2015; pp. 477–518. ISBN 978-3-319-21029-2.

105. Hoskins, J.M.; Goldberg, R.M.; Qu, P.; Ibrahim, J.G.; McLeod, H.L. UGT1A1*28 Genotype and Irinotecan-Induced Neutropenia: Dose Matters. *J. Natl. Cancer Inst.* **2007**, *99*, 1290–1295. [CrossRef] [PubMed]

106. Stoehlmacher, J. Association between Glutathione S-Transferase P1, T1, and M1 Genetic Polymorphism and Survival of Patients with Metastatic Colorectal Cancer. *J. Natl. Cancer Inst.* **2002**, *94*, 936–942. [CrossRef] [PubMed]

107. Ramos, P.; Bentires-Alj, M. Mechanism-Based Cancer Therapy: Resistance to Therapy, Therapy for Resistance. *Oncogene* **2015**, *34*, 3617–3626. [CrossRef]

108. Misale, S.; Yaeger, R.; Hobor, S.; Scala, E.; Janakiraman, M.; Liska, D.; Valtorta, E.; Schiavo, R.; Buscarino, M.; Siravegna, G.; et al. Emergence of KRAS Mutations and Acquired Resistance to Anti-EGFR Therapy in Colorectal Cancer. *Nature* **2012**, *486*, 532–536. [CrossRef]

109. Shirota, Y.; Stoehlmacher, J.; Brabender, J.; Xiong, Y.-P.; Uetake, H.; Danenberg, K.D.; Groshen, S.; Tsao-Wei, D.D.; Danenberg, P.V.; Lenz, H.-J. ERCC1 and Thymidylate Synthase MRNA Levels Predict Survival for Colorectal Cancer Patients Receiving Combination Oxaliplatin and Fluorouracil Chemotherapy. *J. Clin. Oncol.* **2001**, *19*, 4298–4304. [CrossRef] [PubMed]

110. Valeri, N.; Gasparini, P.; Braconi, C.; Paone, A.; Lovat, F.; Fabbri, M.; Sumani, K.M.; Alder, H.; Amadori, D.; Patel, T.; et al. MicroRNA-21 Induces Resistance to 5-Fluorouracil by down-Regulating Human DNA MutS Homolog 2 (HMSH2). *Proc. Natl. Acad. Sci. USA* **2010**, *107*, 21098–21103. [CrossRef]

111. Hosoya, N.; Miyagawa, K. Targeting DNA Damage Response in Cancer Therapy. *Cancer Sci.* **2014**, *105*, 370–388. [CrossRef] [PubMed]

112. Zhao, J. Cancer Stem Cells and Chemoresistance: The Smartest Survives the Raid. *Pharmacol. Ther.* **2016**, *160*, 145–158. [CrossRef]

113. Nehls, O.; Okech, T.; Hsieh, C.-J.; Enzinger, T.; Sarbia, M.; Borchard, F.; Gruenagel, H.-H.; Gaco, V.; Hass, H.G.; Arkenau, H.T.; et al. Studies on P53, BAX and Bcl-2 Protein Expression and Microsatellite Instability in Stage III (UICC) Colon Cancer Treated by Adjuvant Chemotherapy: Major Prognostic Impact of Proapoptotic BAX. *Br. J. Cancer* **2007**, *96*, 1409–1418. [CrossRef]

114. Zhao, Y.; Dong, Q.; Li, J.; Zhang, K.; Qin, J.; Zhao, J.; Sun, Q.; Wang, Z.; Wartmann, T.; Jauch, K.W.; et al. Targeting Cancer Stem Cells and Their Niche: Perspectives for Future Therapeutic Targets and Strategies. *Semin. Cancer Biol.* **2018**, *53*, 139–155. [CrossRef]

115. Vermeulen, L.; De Sousa E Melo, F.; van der Heijden, M.; Cameron, K.; de Jong, J.H.; Borovski, T.; Tuynman, J.B.; Todaro, M.; Merz, C.; Rodermond, H.; et al. Wnt Activity Defines Colon Cancer Stem Cells and Is Regulated by the Microenvironment. *Nat. Cell Biol.* **2010**, *12*, 468–476. [CrossRef]

116. Borovski, T.; De Sousa E Melo, F.; Vermeulen, L.; Medema, J.P. Cancer Stem Cell Niche: The Place to Be. *Cancer Res.* **2011**, *71*, 634–639. [CrossRef]

117. De Sousa e Melo, F.; Kurtova, A.V.; Harnoss, J.M.; Kljavin, N.; Hoeck, J.D.; Hung, J.; Anderson, J.E.; Storm, E.E.; Modrusan, Z.; Koeppen, H.; et al. A Distinct Role for Lgr5+ Stem Cells in Primary and Metastatic Colon Cancer. *Nature* **2017**, *543*, 676–680. [CrossRef] [PubMed]

118. Fumagalli, A.; Oost, K.C.; Kester, L.; Morgner, J.; Bornes, L.; Bruens, L.; Spaargaren, L.; Azkanaz, M.; Schelfhorst, T.; Beerling, E.; et al. Plasticity of Lgr5-Negative Cancer Cells Drives Metastasis in Colorectal Cancer. *Cell Stem Cell* **2020**, *26*, 569–578.e7. [CrossRef]

119. Medema, J.P. Targeting the Colorectal Cancer Stem Cell. *N. Engl. J. Med.* **2017**, *377*, 888–890. [CrossRef]

120. Shimokawa, M.; Ohta, Y.; Nishikori, S.; Matano, M.; Takano, A.; Fujii, M.; Date, S.; Sugimoto, S.; Kanai, T.; Sato, T. Visualization and Targeting of LGR5+ Human Colon Cancer Stem Cells. *Nature* **2017**, *545*, 187–192. [CrossRef] [PubMed]

121. Lenos, K.J.; Miedema, D.M.; Lodestijn, S.C.; Nijman, L.E.; van den Bosch, T.; Romero Ros, X.; Lourenço, F.C.; Lecca, M.C.; van der Heijden, M.; van Neerven, S.M.; et al. Stem Cell Functionality Is Microenvironmentally Defined during Tumour Expansion and Therapy Response in Colon Cancer. *Nat. Cell Biol.* **2018**, *20*, 1193–1202. [CrossRef]

122. De Sousa e Melo, F.; de Sauvage, F.J. Cellular Plasticity in Intestinal Homeostasis and Disease. *Cell Stem Cell* **2019**, *24*, 54–64. [CrossRef]

123. Li, Y.; Rogoff, H.A.; Keates, S.; Gao, Y.; Murikipudi, S.; Mikule, K.; Leggett, D.; Li, W.; Pardee, A.B.; Li, C.J. Suppression of Cancer Relapse and Metastasis by Inhibiting Cancer Stemness. *Proc. Natl. Acad. Sci. USA* **2015**, *112*, 1839–1844. [CrossRef] [PubMed]

124. ClinicalTrials.Gov. Available online: https://clinicaltrials.gov/ (accessed on 7 October 2020).

125. Hubbard, J.M.; Grothey, A. Napabucasin: An Update on the First-in-Class Cancer Stemness Inhibitor. *Drugs* **2017**, *77*, 1091–1103. [CrossRef] [PubMed]

126. Langleben, A.; Supko, J.G.; Hotte, S.J.; Batist, G.; Hirte, H.W.; Rogoff, H.; Li, Y.; Li, W.; Kerstein, D.; Leggett, D.; et al. A Dose-Escalation Phase I Study of a First-in-Class Cancer Stemness Inhibitor in Patients with Advanced Malignancies. *J. Clin. Oncol.* **2013**, *31*, 2542. [CrossRef]

127. Jonker, D.J.; Stephenson, J.; Edenfield, W.J.; Supko, J.G.; Li, Y.; Li, W.; Hitron, M.; Leggett, D.; Kerstein, D.; Li, C. A Phase I Extension Study of BBI608, a First-in-Class Cancer Stem Cell (CSC) Inhibitor, in Patients with Advanced Solid Tumors. *J. Clin. Oncol.* **2014**, *32*, 2546. [CrossRef]

128. Larson, T.; Ortuzar, W.F.; Bekaii-Saab, T.S.; Becerra, C.; Ciombor, K.K.; Hubbard, J.M.; Edenfield, W.J.; Shao, S.H.; Grothey, A.; Borodyansky, L.; et al. BBI608-224: A Phase Ib/II Study of Cancer Stemness Inhibitor Napabucasin (BBI-608) Administered with Panitumumab in KRAS Wild-Type Patients with Metastatic Colorectal Cancer. *J. Clin. Oncol.* **2017**, *35*, 677. [CrossRef]

129. O'Neil, B.H.; Hubbard, J.M.; Starodub, A.; Jonker, D.J.; Edenfield, W.J.; El-Rayes, B.F.; Halfdanarson, T.R.; Ramanathan, R.K.; Pitot, H.C.; Britten, C.D.; et al. Phase 1b Extension Study of Cancer Stemness Inhibitor BB608 (Napabucasin) Administered in Combination with FOLFIRI +/− Bevacizumab (Bev) in Patients (Pts) with Advanced Colorectal Cancer (CRC). *J. Clin. Oncol.* **2016**, *34*, 3564. [CrossRef]

130. Bendell, J.C.; Hubbard, J.M.; O'Neil, B.H.; Jonker, D.J.; Starodub, A.; Peyton, J.D.; Pitot, H.C.; Halfdanarson, T.R.; Nadeau, B.R.; Zubkus, J.D.; et al. Phase 1b/II Study of Cancer Stemness Inhibitor Napabucasin (BBI-608) in Combination with FOLFIRI +/− Bevacizumab (Bev) in Metastatic Colorectal Cancer (MCRC) Patients (Pts). *J. Clin. Oncol.* **2017**, *35*, 3529. [CrossRef]
131. Jonker, D.J.; Nott, L.; Yoshino, T.; Gill, S.; Shapiro, J.; Ohtsu, A.; Zalcberg, J.; Vickers, M.M.; Wei, A.; Gao, Y.; et al. A Randomized Phase III Study of Napabucasin [BBI608] (NAPA) vs Placebo (PBO) in Patients (Pts) with Pretreated Advanced Colorectal Cancer (ACRC): The CCTG/AGITG CO.23 Trial. *Ann. Oncol.* **2016**, *27*, vi150. [CrossRef]
132. Jonker, D.J.; Nott, L.; Yoshino, T.; Gill, S.; Shapiro, J.; Ohtsu, A.; Zalcberg, J.; Vickers, M.M.; Wei, A.C.; Gao, Y.; et al. Napabucasin versus Placebo in Refractory Advanced Colorectal Cancer: A Randomised Phase 3 Trial. *Lancet Gastroenterol. Hepatol.* **2018**, *3*, 263–270. [CrossRef]
133. Grothey, A.; Shah, M.A.; Yoshino, T.; Van Cutsem, E.; Taieb, J.; Xu, R.; Tebbutt, N.C.; Falcone, A.; Cervantes, A.; Borodyansky, L.; et al. CanStem303C Trial: A Phase III Study of Napabucasin (BBI-608) in Combination with 5-Fluorouracil (5-FU), Leucovorin, Irinotecan (FOLFIRI) in Adult Patients with Previously Treated Metastatic Colorectal Cancer (MCRC). *J. Clin. Oncol.* **2017**, *35*, TPS3619. [CrossRef]
134. Sonbol, M.; Ahn, D.; Bekaii-Saab, T. Therapeutic Targeting Strategies of Cancer Stem Cells in Gastrointestinal Malignancies. *Biomedicines* **2019**, *7*, 17. [CrossRef]
135. Fischer, M.; Yen, W.-C.; Kapoun, A.M.; Wang, M.; O'Young, G.; Lewicki, J.; Gurney, A.; Hoey, T. Anti-DLL4 Inhibits Growth and Reduces Tumor-Initiating Cell Frequency in Colorectal Tumors with Oncogenic *KRAS* Mutations. *Cancer Res.* **2011**, *71*, 1520–1525. [CrossRef] [PubMed]
136. Hoey, T.; Yen, W.-C.; Axelrod, F.; Basi, J.; Donigian, L.; Dylla, S.; Fitch-Bruhns, M.; Lazetic, S.; Park, I.-K.; Sato, A.; et al. DLL4 Blockade Inhibits Tumor Growth and Reduces Tumor-Initiating Cell Frequency. *Cell Stem Cell* **2009**, *5*, 168–177. [CrossRef] [PubMed]
137. Ridgway, J.; Zhang, G.; Wu, Y.; Stawicki, S.; Liang, W.-C.; Chanthery, Y.; Kowalski, J.; Watts, R.J.; Callahan, C.; Kasman, I.; et al. Inhibition of Dll4 Signalling Inhibits Tumour Growth by Deregulating Angiogenesis. *Nature* **2006**, *444*, 1083–1087. [CrossRef]
138. Smith, D.C.; Eisenberg, P.D.; Manikhas, G.; Chugh, R.; Gubens, M.A.; Stagg, R.J.; Kapoun, A.M.; Xu, L.; Dupont, J.; Sikic, B. A Phase I Dose Escalation and Expansion Study of the Anticancer Stem Cell Agent Demcizumab (Anti-DLL4) in Patients with Previously Treated Solid Tumors. *Clin. Cancer Res.* **2014**, *20*, 6295–6303. [CrossRef]
139. Johnson, M.; Rasco, D.; Schneider, B.; Shu, C.; Jotte, R.; Parmer, H.; Stagg, R.; Lopez, J. Abstract A081: A Phase 1b, Open-Label, Dose Escalation and Expansion Study of Demcizumab plus Pembrolizumab in Patients with Locally Advanced or Metastatic Solid Tumors. *Mol. Cancer Ther.* **2018**, *17*, A081. [CrossRef]
140. Quarni, W.; Dutta, R.; Green, R.; Katiri, S.; Patel, B.; Mohapatra, S.S.; Mohapatra, S. Mithramycin A Inhibits Colorectal Cancer Growth by Targeting Cancer Stem Cells. *Sci. Rep.* **2019**, *9*, 15202. [CrossRef]
141. Rao, M.; Atay, S.M.; Shukla, V.; Hong, Y.; Upham, T.; Ripley, R.T.; Hong, J.A.; Zhang, M.; Reardon, E.; Fetsch, P.; et al. Mithramycin Depletes Specificity Protein 1 and Activates P53 to Mediate Senescence and Apoptosis of Malignant Pleural Mesothelioma Cells. *Clin. Cancer Res.* **2016**, *22*, 1197–1210. [CrossRef]
142. Zhao, Y.; Zhang, W.; Guo, Z.; Ma, F.; Wu, Y.; Bai, Y.; Gong, W.; Chen, Y.; Cheng, T.; Zhi, F.; et al. Inhibition of the Transcription Factor Sp1 Suppresses Colon Cancer Stem Cell Growth and Induces Apoptosis in Vitro and in Nude Mouse Xenografts. *Oncol. Rep.* **2013**, *30*, 1782–1792. [CrossRef]
143. Grohar, P.J.; Glod, J.; Peer, C.J.; Sissung, T.M.; Arnaldez, F.I.; Long, L.; Figg, W.D.; Whitcomb, P.; Helman, L.J.; Widemann, B.C. A Phase I/II Trial and Pharmacokinetic Study of Mithramycin in Children and Adults with Refractory Ewing Sarcoma and EWS–FLI1 Fusion Transcript. *Cancer Chemother. Pharmacol.* **2017**, *80*, 645–652. [CrossRef] [PubMed]
144. Sissung, T.M.; Huang, P.A.; Hauke, R.J.; McCrea, E.M.; Peer, C.J.; Barbier, R.H.; Strope, J.D.; Ley, A.M.; Zhang, M.; Hong, J.A.; et al. Severe Hepatotoxicity of Mithramycin Therapy Caused by Altered Expression of Hepatocellular Bile Transporters. *Mol. Pharmacol.* **2019**, *96*, 158–167. [CrossRef] [PubMed]
145. Takebe, N.; Miele, L.; Harris, P.J.; Jeong, W.; Bando, H.; Kahn, M.; Yang, S.X.; Ivy, S.P. Targeting Notch, Hedgehog, and Wnt Pathways in Cancer Stem Cells: Clinical Update. *Nat. Rev. Clin. Oncol.* **2015**, *12*, 445–464. [CrossRef] [PubMed]

 *cancers*

Article

Loss of SATB2 Occurs More Frequently Than CDX2 Loss in Colorectal Carcinoma and Identifies Particularly Aggressive Cancers in High-Risk Subgroups

Maxime Schmitt [1,2,†], Miguel Silva [3,†], Björn Konukiewitz [2,4], Corinna Lang [2], Katja Steiger [2], Kathrin Halfter [5], Jutta Engel [5], Paul Jank [1], Nicole Pfarr [2], Dirk Wilhelm [6], Sebastian Foersch [7], Carsten Denkert [1], Markus Tschurtschenthaler [3,8], Wilko Weichert [2,9] and Moritz Jesinghaus [1,2,*]

1   Institute of Pathology, University Hospital Marburg, 35043 Marburg, Germany; maxime-schmitt@gmx.net (M.S.); paul.jank@uni-marburg.de (P.J.); carsten.denkert@uni-marburg.de (C.D.)
2   Institute of Pathology, Technical University Munich, 81675 Munich, Germany; Bjoern.Konukiewitz@uksh.de (B.K.); corinna.maria.lang@icloud.com (C.L.); katja.steiger@tum.de (K.S.); nicole.pfarr@tum.de (N.P.); wilko.weichert@tum.de (W.W.)
3   II Medizinische Klinik, Klinikum Rechts der Isar, Technical University Munich, 81675 Munich, Germany; gsilva.miguel@gmail.com (M.S.); markus.tschurtschenthaler@tum.de (M.T.)
4   Institute of Pathology, Christian-Albrechts-University, 23562 Kiel, Germany
5   Munich Cancer Registry (MCR), University Hospital of Munich, Institute for Medical Information Processing, Biometry, and Epidemiology (IBE), Ludwig-Maximilian-University (LMU), 81377 Munich, Germany; halfter@ibe.med.uni-muenchen.de (K.H.); engel@ibe.med.uni-muenchen.de (J.E.)
6   Department of Surgery, Klinikum Rechts der Isar, Technical University Munich, 81675 Munich, Germany; dirk.wilhelm@tum.de
7   Institute of Pathology, University Hospital Mainz, 55131 Mainz, Germany; sebastian.foersch@unimedizin-mainz.de
8   Institute for Translational Cancer Research, German Cancer Consortium (DKTK), Partner Site Munich, 81675 Munich, Germany
9   German Cancer Consortium (DKTK), Partner Site Munich, 81675 Munich, Germany
*   Correspondence: moritz.jesinghaus@uni-marburg.de
†   These authors contributed equally to this work.

Citation: Schmitt, M.; Silva, M.; Konukiewitz, B.; Lang, C.; Steiger, K.; Halfter, K.; Engel, J.; Jank, P.; Pfarr, N.; Wilhelm, D.; et al. Loss of SATB2 Occurs More Frequently Than CDX2 Loss in Colorectal Carcinoma and Identifies Particularly Aggressive Cancers in High-Risk Subgroups. *Cancers* 2021, 13, 6177. https://doi.org/10.3390/cancers13246177

Academic Editors: Marta Baiocchi and Ann Zeuner

Received: 12 November 2021
Accepted: 5 December 2021
Published: 7 December 2021

Publisher's Note: MDPI stays neutral with regard to jurisdictional claims in published maps and institutional affiliations.

Simple Summary: The immunohistochemical analysis of Special AT-rich sequence-binding protein 2 (SATB2) is increasingly being used to detect colorectal differentiation. Our study aimed to investigate SATB2 expression levels and the prognostic relevance of SATB2 loss in colorectal carcinoma (CRC), especially in comparison with CDX2, the standard marker of colorectal differentiation. We tested SATB2 expression in 1039 CRCs and identified SATB2 as a strong prognosticator in the overall cohort as well as in specific subcohorts, including high-risk subgroups. Compared to CDX2, SATB2 showed a higher prognostic power but was lost at a much higher frequency, generally rendering SATB2 as the less sensitive marker for colorectal differentiation compared to CDX2.

Abstract: Background: Special AT-rich sequence-binding protein 2 (SATB2) has emerged as an alternative immunohistochemical marker to CDX2 for colorectal differentiation. However, the distribution and prognostic relevance of SATB2 expression in colorectal carcinoma (CRC) have to be further elucidated. Methods: SATB2 expression was analysed in 1039 CRCs and correlated with clinicopathological and morphological factors, CDX2 expression as well as survival parameters within the overall cohort and in clinicopathological subgroups. Results: SATB2 loss was a strong prognosticator in univariate analyses of the overall cohort ($p < 0.001$ for all survival comparisons) and in numerous subcohorts including high-risk scenarios (UICC stage III/high tumour budding). SATB2 retained its prognostic relevance in multivariate analyses of these high-risk scenarios (e.g., UICC stage III: DSS: $p = 0.007$, HR: 1.95), but not in the overall cohort (DSS: $p = 0.1$, HR: 1.25). SATB2 loss was more frequent than CDX2 loss (22.2% vs. 10.2%, $p < 0.001$) and of higher prognostic relevance with only moderate overlap between SATB2/CDX2 expression groups. Conclusions: SATB2 loss is able to identify especially aggressive CRCs in high-risk subgroups. While SATB2 is the prognostically superior immunohistochemical parameter compared to CDX2 in univariate analyses, it appears to be the less sensitive marker for colorectal differentiation as it is lost more frequently.

Keywords: SATB2; colorectal carcinoma; prognosis; CDX2

1. Introduction

Considering that colorectal carcinoma (CRC) currently ranks among the three most common cancers in humans concerning incidence and mortality worldwide [1,2], further explorations on potentially relevant biomarkers are warranted in order to characterise these tumours precisely and improve prognostic predictions.

Special AT-rich sequence-binding protein 2 (SATB2), a transcription factor interacting with nuclear matrix attachment regions which is highly expressed in the non-neoplastic colorectal mucosa [3,4], attracted increasing scientific notice for the identification of the colorectal origin of cancers of unknown primary and of CRC metastases [5–10], delineating SATB2 as a valid addition to CDX2, which is still the most established immunohistochemical marker associated with colorectal differentiation [5,11].

Previous immunohistochemical assessments of SATB2 in CRC showed a general association of a diminished SATB2 expression with poorer survival characteristics and microsatellite status [12–16]. However, it remains unclear whether SATB2 is differently expressed within purely morphological (adenocarcinoma NOS vs. specific CRC subtypes, tumour budding subcategories (Bd1/2/3), WHO low- vs. high-grade carcinomas), immunohistochemical (CDX2 expression) and pTNM/UICC stage subgroups. Furthermore, it remains to be elucidated, how frequently the loss of SATB2 occurs in comparison to the loss of CDX2 and whether SATB2 can identify distinct prognostic groups within these colorectal cancer subsets.

To address these questions, we investigated SATB2 expression in a large cohort comprising 1039 CRCs and correlated the results with histomorphologic and immunohistochemical (CRC subtypes, tumour budding activity, WHO grade, CDX2 expression) as well as clinicopathological parameters (pTNM/UICC staging, tumour localisation) and explored the prognostic relevance of SATB2 expression in uni- and multivariate survival analyses in the overall cohort as well as in specific subgroups.

2. Materials and Methods

2.1. Study Population

Our study cohort comprised one thousand and thirty-nine CRC patients that underwent surgical resection between 1997 and 2019 at the University Hospital Klinikum rechts der Isar of the Technical University of Munich, Germany. Only patients with colorectal carcinomas were included in this study. Patients suffering from other colorectal tumours (e.g., neuroendocrine tumours, non-epithelial tumours, etc.), insufficient tissue on the constructed tissue microarray or incomplete clinicopathological/survival data were excluded. The original cohort (1997–2018) was recently extended with cases from 2019 ($n = 36$) [17]. Clinicopathological characteristics as well as survival data were extracted from hospital records or from the Munich Cancer Registry. Definitions of survival parameters, survival endpoints and general treatment modalities were defined as described previously [18,19]. The local ethic committee of the Technical University of Munich approved this study (reference number: 252/16 s).

2.2. Evaluation of SATB2 Expression and Clinicopathological Parameters

SATB2 expression was analysed by SATB2 immunohistochemistry in 1039 CRCs on a tissue microarray with two tumour-carrying cores from each tumour. We used an automated immunostainer (BOND RXm System, Leica Biosystems, Germany) for the immunohistochemical staining of SATB2 (EP281, Cellmarque 384R, Ready-To-Use, Cell Marque, USA), which is the standard SATB2 antibody used in daily clinicopathological routine and which has been used by various previous studies [5,11,15,20]. We stained 2 μm thick sections from our tissue microarray. Antigen retrieval was performed with

Epitope Retrieval 1 after deparaffinisation, (Leica Biosystems, Germany; equivalent to citrate buffer pH 6) for 20 min and antibody binding was detected using a Polymer Refine Detection Kit (Leica Biosystems, Nußloch, Germany) without a postprimary antibody and haematoxylin counterstain. Naturally, pretested positive/negative control-tissues were stained in parallel. Two independent observers (MJ, MS (Maxime Schmitt)) that were blinded to clinicopathological parameters manually performed the evaluation of SATB2 expression.

We assessed the number of positive carcinoma cells for each individual patient. Counting a minimum of 500 tumour cells, the resulting cumulative percentage score for both cores was then assigned for each CRC (range: 0–100%). Only a nuclear staining of SATB2 was considered specific. SATB2 expression patterns (combined from both cores) were classified into three separate groups: diffuse, if the tumours either showed a complete expression or only a very focal loss in singular cells; heterogeneous, if areas with a complete loss of staining were observed; absent, if the tumours were completely negative. A strong staining intensity was defined as an intensity comparable to normal colonic mucosa, a still easily identifiable but slightly weaker staining was rated as medium. A barely visible staining intensity was classified as weak. Cases without any detectable staining were classified as absent.

SATB2-grouping results from the TMA were compared with SATB2 staining from 20 randomly selected full block slides, interobserver variability was probed in 150 cases that were assessed by the two observers in a blinded fashion.

SATB2 expression was correlated with clinicopathological variables including staging data and with Haematoxylin and Eosin-based histopathological parameters defined by the recent WHO classification (CRC subtypes: Adenocarcinoma NOS, Mucinous adenocarcinoma, Signet-ring carcinoma, Medullary carcinoma, Serrated adenocarcinoma, Micropapillary adenocarcinoma, Adenoma-like adenocarcinoma, Adenosquamous carcinoma, Carcinoma with sarcomatoid components, Undifferentiated carcinoma, MANEC/NEC; WHO grade: low grade, formerly G1/G2 vs. high grade, formerly G3 and tumour budding: Bd1 (0-4 Buds in 20×), Bd2 (5-9 Buds in 20×), Bd3 ($\geq$10 Buds in 20×)). The parameters were available from a previous study on the same cohort regarding the distribution and the prevalence of the essential morphologic criteria given in the 2019 WHO classification of colorectal carcinoma, from which also the microsatellite status was extracted [17] (cohort details; Table 1). Furthermore, SATB2 expression was correlated with CDX2 expression, which was analysed in a previous study [18], where a similar methodology regarding the finding of an optimised cutoff for CDX2 expression groups was used [21]. The cases from 2019 that were recently added to the collective (n = 36) were classified regarding the aforementioned parameters (histomorphology, CDX2 expression) as described previously [17,18].

Table 1. Distribution and prognostic impact of SATB2-expression and clinicopathological parameters in the overall cohort.

	Variables	Overall n (%)	Median Overall Survival (SE) (Months)	p-Value	Mean Disease Specific Survival (SE) (Months)	p-Value	Mean Disease Free Survival (SE) (Months)	p-Value
Age				<0.001		0.02		0.98
	below median	504 (48.5%)	86.4 (2.2)		91.2 (2.1)		82.4 (2.3)	
	above median	535 (51.5%)	72.1 (2.2)		84.2 (2.3)		82.7 (2.4)	
Sex				0.33		0.93		0.54
	male	599 (57.7%)	78.1 (2.1)		88 (2.0)		83.6 (2.2)	
	female	440 (42.3%)	80.7 (2.4)		87.4 (2.4)		81 (2.6)	
SATB2 Subgroups				<0.001		<0.001		<0.001
	SATB2-low/absent	231 (22.2%)	68.4 (3.6)		74.2 (3.6)		68 (3.9)	
	SATB2-high	808 (77.8%)	82.2 (1.7)		91.5 (1.7)		86.3 (1.8)	
pT				<0.001		<0.001		<0.001
	1	79 (7.6%)	97.7 (4.8)		115.5 (3.0)		109.8 (3.6)	
	2	187 (18%)	92.8 (3.2)		103.9 (2.8)		100 (3.1)	
	3	578 (55.6%)	79.7 (2.1)		88.3 (2.1)		82.2 (2.2)	
	4	195 (18.8%)	57.2 (3.7)		60.4 (3.8)		55 (4.0)	

Table 1. *Cont.*

Variables		Overall *n* (%)	Median Overall Survival (SE) (Months)	*p*-Value	Mean Disease Specific Survival (SE) (Months)	*p*-Value	Mean Disease Free Survival (SE) (Months)	*p*-Value
pN				<0.001		<0.001		<0.001
	0	580 (55.8%)	89 (1.9)		101.2 (1.7)		99.1 (1.8)	
	1	292 (28.1%)	75.7 (3.0)		81 (3.0)		73.2 (3.2)	
	2	167 (16.1%)	51.4 (4.0)		54.6 (4.2)		42.6 (4.0)	
pM				<0.001		<0.001		<0.001
	0	887 (85.4%)	86.5 (1.6)		96.3 (1.5)		91.1 (1.7)	
	1	152 (14.6%)	40.1 (3.5)		42.7 (3.7)		34.4 (3.6)	
UICC Stage				<0.001		<0.001		<0.001
	1	213 (20.5%)	96.6 (2.9)		111.1 (2.1)		107.8 (2.4)	
	2	350 (33.7%)	86 (2.6)		97 (2.4)		95.7 (2.6)	
	3	318 (30.6%)	81.2 (2.8)		87.2 (2.8)		76.6 (3.1)	
	4	158 (15.2%)	39.3 (3.4)		41.8 (3.6)		33.3 (3.5)	
Tumour type (WHO)				<0.001		<0.001		<0.001
	Adenocarcinoma NOS	650 (62.6%)	83.7 (1.9)		92.5 (1.9)		87.4 (2.0)	
	Mucinous adenocarcinoma	88 (8.5%)	76.5 (5.5)		87 (5.6)		78.1 (6.0)	
	Signet-ring cell carcinoma	9 (0.8%)	54 (22.5)		54 (22.5)		34.4 (18.9)	
	Medullary adenocarcinoma	32 (3.1%)	98.6 (7.2)		116.3 (3.6)		112.8 (4.9)	
	Micropapillary adenocarcinoma	129 (12.4%)	53.6 (4.4)		56.2 (4.6)		47.3 (4.5)	
	Serrated adenocarcinoma	91 (8.7%)	78.4 (5.6)		87.7 (5.4)		84.4 (5.6)	
	Adenoma-like adenocarcinoma	33 (3.2%)	98 (6.4)		115.2 (3.5)		116.4 (3.5)	
	MANEC/NEC	7 (0.7%)	18 (8.2)		18.0 (8.1)		15.8 (8.4)	
WHO grade				<0.001		<0.001		<0.001
	low-grade	708 (68.1%)	86 (1.8)		95.2 (1.7)		89.6 (1.9)	
	high-grade	331 (31.9%)	65.4 (2.8)		72.6 (2.9)		67.8 (3.1)	
Tumour budding				<0.001		<0.001		<0.001
	Bd1	560 (53.9%)	97.8 (1.7)		109.3 (1.3)		107 (1.5)	
	Bd2	270 (26%)	70.5 (3.1)		77.6 (3.1)		66.8 (3.3)	
	Bd3	209 (20.1%)	41 (3.1)		44 (3.4)		36.6 (3.4)	
Resection margin				<0.001		<0.001		<0.001
	R0	960 (92.4%)	83.0 (1.6)		92.4 (1.5)		87.4 (1.7)	
	R1	49 (4.7%)	40.8 (7.2)		42.3 (7.4)		29.2 (6.1)	
	R2	30 (2.9%)	25.0 (4.5)		25 (4.5)		21.5 (3.7)	
Lymphatic invasion				<0.001		<0.001		<0.001
	not present	508 (48.9%)	89.7 (2.0)		101.8 (1.8)		100.6 (1.9)	
	present	531 (51.1%)	69 (1.6)		74.2 (2.3)		65.2 (2.5)	
Venous invasion				<0.001		<0.001		<0.001
	not present	904 (87%)	83.7 (1.6)		93.4 (1.6)		89.2 (1.7)	
	present	135 (13%)	48.7 (4.3)		50.9 (4.4)		38.6 (4.1)	
Microsatellite status				0.01		0.001		<0.001
	Microsatellite stable	877 (84.4%)	77.6 (1.7)		85.5 (1.7)		80 (1.8)	
	Microsatellite instable	162 (15.6%)	88.7 (3.8)		101.4 (3.3)		97.8 (3.7)	
CDX2 subgroups				0.012		0.006		0.012
	CDX2-low/absent	106 (10.2%)	67.9 (5.6)		75.8 (5.7)		70.4 (5.9)	
	CDX2-high	933 (89.8%)	80.4 (1.6)		89.1 (1.6)		83.7 (1.7)	
Tumour localization				0.26		0.83		0.93
	Right (Coec/Asc/Trans)	503 (48.4%)	77.2 (2.3)		87.1 (2.3)		82.7 (2.4)	
	Left (Desc/Sigm/Rect)	536 (51.6%)	81.1 (2.1)		88.4 (2.1)		82.3 (2.3)	

2.3. Statistics

Using SPSS version 26 (SPSS Institute, Chicago, IL, USA) statistical analyses were performed applying X^2 test as well as X^2 test for trends and Fisher's exact test. The Cutoff Finder, a publicly available biostatistical tool that represents a bundle of optimisation and visualisation methods for cutoff determination, was used to define the optimal cutoffs for SATB2 expression groups [21]. Where applicable, the Bonferroni method was used to correct for multiple testing. Univariate survival analyses were performed using the Kaplan–Meier method and significance of survival differences was tested by a log-rank test. The Cox proportional hazard model was used for multivariate analyses. All statistical tests were performed two-sided, *p*-values ≤ 0.05 were considered significant.

3. Results

3.1. Clinicopathological Features and Survival

The median patient age was 69 years. The majority of patients were male (*n* = 599; 58%). Left- (descending colon until rectum; *n* = 536; 52%) and right-sided (coecum until splenic flexure; *n* = 503; 48%) neoplasms showed an almost even distribution. Postoperative UICC staging (eighth edition of the TNM classification of malignant tumours) [22] resulted in 213 stage I (21%), 350 (34%) stage II, 318 (31%) stage III and 158 (15%) stage IV cancers. Three hundred and thirty patients (32%) relapsed, 411 patients (40%) died during follow up, for 301 (29%) patients a tumour-specific death was noted (cohort details: Table 1).

3.1.1. Distribution of SATB2 Expression and Biostatistical Generation of SATB2 Expression Groups

Most CRCs showed a diffuse SATB2 expression (61%, *n* = 639), a heterogeneous staining was noted for 340 cancers (33%), 60 tumours (6%) showed a complete absence of SATB2. A nuclear staining in ≥90% of tumour cells was observed in 65% (679/1039) of cases. In order to transform this continuous variable into dichotomous SATB2 groups (binary variable), we used the Cutoff Finder [21], a publicly available biostatistical tool for cutoff determination, to identify the best cutoff for SATB2 stratification. Following these initial statistical analyses, two SATB2-groups were formed: CRCs that showed an SATB2 expression above the 20th percentile (>70% tumour cells; *n* = 808, 78%) were categorised as SATB2-high, CRC on/below the 20th percentile (range:0–70% of tumour cells; *n* = 231, 22%) were categorised as SATB2-low/absent. Examples of the two SATB2 expression groups among certain CRC subtypes are given in Figure 1. SATB2-low/absent CRCs usually showed a reduced SATB2 staining intensity and a significantly higher rate of a heterogeneous/absent staining pattern (*p* < 0.001, details see Table S1). Only the number of positive tumour cells (regardless of staining pattern or intensity) were used to form the SATB2 expression groups. A comparison of the results of the SATB2-grouping with full block slides showed an excellent concordance with the results from the TMA (95%, *p* < 0.001, Kappa Cohens value: 0.88). Furthermore, an excellent interobserver variance was evident (*p* < 0.001, Kappa Cohens value: 0.95).

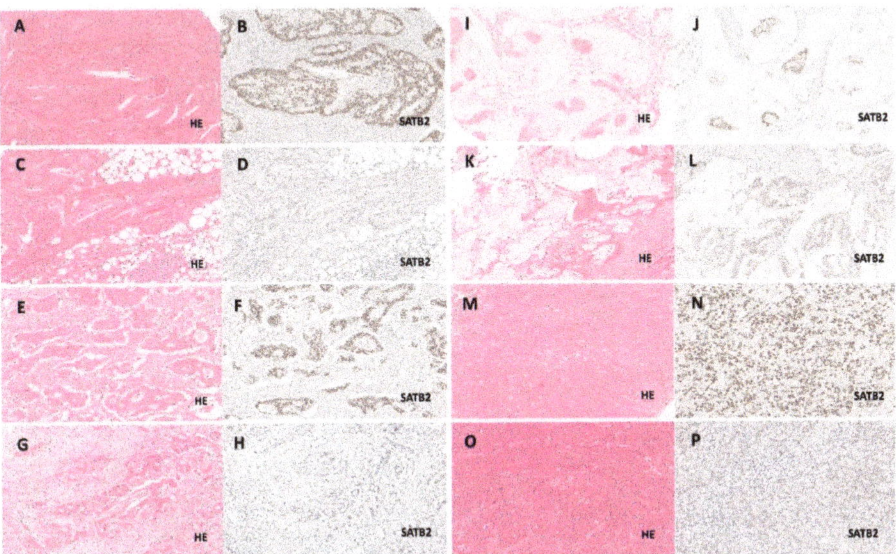

Figure 1. SATB2-low/absent and SATB2-high expression groups in selected colorectal carcinoma subtypes. (**A–D**): Adenocarcinoma NOS from the SATB2-high ((**A**): HE, 20× + (**B**): SATB2, 20×) and from association the SATB2-low/absent

((**C**): HE, 20× + (**D**): SATB2, 20×) expression subgroup. (**E–H**): Micropapillary adenocarcinoma from the SATB2-high ((**E**): HE, 20× + (**F**): SATB2, 20×) and from association the SATB2-low/absent ((**G**: HE, 20× + (**H**: SATB2, 20×) expression subgroup, also shown as an example of carcinomas with a high tumour budding activity from both expression groups. (**I–L**): Mucinous adenocarcinoma from the SATB2-high ((**I**): HE, 20× + (**J**): SATB2, 20×) and from association the SATB2-low/absent ((**K**): HE, 20× + (**L**): SATB2, 20×) expression subgroup. (**M–P**): Medullary carcinoma from the SATB2-high ((**M**): HE, 20× + (**N**): SATB2, 20×) and from association the SATB2-low/absent ((**O**): HE, 20× + (**P**): SATB2, 20×) expression subgroup.

3.1.2. Association of SATB2-Groups with pTNM/UICC Staging, Morphologic Parameters (CRC Subtypes/Tumour Budding/WHO Grade) and Microsatellite Status

As illustrated in Figure 2 and depicted in detail in Table S2, SATB2-low/absent CRCs were significantly enriched in higher pT/pN/pM and combined UICC-stages, right-sided tumours, carcinomas with lymphatic and blood vessel invasion as well as in tumours with positive margins ($p < 0.001$, respectively). Compared to SATB2-high neoplasms, SATB2-low/absent CRCs were significantly increased in CRCs with high (Bd3) tumour budding activity and in poorly differentiated carcinomas according to the WHO grade ($p < 0.001$, respectively). Furthermore, a low/absent SATB2 expression was significantly enriched in the mucinous, micropapillary, medullary and signet-ring CRC subtypes as well as in MANEC/NEC ($p < 0.001$). MSI-H CRCs were also associated with an absent or low SATB2 expression ($p = 0.01$).

3.1.3. Association of SATB2 Expression Groups with CDX2 Expression Groups

Far more CRCs were allocated to the SATB2 low/absent expression group than to the CDX2 low/absent expression group (CDX2: 10.2% vs. SATB2: 22.2%). Although both expression groups were associated with one another ($p < 0.001$), there were many tumours with a discordant SATB2/CDX2 expression status (Kappa Cohens value: 0.30). For example, only 28% (64/231) of CRCs from the SATB2-low/absent subgroup showed a concordant low/absent CDX2 expression, while 40% (42/106) of CDX2-low/absent CRCs showed a high SATB2 expression level. Sixty CRCs (6%) showed a complete absence of SATB2 expression compared to only 13 CRCs (1.3%) that showed a complete negativity for CDX2. Only two CRCs remained negative for both markers, while the rest of the CRCs without any SATB2 expression showed a heterogeneous ($n = 19$, 32%) or diffuse expression of CDX2 ($n= 39$, 65%). Of the 13 completely CDX2 negative cases, the vast majority ($n = 11$, 85%) showed a heterogenous or diffuse SATB2 expression (details: Table 2).

Table 2. Correlation of SATB2 and CDX2 expression groups and CDX2/SATB2 staining patterns.

	Variables			Total	*p*-Value
A	SATB2 expression group				
		low/absent	high		
CDX2 expression group	low/absent	64	42	106	
	high	167	766	933	$p < 0.001$
total		231	808	1039	
B	CDX2 staining pattern				
		Absent	heterogenous	diffuse	
	absent	2	19	39	60	
SATB2 staining pattern	heterogenous	7	70	263	340	$p < 0.001$
	diffuse	4	32	603	639	
total		13	121	905	1039	

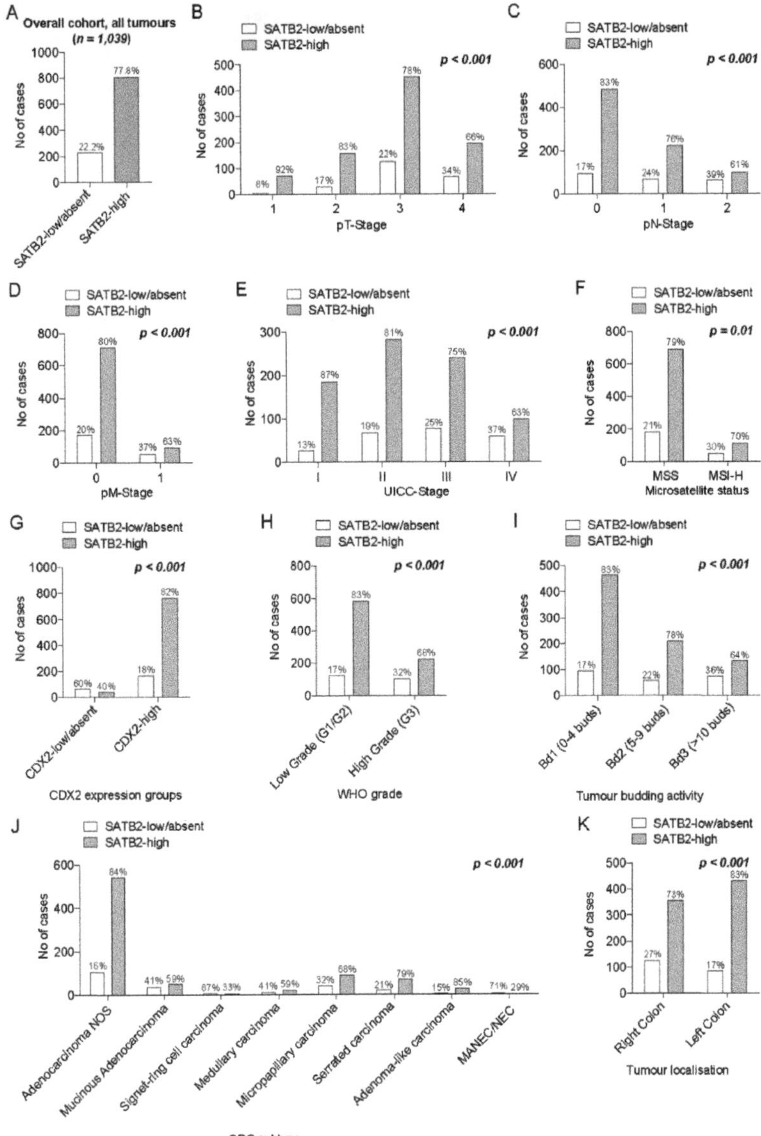

Figure 2. (**A–D**): Prevalence of SATB2 expression groups within the overall cohort (**A**) and in pT (**B**), pN (**C**), pM (**D**) and combined UICC stage subgroups (**E**). (**F,G**): Distribution of SATB2 expression groups among MSS vs. MSI-H (**F**) and CDX2-low/absent vs. CDX2-high CRCs (**G**). (**H–K**): Differential distribution of SATB2 within WHO low-grade vs. WHO high-grade CRCs (**H**), among the different tumour budding subgroups (Bd1, Bd2, Bd3; (**I**)), among colorectal cancer subtypes (**J**) and in right- vs. left-sided CRCs (**K**).

3.2. Prognostic Relevance of SATB2-Groups in the Overall Cohort

As illustrated in Figure 3 and Table 1, compared to SATB2-high CRCs, the SATB2-low/absent group showed a significantly decreased OS (SATB2-high 82.2 months vs. SATB2-low/absent 68.4 months, $p < 0.001$), DSS (SATB2-high 91.5 months vs. SATB2-

low/absent 74.2 months, *p* < 0.001) and DFS (SATB2-high 86.3 months vs. SATB2-low/absent 68 months, *p* < 0.001) in univariate analyses (log-rank test) of the overall cohort of 1039 CRCs.

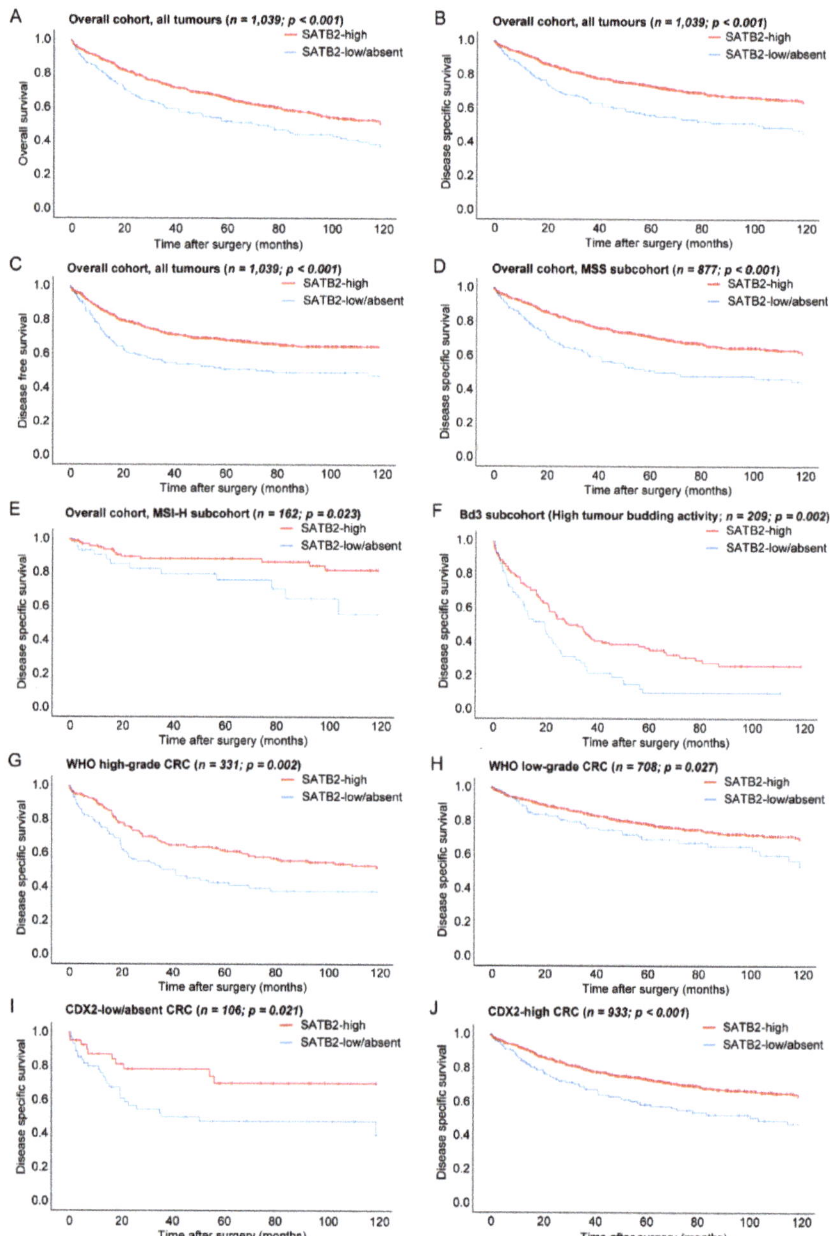

Figure 3. Prognostic relevance of SATB2 expression in univariate analyses on overall-, disease specific- and disease free-survival in the overall cohort (**A–C**) and for disease-specific survival in specific CRC subgroups: MSS (**D**) and MSI-H (**E**) subcohorts, high tumour budding activity subcohort (**F**), WHO high-grade (**G**)/WHO low-grade (**H**) subcohorts, CDX2-low/absent (**I**)/CDX2-high (**J**) subcohorts.

3.2.1. Prognostic Relevance of SATB2 in Microsatellite and CDX2 Expression Subgroups

A strong prognostic impact of the SATB2-low/absent group on OS/DSS/DFS was present in MSS CRCs (65.4 months/70 months/62.8 months, $p < 0.001$, respectively). SATB2 was also prognostic for DSS (SATB2-high 105.8 months vs. SATB2-low/absent 90.6 months, $p = 0.023$), but not for OS (SATB2-high 92.3 months vs. SATB2-low/absent 78.8 months, $p = 0.072$) or DFS (SATB2-high 101 months vs. SATB2-low/absent 89 months, $p = 0.163$) in MSI-H CRC.

When analysed in CDX2 expression subgroups, SATB2 expression groups showed a strong prognostic demarcation in both CDX2-low/absent (e.g., DSS: SATB2-high 91.2 months vs. SATB2-low/absent 64.5 months, $p = 0.021$) as well as in CDX2 high CRCs (e.g., DSS: SATB2-high 91.6 months vs. SATB2-low/absent 77.3 months, $p < 0.001$) (Figure 3, Table S3), while CDX2 expression showed no prognostic relevance in any of the SATB2 expression groups (e.g., $p > 0.05$, data not shown).

3.2.2. Prognostic Relevance of SATB2 in WHO Grade and Tumour Budding Subgroups

SATB2 expression showed a strong prognostic impact in WHO low-grade (e.g., DSS: SATB2-high 96.7 months vs. SATB2-low/absent 88.1 months, $p = 0.027$) and high-grade CRCs (e.g., DSS: SATB2-high 78.8 months vs. SATB2-low/absent 59.5 months, $p = 0.002$).

When analysed within the different tumour budding subgroups (Bd1/2/3), SATB2 showed a weak prognostic relevance in tumours with a low tumour budding activity (e.g., DSS: SATB2-high 110.2 months vs. SATB2-low/absent 104.4 months $p = 0.05$) and especially a strong prognostic impact within CRCs with a high tumour budding activity, where CRCs with a low/absent SATB2 expression showed a significantly worse survival rate compared to SATB2-high carcinomas (e.g., DSS: SATB2-high 51.1 months vs. SATB2-low/absent 29 months, $p = 0.002$) (Figure 3, Table S3).

3.2.3. Prognostic Relevance of SATB2 in UICC Stage Subgroups and Right vs. Left-Sided CRCs

In UICC stage III tumours, SATB2-low/absent showed significantly shortened survival characteristics (OS: 70.3 months vs. 84.4 months, $p = 0.025$; DSS: 74.5 months vs. 91 months, $p = 0.012$; DFS: 62.1 months vs. 81.2 months; $p = 0.004$) (Table S4) in all survival comparisons. In UICC stage I/II and IV, no significant survival impact was visible.

SATB2 expression showed a strong prognostic impact in right-sided (e.g., DSS: SATB2-high 90.8 months vs. SATB2-low/absent 77.4 months, $p = 0.002$) and left-sided CRCs (e.g., DSS: SATB2-high 92.1 months vs. SATB2-low/absent 68.9 months, $p < 0.001$) (Figure S1).

3.3. Multivariate Analyses

In multivariate analyses (including age, gender, resection status, UICC stage, MSI-status, WHO grade, tumour budding, CRC subtypes and SATB2-groups) SATB2-expression was not an independent prognostic factor (e.g., DSS: $p = 0.1$, hazard ratio: 1.25, Table S5) in the overall cohort comprising all CRCs. When the other main histological confounders (WHO grade, tumour budding, CRC subtypes) were excluded from the Cox-regression analysis, SATB2-expression remained a prognostic factor independent of UICC stage, age, gender, resection status and MSI-status (DSS: 0.029, HR:1.32, Table S6).

In a full multivariate analyses (including all parameters mentioned above) of high risk CRC subcohorts (UICC stage III/high tumour budding activity), SATB2 fully retained its prognostic relevance demonstrated in univariate analyses (UICC stage III CRC subcohort: DSS: $p = 0.007$, hazard ratio: 1.95, Table 3; Bd3-CRCs with a high tumour budding activity DSS: $p = 0.01$, hazard ratio: 1.67, Table 4; DFS: $p = 0.02$, Hazard Ratio: 1.79; OS: $p = 0.01$, hazard ratio: 1.58, data not shown).

Table 3. Multivariate disease-specific survival analysis in the UICC stage III subcohort under inclusion of SATB2 expression, age, gender, CRC subtype, tumour budding, WHO grade, resection status and microsatellite status.

Variables		HR (DSS)	Lower CI (95%)	Upper CI (95%)	*p*-Value
SATB2 subgroups					0.007
	SATB2 high	1.00			
	SATB2 Low/absent	1.95	1.20	3.16	
WHO Subtype					0.026
	Adenocarcinoma NOS	1.00			
	Mucinous adenocarcinoma	0.41	0.12	1.34	
	Signet-ring cell carcinoma	1.89	0.53	6.70	
	Medullary carcinoma	0.17	0.02	1.40	
	Micropapillary adenocarcinoma	0.75	0.43	1.29	
	Serrated adenocarcinoma	0.82	0.40	1.68	
	Adenoma-like adenocarcinoma	0.00	0.00	>30	
	MANEC/NEC	5.67	1.23	26.04	
Tumour budding					<0.001
	Bd1	1.00			
	Bd2	3.35	1.76	6.36	
	Bd3	5.78	2.95	11.34	
WHO-grade					0.027
	Low grade	1.00			
	High grade	1.63	1.06	2.51	
Gender					0.723
	female	1.00			
	male	1.08	0.69	1.69	
Resection status					0.001
	R0	1.00			
	R1/2	2.72	1.47	5.04	
Tumour localization					0.812
	Right colon	1.00			
	Left colon	0.94	0.59	1.52	
Age group					0.112
	Below median	1.00			
	Median and above	1.43	0.92	2.34	
Microsatellite status					0.817
	Microsatellite instable	1.00			
	Microsatellite stable	1.09	0.51	2.32	

Table 4. Multivariate disease-specific survival analysis in the high tumour budding (Bd3) subcohort under inclusion of SATB2 expression, age, gender, CRC subtype, UICC stage, WHO grade, resection status and microsatellite status.

Variables		HR (DSS)	Lower CI (95%)	Upper CI (95%)	*p*-Value
SATB2 subgroups					0.01
	SATB2 high	1.00			
	SATB2 Low/absent	1.67	1.13	2.46	
WHO Subtype					0.1
	Adenocarcinoma NOS	1.00			
	Mucinous adenocarcinoma	1.49	0.70	3.16	
	Signet-ring cell carcinoma	1.30	0.44	3.80	
	Micropapillary adenocarcinoma	0.80	0.54	1.19	
	Serrated adenocarcinoma	0.90	0.43	1.90	
	MANEC/NEC	2.31	0.85	6.25	

Table 4. *Cont.*

Variables		HR (DSS)	Lower CI (95%)	Upper CI (95%)	*p*-Value
WHO-grade					0.067
	Low grade	1.00			
	High grade	1.43	0.97	2.09	
UICC Stage					0.006
	I	1.00			
	II	0.67	0.24	1.85	
	III	0.64	0.24	1.70	
	IV	1.41	0.53	3.75	
Gender					0.716
	female	1.00			
	male	1.07	0.74	1.55	
Resection status					0.013
	R0	1.00			
	R1/2	1.44	1.08	1.91	
Tumour localization					0.108
	Right colon	1.00			
	Left colon	1.34	0.94	1.93	
Age group					0.767
	Below median	1.00			
	Median and above	1.06	0.72	1.57	
Microsatellite status					0.159
	Microsatellite instable	1.00			
	Microsatellite stable	1.76	0.80	3.88	

4. Discussion

In this study, we investigated the expression of Special AT-rich sequence-binding protein 2 (SATB2) in one thousand and thirty-nine resected CRCs, correlated the results with histomorphologic parameters (CRC subtypes, tumour budding activity, WHO grade) [17], expression of CDX2 [18] as well as clinicopathological parameters (pTNM/UICC staging, microsatellite status, localisation) and analysed the prognostic relevance of SATB2 in the overall cohort and in specific subcohorts.

In recent years, SATB2 has gained increasing attention as a relatively specific marker of colorectal differentiation [23–26] and functional studies have revealed the tumour-suppressive properties of SATB2 in experimental settings [12,27–30], demonstrating that SATB2 is a complexly regulated tumour suppressor that represses CRC progression by inhibiting the transcription of SNAIL, a master regulator of epithelial-mesenchymal transition. In our cohort of more than one thousand tumours, SATB2 low/absent CRCs were significantly associated with higher UICC stages and massively enriched in tumours with high-risk histomorphological features such as high tumour budding (Bd3) or poorly differentiated carcinomas according to the WHO grade, which is in line with the functional studies postulating the tumour suppressive properties of SATB2 [12,27,28]. Consistent with previous findings, SATB2-low/absent CRCs were also associated with strongly reduced survival parameters in univariate analyses (log-rank test) of the overall cohort. Interestingly, although SATB2 expression retained its statistical significance in multivariate analyses when UICC stage, resection status, MSI-status, age and gender were incorporated, this independent prognostic power vanished when we added the most common histomorphological parameters of CRC, tumour budding, WHO grade and the different CRC subtypes to our multivariate analyses. These findings argue towards the fact that the massive enrichment of SATB2 low expressing tumours in high-grade categories of these parameters probably washes out the strong prognostic effect of SATB2 that is present in univariate analyses.

Nevertheless, we wanted to know whether SATB2 can identify any prognostic subgroup that is not identified by either UICC staging or classical histomorphological parameters and which is also retained in multivariate analyses incorporating all factors.

Interestingly, SATB2 low/absent CRCs were associated with an especially aggressive disease course in CRCs with high tumour budding activity and in UICC stage III cancers and showed highly reduced survival times in these high-risk subgroups. In subsequent multivariate analyses incorporating all prognostic factors, SATB2 retained its prognostic relevance in both UICC stage III carcinomas and dissociative cancers with a high tumour budding activity. These findings delineate SATB2 as a worthwhile immunohistochemical biomarker in CRC that can identify especially aggressive cancers in these high-risk subgroups of CRC and delivers valuable additional prognostic information in addition to standard histomorphological factors and UICC staging. Regarding the translation of these results into clinicopathological routine, we suggest that the responsible pathologist reports SATB2 loss by stating the percentage of SATB2 expressing cancer cells, ideally combined with the notion that a significantly reduced expression of SATB2 has been associated with a poorer clinical outcome.

As SATB2 has emerged as a considerable alternative to CDX2 to verify or rule out colorectal differentiation, another aim of our study was to compare the incidence and overlap of SATB2 loss with loss of CDX2 and also to compare the prognostic relevance of these two markers with one another [18]. The findings of these aspects of our study are also particularly interesting, because they highlight possible strengths and weaknesses of the most commonly used colorectal markers. Notably, the frequency of a reduced or completely lost SATB2 expression is much higher compared to CDX2 and especially a completely absent SATB2 expression was far more frequent than a complete negativity for CDX2. This implies, that when both CDX2 and SATB2 (as singular markers) are assessed regarding their ability to detect a colorectal origin in neoplastic tissues, SATB2 has to be ranked as the less sensitive marker compared to CDX2. However, as only two out of 1039 CRCs (0.2%) showed a complete loss of both SATB2 and CDX2, and the majority of SATB2 negative CRCs showed a strong and diffuse expression of CDX2 (and vice versa), a combined panel of both markers appears to be able to identify the overwhelming majority of colorectal cancers and is probably the most expedient approach for routine diagnostic settings.

When we then moved on to compare the prognostic impact of SATB2 and CDX2, we observed an opposing picture. Compared to CDX2, the loss of SATB2 showed a considerably higher prognostic impact in univariate analyses (log-rank test) of the overall cohort and in nearly all clinicopathological subscenarios of CRC, in which CDX2 mostly showed at best minimal prognostic impact in our cohort. A crucial difference between the prognostic difference of SATB2 and CDX2 was also that SATB2 retained its prognostic power in right- and left-sided CRCs, while CDX2 did not show any prognostic significance in right-sided tumours although it is more frequently lost in the right colon [18]. In line with these findings, we additionally observed that SATB2-low/absent CRCs were able to identify patients with a poor prognosis in both CDX2 expression groups (CDX2-low/absent vs. CDX2- high), while CDX2 showed no prognostic relevance in SATB2 expression subgroups, rendering SATB2 as the prognostically superior immunohistochemical biomarker in CRC compared to CDX2.

5. Conclusions

Our study has five major findings: (1) a low/absent SATB2 expression is significantly enriched in advanced stage CRCs that have an aggressive histomorphological phenotype with high tumour budding activity and/or a poor differentiation according to the WHO grade. (2) Loss of SATB2 is of high prognostic relevance in uni- and multivariate analyses (including UICC stage) in the overall cohort, but shows no independent prognostic value in the overall cohort when the main histomorphological parameters of CRC (tumour budding, WHO grade, CRC subtypes) are added to the multivariate analyses. (3) SATB2 shows an especially high prognostic relevance in uni- and multivariate analyses of high-risk clinicopathological subgroups (high tumour budding/UICC stage III) and identifies CRCs with a particularly aggressive disease course in these high-risk scenarios. (4) SATB2 loss occurs much more frequently than loss of CDX2, with a substantial portion of SATB2-negative CRCs showing a diffuse or at least heterogeneous CDX2 positivity, generally delineating CDX2 as the more sensitive marker of colorectal differentiation in carcinomas. (5) SATB2, in general, showed a vastly better prediction of survival outcome compared to CDX2, with SATB2 retaining its prognostic impact in CDX2 expression subgroups (CDX2 low/absent vs. high), rendering SATB2 as the superior prognostic biomarker compared to CDX2.

In conclusion, our study identifies SATB2 as a potentially valuable additional prognostic biomarker in CRC. Further studies are warranted to explore the possible therapeutic implications of a diminished or completely lost SATB2 expression. Both SATB2 and CDX2 can individually be completely lost in CRCs, while a total absence of both markers is almost never observed. Therefore, a combined panel of both markers appears to be the most solid approach to pinpoint or rule out colorectal differentiation.

Supplementary Materials: The following are available online at https://www.mdpi.com/article/10.3390/cancers13246177/s1. Figure S1: Prognostic relevance of SATB2 expression for disease-specific survival in right-sided (A, coecum until splenic flexure) and left-sided (descending colon until rectum) CRCs, Table S1: Correlation between SATB2 expression groups with SATB2 staining pattern (A) and SATB2 staining intensity (B), Table S2: Distribution of SATB2 expression groups with clinicopathological and morphological parameters in the overall cohort, Table S3: Impact of SATB2 expression on overall, disease-specific and disease-free survival in tumour budding—(Bd1, Bd2, Bd3), WHO grade (low, high), microsatellite (MSS, MSI-H) and CDX2 expression subcohorts (low/absent, high), Table S4: Impact of SATB2 expression on overall, disease-specific and disease-free survival in the different UICC stage groups, Table S5: Multivariate overall survival analysis in the overall cohort under inclusion of SATB2 expression, age, gender, CRC subtype, tumour budding, WHO grade and microsatellite status, Table S6: Multivariate overall survival analysis in the overall cohort under exclusion of CRC subtype, tumour budding and WHO grade.

Author Contributions: M.J. designed this study, M.J., M.S. (Miguel Silva) and M.S. (Maxime Schmitt) wrote the manuscript with assistance from C.D., B.K., S.F., P.J., D.W., W.W., M.J., B.K. and M.S. (Maxime Schmitt) performed histopathological analyses. M.J., C.D., S.F., W.W., P.J., M.S. (Miguel Silva) and M.T. performed statistical analyses. M.J., K.S., C.L., N.P., D.W., K.H., J.E. collected clinicopathological data. All authors have read and agreed to the published version of the manuscript.

Funding: This research was funded by grants of the Deutsche Forschungsgemeinschaft (DFG, German Research Foundation): Project-ID 360372040—SFB 1335 to WW and Project-ID 39535707—SFB 1371 to KS and MT. The authors have disclosed that they have no significant relationships with, or financial interest in, any commercial companies pertaining to this article.

Institutional Review Board Statement: The study was conducted according to the guidelines of the Declaration of Helsinki. The local ethic committee of the Technical University of Munich approved this study (reference number: 252/16 s).

Informed Consent Statement: Patients signed a general informed consent protocol during admission to the hospital.

Data Availability Statement: All relevant data are within the paper and its supporting information files. The data underlying the results presented in the study are available from the study group upon reasonable request, some restrictions apply due to confidentiality of patient data. Since these data are derived from a research trial with ongoing follow up there are legal and ethical restrictions to share sensitive patient related data publicly. Data can be requested in the context of a translational research project by sending a request to the corresponding author.

Acknowledgments: The authors thank the Comparative Experimental Pathology Unit of the Institute of Pathology and especially Marion Mielke and Olga Seelbach for excellent technical assistance. Furthermore, we thank Simone Köppel and Christina Schott from the Biobank of the Klinikum rechts der Isar (MTBio) for their excellent support.

Conflicts of Interest: The authors declare no conflict of interest.

References

1. Siegel, R.L.; Miller, K.D.; Goding Sauer, A.; Fedewa, S.A.; Butterly, L.F.; Anderson, J.C.; Cercek, A.; Smith, R.A.; Jemal, A. Colorectal cancer statistics, 2020. *CA Cancer J. Clin.* **2020**, *70*, 145–164. [CrossRef]
2. Sung, H.; Ferlay, J.; Siegel, R.L.; Laversanne, M.; Soerjomataram, I.; Jemal, A.; Bray, F. Global cancer statistics 2020: Globocan estimates of incidence and mortality worldwide for 36 cancers in 185 countries. *CA Cancer J. Clin.* **2021**, *71*, 209–249. [CrossRef]
3. Zarate, Y.A.; Kaylor, J.; Fish, J. Satb2-associated syndrome. In *Genereviews((r))*; Adam, M.P., Ardinger, H.H., Pagon, R.A., Wallace, S.E., Bean, L.J.H., Mirzaa, G., Amemiya, A., Eds.; National Center for Biotechnology Information Bookshelf Site: Seattle, WA, USA, 1993.
4. FitzPatrick, D.R.; Carr, I.M.; McLaren, L.; Leek, J.P.; Wightman, P.; Williamson, K.; Gautier, P.; McGill, N.; Hayward, C.; Firth, H.; et al. Identification of satb2 as the cleft palate gene on 2q32-q33. *Hum. Mol. Genet.* **2003**, *12*, 2491–2501. [CrossRef]
5. Magnusson, K.; de Wit, M.; Brennan, D.J.; Johnson, L.B.; McGee, S.F.; Lundberg, E.; Naicker, K.; Klinger, R.; Kampf, C.; Asplund, A.; et al. Satb2 in combination with cytokeratin 20 identifies over 95% of all colorectal carcinomas. *Am. J. Surg. Pathol.* **2011**, *35*, 937–948. [CrossRef]
6. Ordonez, N.G. Satb2 is a novel marker of osteoblastic differentiation and colorectal adenocarcinoma. *Adv. Anat. Pathol.* **2014**, *21*, 63–67. [CrossRef]
7. Moh, M.; Krings, G.; Ates, D.; Aysal, A.; Kim, G.E.; Rabban, J.T. Satb2 expression distinguishes ovarian metastases of colorectal and appendiceal origin from primary ovarian tumors of mucinous or endometrioid type. *Am. J. Surg. Pathol.* **2016**, *40*, 419–432. [CrossRef] [PubMed]
8. Conner, J.R.; Hornick, J.L. Metastatic carcinoma of unknown primary: Diagnostic approach using immunohistochemistry. *Adv. Anat. Pathol.* **2015**, *22*, 149–167. [CrossRef] [PubMed]
9. Zhang, Y.J.; Chen, J.W.; He, X.S.; Zhang, H.Z.; Ling, Y.H.; Wen, J.H.; Deng, W.H.; Li, P.; Yun, J.P.; Xie, D.; et al. Satb2 is a promising biomarker for identifying a colorectal origin for liver metastatic adenocarcinomas. *EBioMedicine* **2018**, *28*, 62–69. [CrossRef]
10. Kriegsmann, M.; Harms, A.; Longuespee, R.; Muley, T.; Winter, H.; Kriegsmann, K.; Kazdal, D.; Goeppert, B.; Pathil, A.; Warth, A. Role of conventional immunomarkers, hnf4-alpha and satb2, in the differential diagnosis of pulmonary and colorectal adenocarcinomas. *Histopathology* **2018**, *72*, 997–1006. [CrossRef] [PubMed]
11. Dragomir, A.; de Wit, M.; Johansson, C.; Uhlen, M.; Ponten, F. The role of satb2 as a diagnostic marker for tumors of colorectal origin: Results of a pathology-based clinical prospective study. *Am. J. Clin. Pathol.* **2014**, *141*, 630–638. [CrossRef]
12. Wang, S.; Zhou, J.; Wang, X.Y.; Hao, J.M.; Chen, J.Z.; Zhang, X.M.; Jin, H.; Liu, L.; Zhang, Y.F.; Liu, J.; et al. Down-regulated expression of satb2 is associated with metastasis and poor prognosis in colorectal cancer. *J. Pathol.* **2009**, *219*, 114–122. [CrossRef] [PubMed]
13. Eberhard, J.; Gaber, A.; Wangefjord, S.; Nodin, B.; Uhlen, M.; Ericson Lindquist, K.; Jirstrom, K. A cohort study of the prognostic and treatment predictive value of satb2 expression in colorectal cancer. *Br. J. Cancer* **2012**, *106*, 931–938. [CrossRef] [PubMed]
14. Mezheyeuski, A.; Ponten, F.; Edqvist, P.H.; Sundstrom, M.; Thunberg, U.; Qvortrup, C.; Pfeiffer, P.; Sorbye, H.; Glimelius, B.; Dragomir, A. Metastatic colorectal carcinomas with high satb2 expression are associated with better prognosis and response to chemotherapy: A population-based scandinavian study. *Acta Oncol.* **2020**, *59*, 284–290. [CrossRef]
15. Lin, F.; Shi, J.; Zhu, S.; Chen, Z.; Li, A.; Chen, T.; Wang, H.L.; Liu, H. Cadherin-17 and satb2 are sensitive and specific immunomarkers for medullary carcinoma of the large intestine. *Arch. Pathol. Lab. Med.* **2014**, *138*, 1015–1026. [CrossRef]
16. Ma, C.; Lowenthal, B.M.; Pai, R.K. Satb2 is superior to cdx2 in distinguishing signet ring cell carcinoma of the upper gastrointestinal tract and lower gastrointestinal tract. *Am. J. Surg. Pathol.* **2018**, *42*, 1715–1722. [CrossRef]
17. Jesinghaus, M.; Schmitt, M.; Lang, C.; Reiser, M.; Scheiter, A.; Konukiewitz, B.; Steiger, K.; Silva, M.; Tschurtschenthaler, M.; Lange, S.; et al. Morphology matters: A critical reappraisal of the clinical relevance of morphologic criteria from the 2019 who classification in a large colorectal cancer cohort comprising 1004 cases. *Am. J. Surg. Pathol.* **2021**, *45*, 969–978. [CrossRef]
18. Konukiewitz, B.; Schmitt, S.; Silva, M.; Pohl, J.; Lang, C.; Steiger, K.; Halfter, K.; Engel, J.; Schlitter, A.M.; Boxberg, M.; et al. Loss of cdx2 in colorectal cancer is associated with histopathologic subtypes and microsatellite instability but is prognostically inferior to hematoxylin-eosin based morphologic parameters from the who classification. *Br. J. Cancer* **2021**, in press. [CrossRef]
19. Konukiewitz, B.; Kasajima, A.; Schmitt, M.; Schwamborn, K.; Groll, T.; Schicktanz, F.; Delbridge, C.; Schutze, L.M.; Wilhelm, D.; Lang, C.; et al. Neuroendocrine differentiation in conventional colorectal adenocarcinomas: Incidental finding or prognostic biomarker? *Cancers* **2021**, *13*, 5111. [CrossRef] [PubMed]
20. Li, Z.; Yuan, J.; Wei, L.; Zhou, L.; Mei, K.; Yue, J.; Gao, H.; Zhang, M.; Jia, L.; Kang, Q.; et al. Satb2 is a sensitive marker for lower gastrointestinal well-differentiated neuroendocrine tumors. *Int. J. Clin. Exp. Pathol.* **2015**, *8*, 7072–7082.
21. Budczies, J.; Klauschen, F.; Sinn, B.V.; Gyorffy, B.; Schmitt, W.D.; Darb-Esfahani, S.; Denkert, C. Cutoff finder: A comprehensive and straightforward web application enabling rapid biomarker cutoff optimization. *PLoS ONE* **2012**, *7*, e51862. [CrossRef]
22. Brierley, J.D.; Gospodarowicz, M.K.; Wittekind, C. *Tnm Classification of Malignant Tumours*; John Wiley & Sons: Hoboken, NJ, USA, 2017.
23. Berg, K.B.; Schaeffer, D.F. Satb2 as an immunohistochemical marker for colorectal adenocarcinoma: A concise review of benefits and pitfalls. *Arch. Pathol. Lab. Med.* **2017**, *141*, 1428–1433. [CrossRef]
24. Cigerova, V.; Adamkov, M.; Drahosova, S.; Grendar, M. Immunohistochemical expression and significance of satb2 protein in colorectal cancer. *Ann. Diagn. Pathol.* **2021**, *52*, 151731. [CrossRef] [PubMed]

25. Ma, C.; Olevian, D.; Miller, C.; Herbst, C.; Jayachandran, P.; Kozak, M.M.; Chang, D.T.; Pai, R.K. Satb2 and cdx2 are prognostic biomarkers in DNA mismatch repair protein deficient colon cancer. *Mod. Pathol.* **2019**, *32*, 1217–1231. [CrossRef]
26. Ma, C.; Olevian, D.C.; Lowenthal, B.M.; Jayachandran, P.; Kozak, M.M.; Chang, D.T.; Pai, R.K. Loss of satb2 expression in colorectal carcinoma is associated with DNA mismatch repair protein deficiency and braf mutation. *Am. J. Surg. Pathol.* **2018**, *42*, 1409–1417. [CrossRef]
27. Wang, Y.Q.; Jiang, D.M.; Hu, S.S.; Zhao, L.; Wang, L.; Yang, M.H.; Ai, M.L.; Jiang, H.J.; Han, Y.; Ding, Y.Q.; et al. Satb2-as1 suppresses colorectal carcinoma aggressiveness by inhibiting satb2-dependent snail transcription and epithelial-mesenchymal transition. *Cancer Res.* **2019**, *79*, 3542–3556. [CrossRef]
28. Xu, M.; Xu, X.; Pan, B.; Chen, X.; Lin, K.; Zeng, K.; Liu, X.; Xu, T.; Sun, L.; Qin, J.; et al. Lncrna satb2-as1 inhibits tumor metastasis and affects the tumor immune cell microenvironment in colorectal cancer by regulating satb2. *Mol. Cancer* **2019**, *18*, 135. [CrossRef] [PubMed]
29. Gu, J.; Wang, G.; Liu, H.; Xiong, C. Satb2 targeted by methylated mir-34c-5p suppresses proliferation and metastasis attenuating the epithelial-mesenchymal transition in colorectal cancer. *Cell Prolif.* **2018**, *51*, e12455. [CrossRef] [PubMed]
30. Yang, M.H.; Yu, J.; Jiang, D.M.; Li, W.L.; Wang, S.; Ding, Y.Q. Microrna-182 targets special at-rich sequence-binding protein 2 to promote colorectal cancer proliferation and metastasis. *J. Transl. Med.* **2014**, *12*, 109. [CrossRef]

Article

The Targeting of MRE11 or RAD51 Sensitizes Colorectal Cancer Stem Cells to CHK1 Inhibition

Luca Mattiello [1,2,†], Sara Soliman Abdel Rehim [1,3,†], Martina Musella [4,†], Antonella Sistigu [4,5], Andrea Guarracino [1,3], Sara Vitale [4], Francesca Corradi [3], Claudia Galassi [4], Francesca Sperati [6], Gwenola Manic [1,2,*,‡], Ruggero De Maria [4,7,*,‡] and Ilio Vitale [1,2,*,‡]

[1] Italian Institute for Genomic Medicine (IIGM), c/o IRCSS, 10060 Candiolo, Italy; mattiello.borsisti@iigm.it (L.M.); soliman.borsisti@iigm.it (S.S.A.R.); andrea.guarracino@uniroma2.it (A.G.)
[2] Candiolo Cancer Institute, FPO—IRCCS, 10060 Candiolo, Italy
[3] Department of Biology, University of Rome "Tor Vergata", 00133 Rome, Italy; francesca.corradi89@gmail.com
[4] Dipartimento di Medicina e Chirurgia Traslazionale, Università Cattolica del Sacro Cuore, 00168 Rome, Italy; martina.musella@unicatt.it (M.M.); antonella.sistigu@unicatt.it (A.S.); sara.vitale@unicatt.it (S.V.); claudia.galassi@unicatt.it (C.G.)
[5] UOSD Tumor Immunology and Immunotherapy Unit, IRCCS Regina Elena National Cancer Institute, 00144 Rome, Italy
[6] UOSD Biostatistics, Bioinformatics and Clinical Trial Center, San Gallicano Dermatological Institute IRCCS, 00144 Rome, Italy; francesca.sperati@ifo.gov.it
[7] Fondazione Policlinico Universitario "A. Gemelli"—IRCCS, 00168 Rome, Italy
[*] Correspondence: gwenola.manic@gmail.com or manic.esterni@iigm.it (G.M.); ruggero.demaria@unicatt.it (R.D.M.); iliovit@gmail.com or ilio.vitale@iigm.it (I.V.)
[†] Shared first authorship.
[‡] Shared senior co-authorship.

Citation: Mattiello, L.; Soliman Abdel Rehim, S.; Musella, M.; Sistigu, A.; Guarracino, A.; Vitale, S.; Corradi, F.; Galassi, C.; Sperati, F.; Manic, G.; et al. The Targeting of MRE11 or RAD51 Sensitizes Colorectal Cancer Stem Cells to CHK1 Inhibition. *Cancers* **2021**, *13*, 1957. https://doi.org/10.3390/cancers13081957

Academic Editor: David Wong

Received: 15 March 2021
Accepted: 11 April 2021
Published: 19 April 2021

Publisher's Note: MDPI stays neutral with regard to jurisdictional claims in published maps and institutional affiliations.

Simple Summary: The ATR-CHK1 axis of the DNA damage response is crucial for the survival of most colorectal cancer stem cells (CRC-SCs), but a significant fraction of primary CRC-SCs either is resistant to ATR or CHK1 inhibitors or survives the abrogation of the ATR-CHK1 cascade despite an initial response. Here, we demonstrate that the targeting of RAD51 or MRE11 improves the sensitivity of primary CRC-SCs to the CHK1/2 inhibitor prexasertib by sequentially inducing replication stress, the abrogation of cell cycle checkpoints, and the emergence of mitotic defects. This results in the induction of mitotic catastrophe and CRC-SC killing via a caspase-dependent apoptosis.

Abstract: Cancer stem cells (CSCs) drive not only tumor initiation and expansion, but also therapeutic resistance and tumor relapse. Therefore, CSC eradication is required for effective cancer therapy. In preclinical models, CSCs demonstrated high capability to tolerate even extensive genotoxic stress, including replication stress, because they are endowed with a very robust DNA damage response (DDR). This favors the survival of DNA-damaged CSCs instead of their inhibition via apoptosis or senescence. The DDR represents a unique CSC vulnerability, but the abrogation of the DDR through the inhibition of the ATR-CHK1 axis is effective only against some subtypes of CSCs, and resistance often emerges. Here, we analyzed the impact of druggable DDR players in the response of patient-derived colorectal CSCs (CRC-SCs) to CHK1/2 inhibitor prexasertib, identifying RAD51 and MRE11 as sensitizing targets enhancing prexasertib efficacy. We showed that combined inhibition of RAD51 and CHK1 (via B02+prexasertib) or MRE11 and CHK1 (via mirin+prexasertib) kills CSCs by affecting multiple genoprotective processes. In more detail, these two prexasertib-based regimens promote CSC eradication through a sequential mechanism involving the induction of elevated replication stress in a context in which cell cycle checkpoints usually activated during the replication stress response are abrogated. This leads to uncontrolled proliferation and premature entry into mitosis of replication-stressed cells, followed by the induction of mitotic catastrophe. CRC-SCs subjected to RAD51+CHK1 inhibitors or MRE11+CHK1 inhibitors are eventually eliminated, and CRC-SC tumorspheres inhibited or disaggregated, via a caspase-dependent apoptosis. These results support further clinical development of these prexasertib-based regimens in colorectal cancer patients.

Keywords: colorectal cancer; chromosomal instability; DNA damage; targeted therapy; tumor-initiating cells

1. Introduction

Robust experimental evidence indicates that human tumors often exhibit a hierarchical organization, whose apex is occupied by a subpopulation of cancer cells known as cancer stem cells (CSCs) because of the ability to self-renew and generate a progeny with different degree of differentiation (reviewed in [1]). CSCs have been identified and prospectively isolated from colorectal and other cancers, where they are believed to promote tumor development and to contribute to disease expansion, evolution and dissemination [2–6]; for these reasons, they are also known as tumor-initiating cells or tumor-propagating cells. In addition, CSCs have been shown to act as a main source of tumor heterogeneity [7], which is in turn linked to dismal prognosis and therapy resistance, as well as of tumor relapse [1,8]. At least in part, these features are linked to the relative low proneness of CSCs to undergo regulated cell death under stress conditions. In particular, in preclinical patient-derived models, CSCs have demonstrated elevated resistance to DNA damages, making them able to tolerate constitutive replication stress–defined as the slowing or stalling of replication fork progression and/or DNA synthesis [9]–or survive conventional genotoxic agents, including ionizing radiation and chemotherapeutics (reviewed in [10]).

The current view is that the resistance of CSCs derives from the orientation of the DNA damage response (DDR) towards cytoprotection. The DDR is a multipronged mechanism specifically activated in cells experiencing DNA lesions, operating through a two-step sensing-signal transduction cascade. This culminates either in the activation of cytoprotective signaling modules favoring the repair of or tolerance to DNA lesions, or in the activation of cytotoxic signaling modules leading to proliferation arrest or demise of irreversibly damaged cells upon induction of cell senescence or regulated cell death [10,11]. The activation of these pathways depends on the severity of the insults as well as on the efficiency of the machineries responsible for sensing, repairing, or tolerating DNA lesions [10]. Other crucial factors orienting the DDR towards cytosurvival or cytotoxicity encompass the proficiency, robustness, and activation kinetics of cytoprotective versus cytosurvival pathways. Collectively, these features dictate the outcome (i.e., survival or death) of DNA-damaged cells.

There is evidence that CSCs survive extensive DNA damage because of their capability not only to tolerate and repair DNA lesion, but also to detoxify reactive oxygen species and/or extrude DNA-damaging drugs [12–14]. Other factors tipping the balance toward the survival of CSC experiencing extensive DNA lesions include: (i) limited activation (or silencing) of cytotoxic mechanisms, which is due to the intrinsic deregulation of apoptotic pathways and evasion of regulated cell death or senescence [15,16], together with (ii) high proneness to activate cytosurvival cascades, originating from the constitutive activation of the DDR signaling [17–19]. In particular, a basal overactivaton of the ataxia telangiectasia mutated serine/threonine kinase (ATM)-checkpoint kinase 2 (CHEK2, best known as CHK2) axis, which is primarily activated by double-strand breaks (DSBs), or the ataxia telangiectasia mutated and Rad3 related serine/threonine kinase (ATR)-checkpoint kinase 1 (CHEK1, best known as CHK1) axis, which is primarily activated by long stretches of single-stranded DNA (ssDNA) adjacent to double-stranded (ds)/ssDNA junctions during replication stress, has been observed in multiple experimental models of CSC enrichment, including patient-derived models [20–22].

Importantly, the presence of a robust DDR constitutes a vulnerability of CSCs. On the one hand, the targeting of these DDR kinases in combination with DNA damaging agents de facto silences the cytosurvival modules of the DDR, thus reestablishing tumor sensitivity to DNA damage. On the other hand, CSCs can be particularly dependent on DDR kinase(s) for survival even in the absence of exogenous sources of DNA damage, making them targetable by the abrogation of the DDR signaling. Therefore, the DDR can be harnessed to revert the intrinsic capacity of CSCs to evade apoptosis. Accordingly, inhibitors of the DDR kinase ATM, ATR and CHK1 are reported to kill CSCs when administered either alone [20,23] or in combination with other DNA damaging agents to which CSCs were formerly resistant [17,19,24,25].

However, one emerging concept is that the response of CSCs to inhibitors of DDR kinases is heterogeneous, restricting the therapeutic use of these drugs to specific subtypes. Although most colorectal CSCs (CRC-SCs) are sensitive to inhibitors of the ATR-CHK1 cascade ultimately succumbing via replication catastrophe subsequent to the induction of intolerable or lethal levels of replication stress, a significant fraction of CSCs survives this regimen [20]. As a further limitation of these strategies, we recently revealed a mechanism of resistance to ATR-CHK1 inhibitors in CRC-SCs based on upregulation of poly(ADP-ribose) polymerase 1 (PARP1) [26], a DDR player with pleiotropic roles in DNA damage repair, the response to replication stress, and regulated cell death [27].

Based on these considerations, in this study, we investigated the impact of relevant druggable DDR players on the survival of patient-derived CRC-SCs, identifying MRE11 homolog, double-strand break repair nuclease (MRE11) and RAD51 recombinase (RAD51) as targets for sensitizing CRC-SCs to CHK1 inhibitors.

2. Results

2.1. Identification of RAD51 and MRE11 as Sensitizing Targets to Enhance the Efficacy of the CHK1 Inhibitor Prexasertib in CRC-SCs

We recently established distinct pairs of primary CRC-SCs sensitive (SENS) and resistant (RES) to pharmacological inhibitors of ATR and CHK1 kinases [26], the principal transducers of the cellular response to replication stress. From this panel, we selected two pairs of CRC-SCs: #1SENS/#1RES and #19SENS/#19RES, and used them to identify novel actionable targets improving the sensitivity and/or reverting the resistance to CHK1 inhibitors. We first analyzed the impact of crucial druggable DDR players on RES-CRC-SC survival. To this aim, RES-CRC-SCs were treated or not with the CHK1 inhibitor prexasertib together with inhibitors of ATR (VE-821), ATM (KU-60019), DNA-PK (NU-7026), MRE11 (mirin), and RAD51 (B02). After (co)treatments, cells were assessed for their proliferation/survival with a CellTiter-Glo® Luminescent Cell Viability assay. As internal comparison, RES-CRC-SCs were exposed to two representative pharmacological agents we previously demonstrated to potently sensitize to ATR/CHK1 inhibitors: triapine and adavosertib, which inhibit, respectively, the ribonucleotide reductase regulatory subunit M2 (RRM2) and WEE1 G2 checkpoint kinase (WEE1). This analysis led to the identification of the inhibitors of MRE11 (mirin) and of RAD51 (B02) as novel sensitizers of CRC-SCs to prexasertib (Figure 1A, Table S1). As opposed to mirin and B02, pharmacological inhibitors of the DDR kinases ATM, ATR and DNA-PK were ineffective or mildly effective in sensitizing CRC-SCs to prexasertib (Figure 1A, Table S1). This result rules out redundant or parallel roles of DDR kinases in the replication stress of CSCs, also excluding a potential contribution of ATR in resolving non-stringent RS independently of CHK1, as previously reported [28].

In following dose-response studies, we confirmed that the inhibition of RAD51 or MRE11 mildly affected the survival of RES-CRC-SCs when administered alone (Figure 1B,C). On the contrary, both mirin and B02 sensitized RES-CRC-SCs to CHK1 inhibitors (Figure 1B,C), thus validating our previous result. Along similar lines, B02 or mirin administered as monotherapies had minimal effect also on the survival of SENS-CRC-SCs (Figure 1D) and on that of CRC-SCs of our panel intrinsically resistant to CHK1 inhibitors (innRES-CRC-SCs) (Figure 1E). However, these agents increased the responsiveness of all these cells to CHK1 inhibition. Indeed, the inhibition of RAD51 or of MRE11 promoted sensitization of SENS-CRC-SCs (Figure 1D) and of innRES-CRC-SCs (Figure 1E) to sublethal doses of prexasertib. In conclusion, these findings demonstrate that the targeting of MRE11 or RAD51 sensitizes CRC-SCs to the inhibition of CHK1.

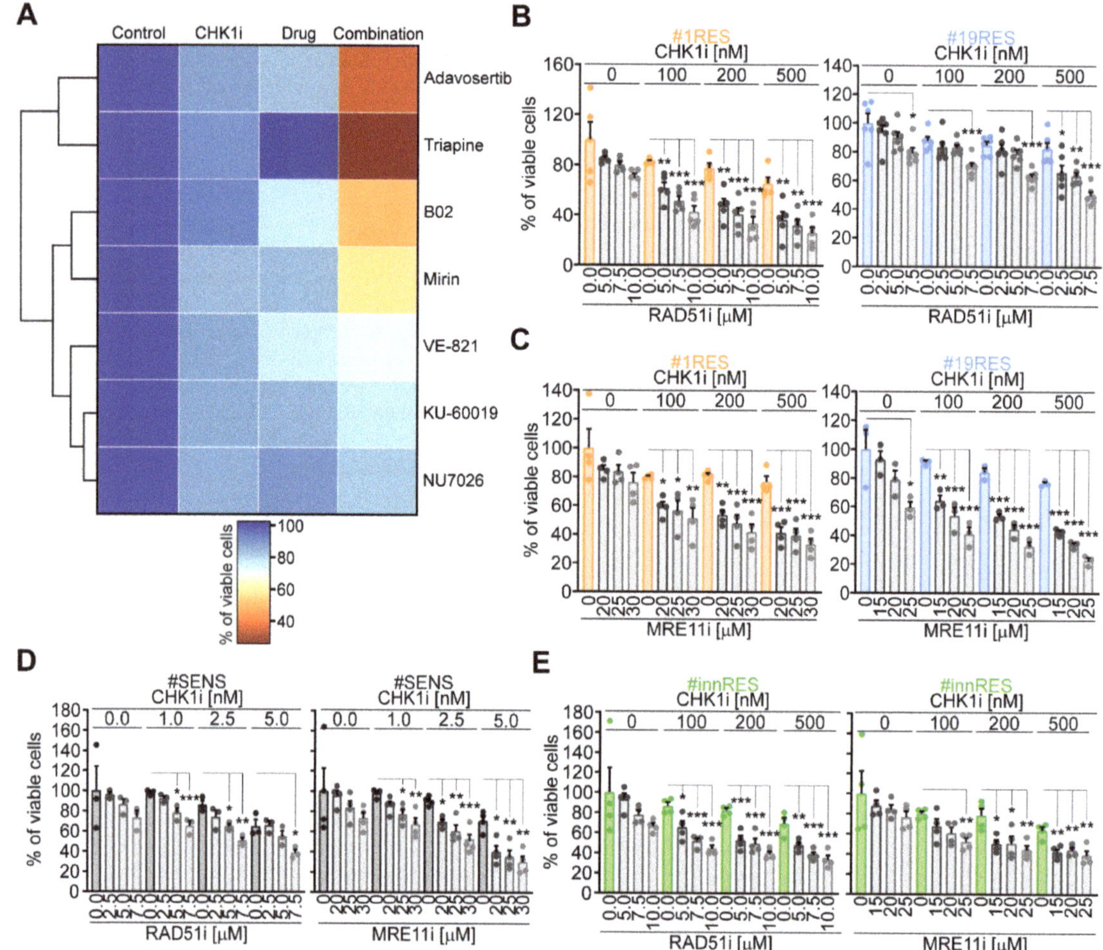

Figure 1. Identification of RAD51 and MRE11 inhibitors as prexasertib-sensitizing agents in CRC-SCs. (**A**) CRC-SCs previously characterized as resistant to ATR-CHK1 abrogation (RES-CRC-SCs) were left untreated or treated for 72 h with the CHK1/2 inhibitor prexasertib (CHK1i), and/or a set of modulators of the DNA damage response (DDR) with known prexasertib-sensitizing effect (i.e., adavosertib and triapine) or with unknown impact on prexasertib CSC toxicity (i.e., B02, KU-60019, mirin, NU7026, and VE-821), as indicated. Cell proliferation and viability were assessed by CellTiter-Glo® assay. The heatmap shows prexasertib-sensitizing effects of DDR modulators, with values corresponding to the percentage of viable cells upon normalization on control conditions. Data are means of three independent experiments, with values reported in Supplementary Table S1. The heatmap and clusterization were generated with Python. (**B,C**) Cell proliferation/viability (assessed by CellTiter-Glo® assay) of distinct RES-CRC-SCs left untreated or exposed for 96 h to CHK1i alone or in combination with RAD51i (B02) (**B**) or MRE11i (mirin) (**C**), as indicated. Results are means±SEM and individual data points of six (RAD51i-treated #19RES), five (RAD51i-treated #1RES), four (MRE11i-treated #1RES), or three (MRE11i-treated #19RES) independent experiments. * $p < 0.05$, ** $p < 0.01$, *** $p < 0.001$ (one-way ANOVA and Bonferroni or Dunnett's T3 post-hoc test) as reported. (**D,E**) Cell proliferation/viability (evaluated by CellTiter-Glo® assay) of representative CRC-SCs sensitive to CHK1i (SENS-CRC-SCs) (**D**) or intrinsically resistant to CHK1i (innRES-CRC-SCs) (**E**) left untreated or subjected for 96 h to CHK1i alone or in combination with MRE11i or RAD51i, as indicated. Results are means ± SEM and individual data points of four (MRE11i-treated SENS-CRC-SCs and #innRES) or three (RAD511i-treated SENS-CRC-SCs) independent experiments. * $p < 0.05$, ** $p < 0.01$, *** $p < 0.001$ (one-way ANOVA and Bonferroni or Dunnett's T3 post-hoc test) as reported. Associated supplementary table: Supplementary Table S1.

2.2. The Targeting of RAD51 or MRE11 Sensitize to CHK1 Inhibitors by Boosting Replication Stress

We then analyzed the impact of combined inhibition of CHK1 and RAD51 (by prexasertib+B02) and CHK1 and MRE11 (by prexasertib+mirin) on DNA replication and DNA damage in CRC-SCs. Through western-blot studies, we provided evidence that these two prexasertib-based regimens promoted an elevated increase in the phosphorylation of RPA32 (pRPA32) (Figure 2A), a marker of ongoing replication stress [20,26,29], as compared to prexasertib alone. As expected, replication stress was not augmented by the administration of mirin or B02 alone (Figure 2A; Figure S1). Flow cytometry and fluorescence microscopy studies confirmed the induction of extensive DNA damage upon prexasertib-based combinatorial regimens in RES-CRC-SCs. Indeed, combined inhibition of CHK1+RAD51 and of CHK1+MRE11 significantly promoted the phosphorylation of H2AX (Figure 2B,C), a post-translational modification sensing DNA lesions best known as γH2AX [30]. In line with the induction of replication stress, in RES-CRC-SCs the two prexasertib-based combinations significantly increased the percentage of S-phase cells positive to γH2AX (Figure 2B), a marker of fork breakage occurring during elevated replication stress. Moreover, fluorescence microscopy studies revealed that a significant percentage of CRC-SCs cotreated either with CHK1+RAD51 inhibitors or with CHK1+MRE11 inhibitors displayed a diffuse γH2AX nuclear staining instead of classical nuclear foci, covering either all or a portion of the nuclei (Figure 2C). Such diffuse nuclear staining confirms the presence of excessive RS, indicating the induction of replication catastrophe.

Collectively, these findings indicate that the targeting of RAD51 or MRE11 sensitizes to CHK1 inhibitors by boosting replication stress to lethal levels.

2.3. The Administration of Prexasertib Together with Mirin or B02 Alters Mitotic Timing

We then performed an extensive flow cytometry-mediated characterization of cell cycle progression. We provided evidence that elevated replication stress induced by MRE11+CHK1 inhibitors in RES-CRC-SCs was accompanied by changes in cell cycle profiles, manifested with a significant accumulation of cells with a DNA content between $2n$ and $4n$ (presumably S-phase cells) (Figure 3A,B). A similar accumulation (though to a lesser extent) was observed in RES-CRC-SCs treated with CHK1+RAD51 inhibitors, while prexasertib, mirin and B02 monotherapies did not significantly affect cell cycle progression (Figure 3A,B). To enter more in-depth into such an effect, we extended cell cycle analysis focusing on phosphorylated histone 3 (pH3), a marker of mitosis. By cytofluorimetry, we observed that, when combined with prexasertib, mirin and B02 significantly increased the mitotic (pH3$^+$) fraction (Figure 3A,C). Intriguingly, upon exposure to prexasertib+B02 or prexasertib+mirin around 30% of cell positive for pH3$^+$ did not present the classic $4n$ DNA content, but bore a DNA content between $2n$ and $4n$ (Figure 3A,C). The presence of such a high number of pH3$^+$ cells with a DNA content lower than $4n$ indicates premature entry into mitosis upon cotreatment. This evidence demonstrates that combined inhibitions of CHK1 and RAD51 and of CHK1 and MRE11 in RES-CRC-SCs not only affect DNA replication but also deregulate cell cycle proliferation and mitotic timing. In particular, it suggests that prexasertib-based combinations push RES-CRC-SCs with unreplicated and/or damaged DNA into aberrant mitosis rather than causing an S-phase blockade. Again, this effect was absent or less evident in cells treated with prexasertib, B02 or mirin alone (Figure 3A–C). To further confirm this result, we performed a biparametric analysis of pH3 and γH2AX, observing a significant augmentation of the percentage of cells with double positivity for pH3 and γH2AX in RES-CRC-SCs subjected to prexasertib-based cotreatments (Figure 3D). Finally, by fluorescence microscopy studies we observed that, upon prexasertib+B02 or prexasertib+mirin cotreatment, almost half of the analyzed (pro)metaphases showed γH2AX foci (Figure 3E). This evidence confirms the presence of DNA damage in mitosis.

Altogether, these findings indicate that the induction of replication stress by combined inhibition of CHK1 and RAD51 or of CHK1 and MRE11 culminates in premature mitosis entry and aberrant mitotic execution, thus supporting the occurrence of mitotic catastrophe [31].

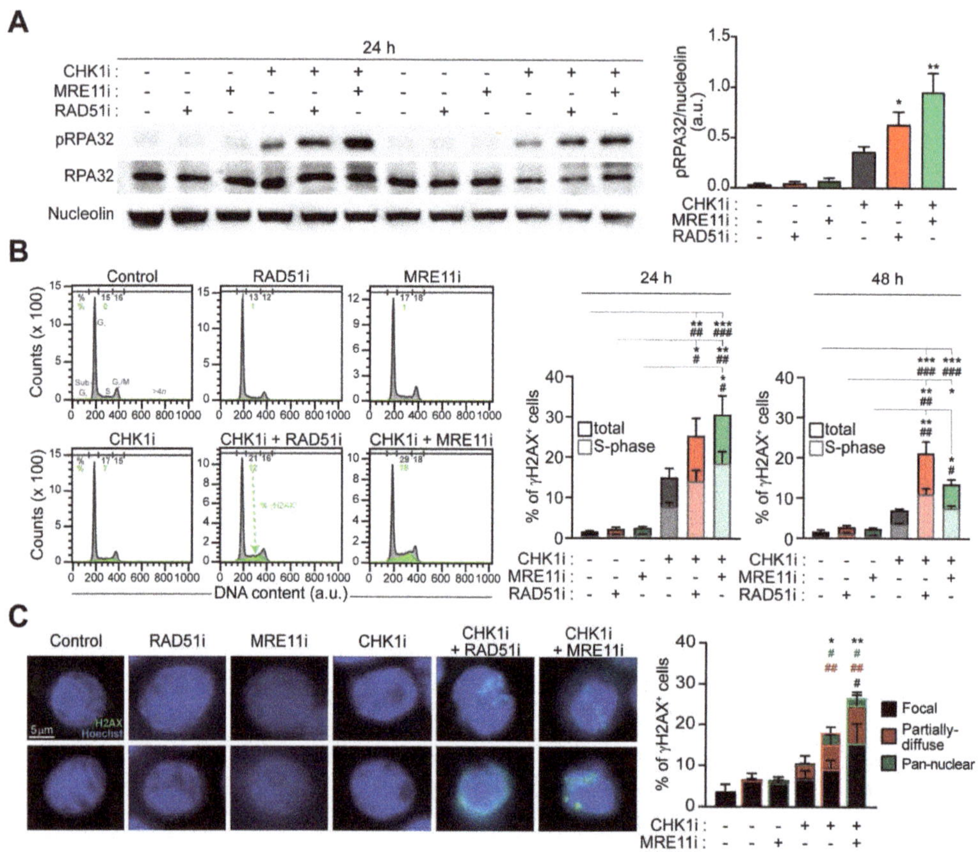

Figure 2. Combined inhibition of CHK1 and MRE11 or CHK1 and RAD51 induces replication stress in prexasertib-resistant CRC-SCs. (**A**) Western-blot analysis of representative RES-CRC-SCs left untreated or administrated for 24 h with prexasertib (CHK1i), either alone or in combination with MRE11i (mirin) or RAD51i (B02), and then stained with antibodies recognizing phospho(p)RPA32 (S4/S8) and RPA32 (as markers of replication stress, RS) and nucleolin (to ensure equal lane loading). One representative western-blot and the quantification of the ratio pRPA32/nucleolin are shown (see also Supplementary Figure S1). Results are expressed as means±SEM and individual data points of five independent experiments. * $p < 0.05$, ** $p < 0.01$, *** $p < 0.001$ (Kruskal-Wallis test and Dunn's post-hoc test) compared to untreated conditions. (**B**) Flow cytometry analysis in representative RES-CRC-SCs left untreated or treated with CHK1i, either alone or combined with MRE11i or RAD51i, and then stained with a DNA intercalant (DAPI) together with an anti-γH2AX antibody. Cell cycle profiles and quantitative data (means ± SEM; six independent experiments at 24 h and seven independent experiments at 48 h) are reported. In cell cycle profiles, cells positive for γH2AX in S-phase are in green. Numbers indicate the percentage of corresponding events. In the histograms, the percentage of γH2AX+ cells in all cell cycle phases (total) are in dark color, while the percentage of γH2AX+ cells in S-phase are in pale color. * $p < 0.05$, ** $p < 0.01$, *** $p < 0.001$ (Kruskal-Wallis test and Dunn's post-hoc test), as indicated (for γH2AX+ cells in all cell cycle phases). # $p < 0.05$, ## $p < 0.01$, ### $p < 0.001$ (Kruskal-Wallis test and Dunn's post-hoc test), as indicated (for γH2AX+ cells in S-phase). (**C**) Immunofluorescence analysis in representative RES-CRC-SCs left untreated or exposed for 24 h to CHK1i, either alone or in combination with MRE11i or RAD51i, and then strained with an antibody recognizing γH2AX. Representative images and quantification of percentages of γH2AX+ cells presenting "focal", "partially diffuse" or "pan-nuclear" γH2AX positivity are shown. For more information about the category of γH2AX positivity, see Materials and Methods. Data are expressed as means±SEM and individual

data points of five independent experiments. * $p < 0.05$, ** $p < 0.01$, *** $p < 0.001$ (Kruskal-Wallis test and Dunn's post-hoc test) compared to untreated conditions (for all γH2AX$^+$ cells); # $p < 0.05$, ## $p < 0.01$, ### $p < 0.001$ (Kruskal-Wallis test and Dunn's post-hoc test) compared to untreated conditions (for each distinct category of γH2AX positivity, depicted with the indicated color code). Dose range in (**A–C**): 100 nM CHK1i, 20 µM MRE11i for #19RES or 30 µM MRE11i for #1RES, 7.5 µM RAD51i; a.u., arbitrary units. Associated supplementary figure: Supplementary Figure S1.

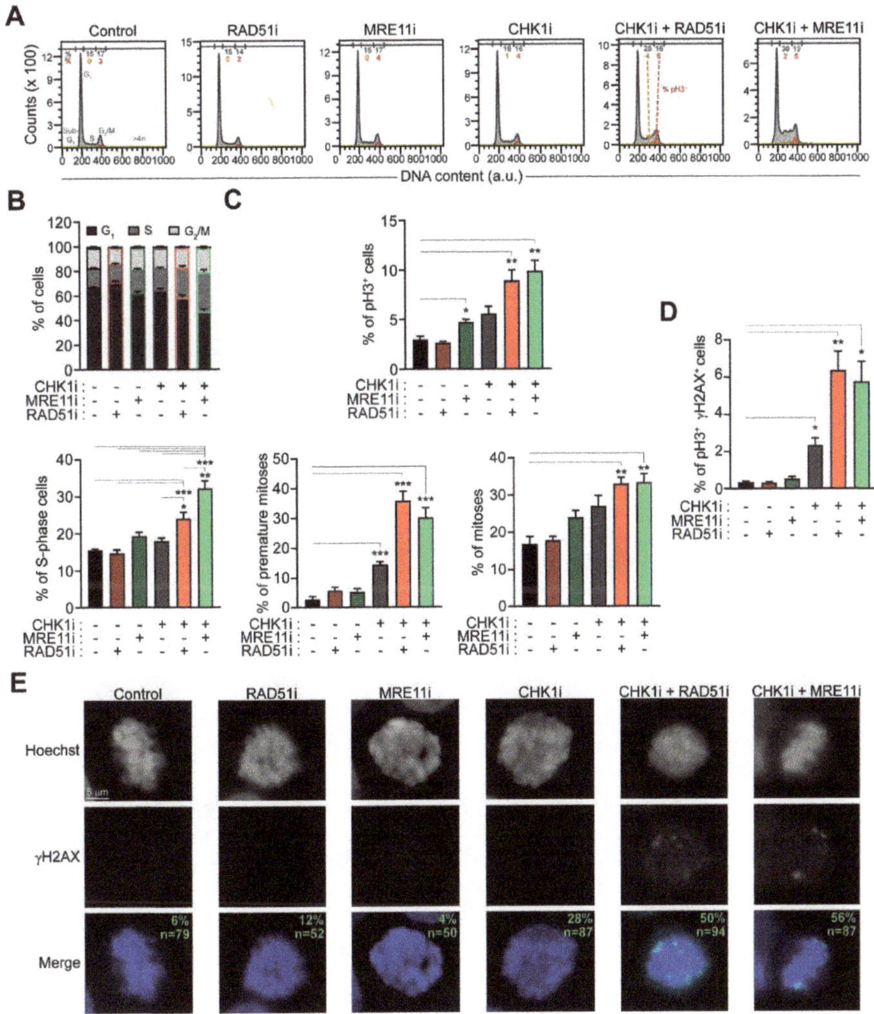

Figure 3. Combined inhibition of MRE11 and CHK1 or of RAD51 and CHK1 causes an accumulation of premature and damaged mitoses in CRC-SCs resistant to prexasertib. (**A–D**) Cytofluorimetric assessment of cell cycle profiles (**A,B**) or the levels of phospho(p)H3 (S10) (**A,C**) and/or pH3 and γH2AX (**D**) in representative RES-CRC-SC left untreated or treated for 24 h with prexasertib (CHK1i) alone or in combination with MRE11i (mirin) or RAD51i (B02), and then stained with DAPI and the appropriates antibodies. Cell cycle profiles and quantitative data (means $\pm$ SEM from six independent experiments) are reported. In (**A**), cells positive for pH3 in S-phase and G$_2$/M-phase are in orange and red, respectively. Numbers indicate the percentage of corresponding events. In (**C**), the percentage of premature mitosis corresponds to the percentage of pH3$^+$ cells with a DNA content lower than $4n$ among all pH3$^+$ cells, while the percentage of normal mitoses corresponds to the percentage of pH3$^+$ cells with a $4n$ DNA content among all cells. * $p < 0.05$, ** $p < 0.01$, *** $p < 0.001$ (one-way ANOVA and Bonferroni or Dunnett's T3 post-hoc test in (**B**), in (**C**) for the analysis of the percentages of pH3$^+$ cells and of premature

mitoses, and in (**D**); Kruskal-Wallis test and Dunn's post-hoc test in (**C**), for the analysis of the percentages of mitoses), as indicated. (**E**) Immunofluorescence detection of DNA damage in mitosis in RES-CRC-SCs left untreated or exposed for 24 h to CHK1i, alone or in combination with MRE11i or RAD51i, and then strained with DAPI and an antibody recognizing γH2AX. The panel shows representative images of (pro)metaphases (as demonstrated by the classical chromosome condensation in gray scale images) with DNA damage foci. Green numbers refer to numbers of analyzed mitoses (n) and the percentages of γH2AX$^+$ mitoses pooled from five independent experiments. Dose range in (**A–E**): 100 nM CHK1i, 20 μM MRE11i for #19RES or 30 μM MRE11i for #1RES, 7.5 μM RAD51i; a.u., arbitrary units.

2.4. Mitotic Catastrophe by CHK1-Based Regimens Is Executed via Caspase-Dependent Apoptosis

We then explored the mechanism of cell killing by prexasertib+mirin and prexasertib+B02. Through western-blot, we observed similar constitutive levels of MRE11 and RAD51 in both SENS- and RES-CRC-SCs (Figure 4A; Figure S2). Moreover, no significant modulation of the level of these proteins was detected upon administration of CHK1 inhibitor, alone or in combination with MRE11 and/or RAD51 inhibitors (Figure 4B; Figure S3), ruling out a mechanism of sensitization mediated by the upregulation of MRE11 and/or RAD51. We then investigated whether the demise of RES-CRC-SCs undergoing mitotic catastrophe occurred via apoptosis, focusing on the involvement of caspases. Immunofluorescence microscopy studies of caspase 3 (CASP3) activation revealed that both prexasertib+B02 and prexasertib+mirin induced a significant increase in the level of cleaved CASP3 (Figure 4C), the activated form of this protease during apoptosis. On the contrary, monotherapies with these drugs did not significantly activate CASP3 (Figure 4C). Similar findings were obtained through flow cytometry analyses. In these studies, we observed a global increase of CASP3 activation in CRC-SCs subjected to prexasertib-based regimens, and in particular of CASP3A$^+$ cells displaying a DNA content between $2n$ and $4n$ (which include premature mitoses) and a $4n$ DNA content (which include apparent normal mitoses) (Figure 4D). In line with CASP3 involvement, combined inhibitions of CHK1 with RAD51 or CHK1 with MRE11 induced the cleavage of PARP1 (Figure 4E; Figure S4), a downstream target of activated CASP3. Consistently, PARP1 cleavage was completely abolished in CRC-SCs subjected to these prexasertib-based combinations by the administration of the caspase inhibitor Q-VD-Oph (Figure 4E; Figure S4).

Collectively, these results indicate that combined inhibition of CHK1 and MRE11 or RAD51 ultimately kills CRC-SCs via caspase-dependent apoptosis.

2.5. Prexasertib in Combination with RAD51 or MRE11 Inhibitors Disrupts 3D Tumorsphere Organization and Growth

We then analyzed the impact of the coinhibition of CHK1 and RAD51 or MRE11 on the organization and survival of CRC-SCs grown in vitro as 3D tumorspheres. We first performed live fluorescence microscopy in RES-CRC-SCs left untreated or administered with CHK1, RAD51 and/or MRE11 inhibitors, either alone or in combination. After treatment, RES-CRC-SCs were co-incubated with SYTOX, a vital dye incorporated by cells undergoing regulated cell death due to loss of membrane integrity, together with Hoechst 33342 to visualize the nuclei (Figure 5A). Image analysis of SYTOX incorporation confirmed that combined treatment of prexasertib with either B02 or mirin induced a high level of apoptosis in tumorspheres as compared to prexasertib, B02 or mirin monotherapies (Figure 5A). This evidence demonstrates the occurrence of apoptosis in CRC-SC tumorspheres. To explore more in-depth this phenomenon, we performed live videomicroscopy studies monitoring RES-CRC-SCs grown as spheres for approximately 67 h upon drug administration. In this analysis, we provided evidence that the two prexasertib-based combinations affected 3D tumorsphere growth and organization. Indeed, prexasertib+mirin (and to a lesser extent) prexasertib+B02 induced disaggregation of a significant number of spheres, manifested with sphere demise or dissolution (Figure 5B). Moreover, both combinations (and in particular prexasertib+B02) decelerated the growth of viable spheres (i.e., those that did not undergo disaggregation) as shown by the mild increase in sphere diameter at the end of the experimentation in these conditions, a phenomenon also observed with prexasertib (Figure 5B). Further confirming the effect on sphere organization, combined administration of prexasertib with B02 or mirin (but not or less monotherapies) promoted the occurrence of multiple rounds of expulsion of cells/spheres displaying an apoptotic morphology (Figure 5B).

Collectively, these results demonstrate that mirin and B02 sensitize CRC-SCs to prexasertib by disrupting tumorsphere organization and growth.

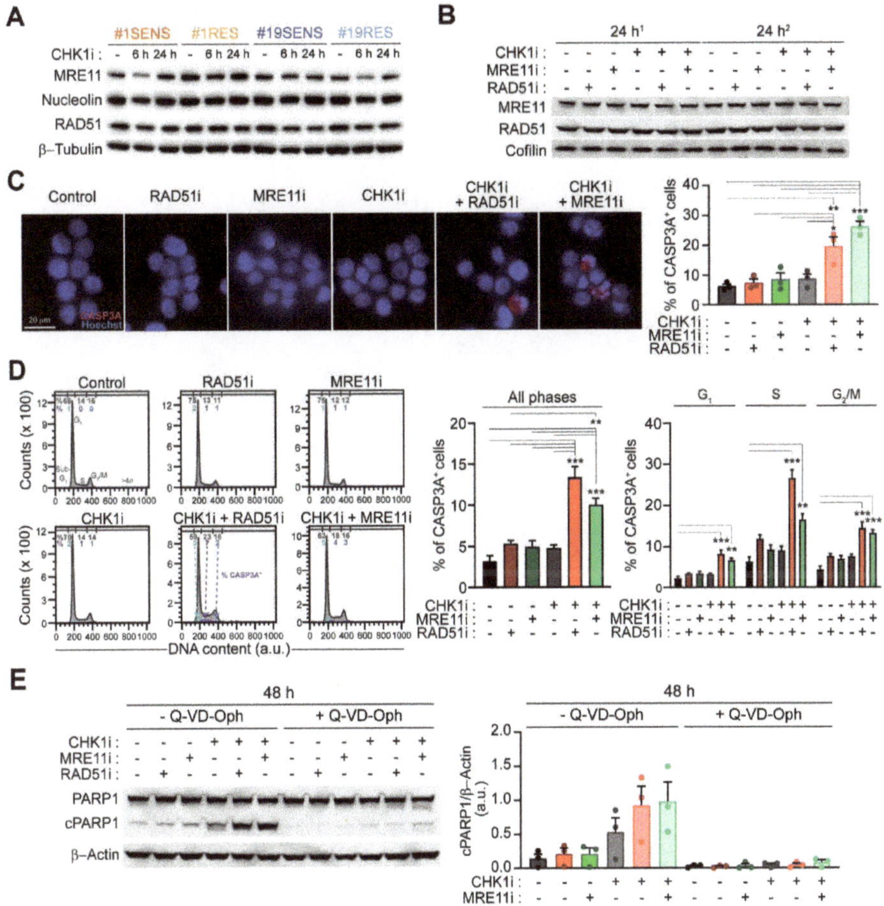

Figure 4. The inhibition of MRE11 or RAD51 sensitizes CRC-SCs to prexasertib by inducing a caspase-dependent mitotic catastrophe. (**A,B**) Western-blot analysis of MRE11 and RAD51 levels in SENS-CRC-SCs (**A**) and RES-CRC-SCs (**A,B**) left untreated or administrated with prexasertib (CHK1i), alone (**A,B**) or in combination with MRE11i (mirin) or RAD51i (B02) (**B**) (as indicated), and then stained with the appropriates antibodies. In (**B**), two independent experiments for #1RES ([1] and [2]) are shown. Cofilin, nucleolin and β-tubulin were used to ensure equal lane loading. Representative western-blots are reported. Quantifications are shown in Supplementary Figures S2 and S3. (**C,D**) Immunofluorescence- and flow cytometry-mediated detection of the activation of caspase 3 (CASP3A) in RES-CRC-SCs upon treatment for 24 h (**C**) or 48 h (**D**) with CHK1i, MRE11i, and/or RAD51i as indicated, followed by co-staining with the DNA intercalant Hoechst (**C**) or DAPI (**D**) and an anti-CASP3a antibody. In (**C**), representative images and data quantification are shown. In (**D**), cell cycle profiles (left) and quantification of the percentage of CASP3A$^+$ cells among all cells (center) and the relative percentage of CASP3A$^+$ cells in G_1-, S-, G_2/M-phase (right) are illustrated. In cell cycle profiles, positivity for CASP3A in G_1-, S-, G_2/M-phase is depicted in pale blue, violet and dark blue, respectively. Numbers indicate the percentages of corresponding events. In the histograms, results are expressed as means ± SEM from three or seven independent experiments in (**C,D**),

respectively. * $p < 0.05$, ** $p < 0.01$, *** $p < 0.001$ (one-way ANOVA and Bonferroni post-hoc test) compared to untreated conditions. (**E**) Western-blot analysis in RES-CRC-SCs treated for 48 h as indicated, using an antibody directed against PARP1, also recognizing the cleaved (c) form. β-Actin was used to monitor equal lane loading. Representative western-blots are shown (see also Supplementary Figure S4). Quantification of data, expressed as means ± SEM, and individual data points are from three independent experiments. cPARP1, cleaved PARP1. Dose range in **A–E**: 100 nM CHK1i, 20 μM MRE11i for #19RES or 30 μM MRE11i for #1RES, 15 μM Q-VD-Oph, 7.5 μM RAD51i; a.u., arbitrary units. Associated supplementary figures: Supplementary Figures S2–S4.

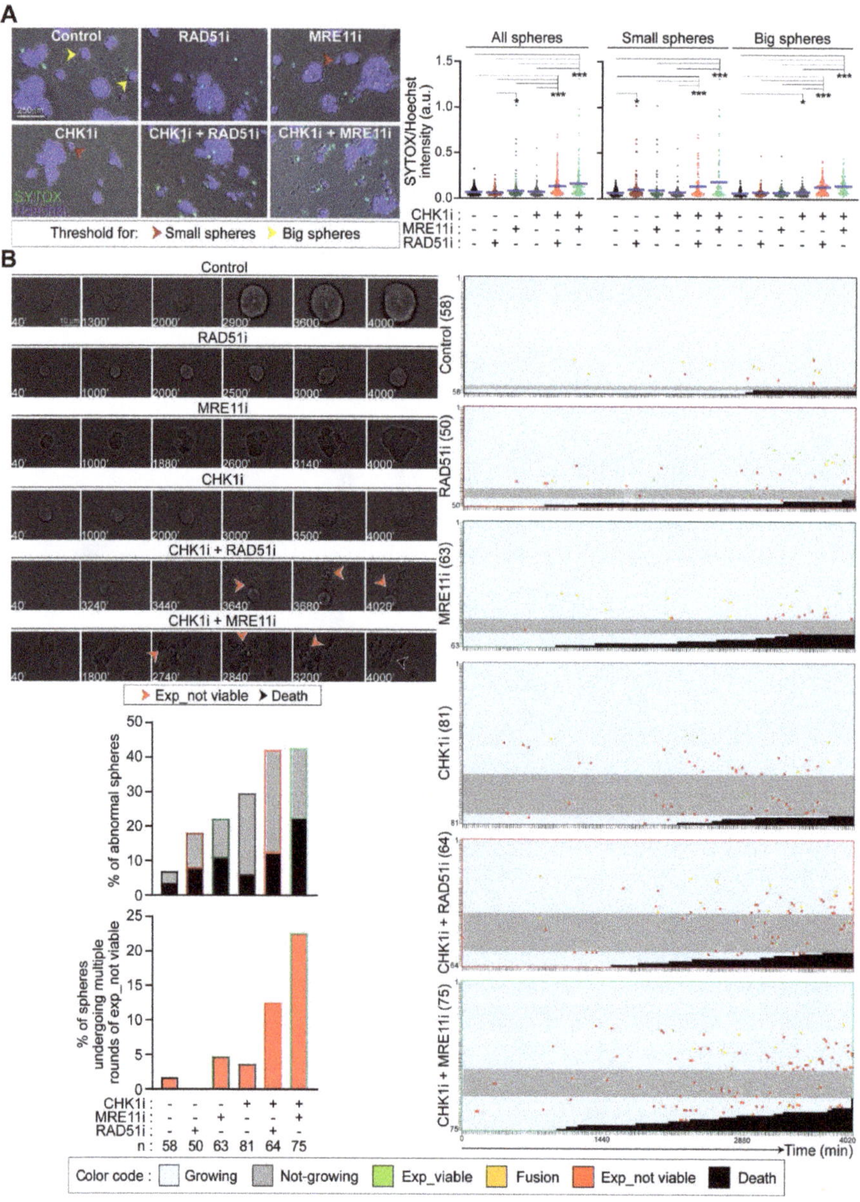

Figure 5. The cooperation between MRE11+CHK1 and RAD51+CHK1 is essential for the survival of CRC-SCs resistant to prexasertib. (**A**) Live cell microscopy assessment of apoptosis induction in representative RES-CRC-SCs left untreated or

exposed for 48 h to prexasertib (CHK1i), MRE11i (mirin) and/or RAD51i (B02) as indicated, and then incubated for 10 min with SYTOX (which incorporates only dead cells) and the DNA dye Hoechst. SYTOX incorporation (as a parameter of regulated cell death activation) was evaluated by live fluorescence microscopy and image analysis. Nuclear fluorescence was used to discriminate spheres (area higher than 1000 pixels) from single cells or small cell aggregates (area lower than 1000 pixels) on the basis of the threshold indicated with a red arrow point (1000 pixels). Spheres were further classified in small spheres (area comprised between 1000 and 3000 pixels) and big spheres (area > 3000 pixels) on the basis of the threshold indicated with a yellow arrow point (approximately 3000 pixels). Green intensity means (i.e., SYTOX incorporation) were quantified in such spheres. Data, expressed as ratios of SYTOX/Hoechst intensity, are a pool of two independent experiments performed on distinct RES-CRC-SCs, and are shown as box-plots with means and individual data points. In the box-plot on the right, spheres are divided in small and big spheres. * $p < 0.05$, *** $p < 0.001$ (Kruskal-Wallis ANOVA and Dunn's post-hoc test). (**B**) Live cell videomicroscopy analysis of one representative RES-CRC-SCs (#19RES) grown as 3D tumorspheres, left untreated or exposed to CHK1i, either alone or in combination with MRE11i or RAD51i. Images were taken every 20 min for up to 67 h (see Materials and Methods). Representative frames of the fate of one sphere per condition are shown, with numbers referring to the time passed from the beginning of the recording. Expulsion of one or more dead/inert cells or of one or more dead/inert cell aggregates are depicted with a red arrow point (**Exp_not viable**), while sphere disaggregation and/or death with a dark arrow point (**Death**). The fate of all spheres analyzed (at least 50 spheres per condition) are represented on the right using the indicated color code. In such "sphere fate profile", each single sphere is depicted by a hyphen, with the first and last spheres analyzed also illustrated by a number. The following events are included: (i) expulsion of one or more viable cells or cell aggregates (**Exp_viable**; in green), (ii) sphere fusion (**Fusion**; in orange), (iii) expulsion of one or more cells or cell aggregates with an apoptotic or inert morphology (**Exp_not viable**; in red), (iv) sphere disaggregation and/or death (**Death**; in dark). The growth of each sphere was determined by measuring the area of the sphere at the beginning and at the end of the recording, and then calculating the ratio of the latter on the former. We considered as growing spheres only those with a growth ratio higher than 1.3. **Growing** and **not-growing** spheres are respectively colored in blue and grey. Numbers indicate the total number of spheres for each condition, counted in two separate videos. In the histogram in the panel, results are expressed as percentages of abnormal spheres (comprising spheres not growing and/or disaggregated/dead spheres) on the top and of spheres undergoing multiple (i.e., more than three) rounds of expulsion of inert/non-viable cells or cell aggregates on the bottom. For more information about the experiments, categories, and exclusion criteria see Materials and Methods. See also Supplementary Videos. Dose range in A–D: 100 nM CHK1i, 20 μM MRE11i for #19RES or 30 μM MRE11i for #1RES, 7.5 μM RAD51i; a.u., arbitrary units. Associated supplementary figure: Supplementary Figure S5.

3. Discussion

The ATR-CHK1 axis is particularly relevant for the survival of CSCs, which often display high levels of replication stress, and, accordingly, ATR and CHK1 inhibitors are being explored for the design of anti-CSC therapies [20,22,23]. Moreover, the antineoplastic activity of ATR and CHK1 inhibitors, including prexasertib, is currently investigated in ongoing clinical trials (https://www.clinicaltrials.gov/). However, a significant fraction of CSCs is intrinsically resistant to these therapies, and resistance mechanisms also emerge. Here, we provided evidence that the targeting of either MRE11 or RAD51 sensitizes primary CRC-SCs to the CHK1 inhibitor prexasertib by inducing a mitotic catastrophe process culminating in caspase-dependent cell death.

Results from this study indicate that sensitization to prexasertib by RAD51 or MRE11 inhibitors involves the boost of replication stress, which reinstates CRC-SC dependence on the function of the ATR-CHK1 pathway of the replication stress response. This is in line with previous evidence showing enhanced anti-CSC activity of ATR-CHK1 inhibitors due to the induction of replication stress by agents including (but not limited to) gemcitabine, irinotecan and inhibitors of PARP1, RRM2 or WEE1 [20,25,26,32–34]. However, here, we showed that the two identified prexasertib-based regimens exerted a rather broad impact on the DDR of CRC-SCs, simultaneously impairing several cellular processes for the preservation of genomic stability. Indeed, concomitant inhibition of CHK1 and RAD51 or MRE11 deregulated not only the DNA replication process but also cell cycle progression and ultimately cell division, resulting in a general reorientation of the DDR toward cytotoxicity. This evidence is relevant for cancer therapy, as it suggests that prexasertib-based combinations can potentially overcome the reported heterogeneity in the therapeutic response of CSCs, which constitutes one major challenge in cancer therapy. In particular, CSC sensitivity to ATR-CHK1 inhibitors could be limited by a variety of cytoprotective mechanisms characterizing CSCs, encompassing (among others) strong DNA damage signaling and repair, cell cycle checkpoint proficiency, elevated tolerance to DNA damage, and limited apoptosis induction (reviewed in [10]). As a consequence, therapeutic regimens that simultaneously target multiple of these aspects—like those based on the combination of CHK1+MRE11 or CHK1+RAD51 inhibitors—can be particularly effective in eradicating CSCs, also limiting the potential development of acquired resistance.

At the mechanistic level, in this study, we demonstrated that combined inhibition of CHK1 and either MRE11 or RAD51 leads to uncontrolled cell cycle progression and untimely mitotic entry of cells with ongoing replication stress. Usually, the proliferation of replication stressed cells is limited by the activation of the intra-S and G_2/M checkpoint, both of which depend on the ATR-CHK1 axis [35,36] and, so, are silenced on its abrogation. Consistently, under prexasertib-based regimens, we found illicit proliferation and mitotic entry of CRC-SCs with unreplicated and damaged DNA, as shown by the presence of premature mitoses (pH3$^+$ cells with DNA content between $2n$ and $4n$) and of mitotic damage (mitotic cells displaying γH2AX foci). Hence, we propose a model in which pre-mitotic defects linked to replication stress emerging upon CHK1+MRE11 or CHK1+RAD51 inhibition are eventually transmitted into mitosis, threatening genomic stability, affecting sister chromatid segregation, and resulting in the activation of mitotic catastrophe.

At this regard, the inhibition of RAD51 or MRE11 may constitute the initiating factor of prexasertib sensitization. Indeed, RAD51 is a homologous recombination (HR) player also contributing to the stabilization, regression and restart of stressed DNA replication forks [37–41], meaning that its inhibition can impair the replication stress response. Similar considerations apply to MRE11, which has a pleiotropic impact on the DDR. In particular, a controlled and limited resection by MRE11 is crucial for the efficient restart or repriming of stalled forks [42,43]. In this context, uncontrolled fork degradation by MRE11, which occurs in conditions of fork instability or unprotection [44,45], is restricted by factors including RAD51 [46]. Moreover, MRE11 as a component of the MRE11-RAD50-NBS1 (MRN) complex has roles in the G_2/M checkpoint, regulating the interphase-prophase transition, and also acts in sister chromatid segregation during mitosis [47–49]. Further studies will uncover the precise mechanisms through which RAD51 or MRE11 inhibitors affect mitosis entry and execution upon concurrent CHK1 inhibition. Irrespective of these mechanistic unknowns, we demonstrated that mitotic catastrophe by RAD51+CHK1 or MRE11+CHK1 inhibitors resulted in tumorsphere disaggregation and CRC-SC death via caspase-dependent apoptosis. The particular proneness of CRC-SCs to mitotic catastrophe suggested by this and our previous study [26] can be potentially exploited for the design of effective colorectal cancer therapy.

4. Materials and Methods

4.1. Cell Lines, Culture Conditions and Chemicals

Media, supplements and plasticware for cell culture were supplied by Thermo Fisher Scientific (Waltham, MA, USA), Sigma-Aldrich (Millipore-Sigma, Merck Group, St. Louis, MO, USA) and Corning Life Sciences (Corning, NY, USA). Colorectal cancer cells (CRC-SCs) used in this study were previously isolated and established from patient sample following the standards of the institutional ethics committee on human experimentation (authorization no. CE5ISS09/282) as reported in [50]. Informed consent was requested in this previous study. The methods for CSC authentication, validation and cultivation are reported in [20,50], while generation of CRC-SCs resistant to CHK1 inhibitors (RES-CRC-SCs) is described in [26]. Adavosertib (#S1525), B02 (#S8434), KU-60019 (#S1570), mirin (#S8096), NU7026 (#S2893), prexasertib (#S7178), Q-VD-Oph (#S7311), triapine (#S7470), and VE-821 (#S8007) were supplied by Selleck Chemicals (Houston, TX, USA), while dimethyl sulfoxide (DMSO, #5879) by Sigma-Aldrich. Twenty-four hours before each experimentation, CRC-SCs were dissociated at single cells, then counted and seeded in specific multi-well plates. For live microscopy and videomicroscopy studies, cell seeding was carried out 48 h before the treatment to allow tumorsphere formation.

4.2. Cell Proliferation and Viability

To assess proliferation and survival, CRC-SCs were dissociated at single cells and seeded at a density of 6–8 × 10^3 cells per well in 96-well plates, in a volume of 100 µL of medium per well. After 24 h, CRC-SCs were treated according to specific experimentations and cell viability/proliferation was determined by assessing ATP levels via CellTiter-Glo® Luminescent Cell Viability Assay (#G7572, Promega, Madison, WI, USA) with a multimode microplate reader (DTX-880; Beckman Coulter, Brea, CA, USA). Experiments were carried out in triplicate and repeated at least three independent times, with results expressed as means of triplicate values or of duplicate values when we encountered technical problems with one replicate, as reported in the statistical procedures. The heatmap in Figure 1A illustrates the percentage of viable RES-CRC-SCs determined via CellTiter-Glo® Assay upon normalization of treated on untreated conditions using data reported in Supplementary Table S1. Drug sensitivity heatmap was generated using a Python script (Python Software Foundation, Wilmington, DE, USA; https://www.python.org/) and the Seaborn library (10.5281/zenodo.592845). The heatmap scale goes from 100 (blue; i.e., RES-CRC-SCs are all alive) to 25 (dark red) in consideration of the fact that 25% is the minimum of viability obtained for RES-CRC-SCs using the combination prexasertib+triapine.

4.3. Cytofluorometric Studies

Flow cytometry studies aimed at determining cell cycle profile, mitosis fraction, DNA damage and/or apoptosis of CRC-SCs were performed as previously reported [26]. The following primary antibodies were used: cleaved caspase 3 (CASP3; #9661) and phospho-histone H3 (pH3; #3377) from Cell Signaling Technology (Danvers, MA, USA), and γH2AX (#05-636) from Merck Millipore. Secondary antibodies used were: Alexa Fluor® 488-goat anti-mouse and 647-donkey anti-rabbit ((#A-21121 and #A-31573). These antibodies were provided by Thermo Scientific. Antibody dilutions and the Research Resource Identifiers (RRIDs) are in [26]. The DNA intercalant DAPI (#D1306, from Thermo Scientific) at 10 µM was employed to stain the DNA. The samples were acquired by a BD FACSCelesta™ flow cytometer (BD Biosciences, BD, Franklin Lakes, NJ). Data and statistical analyses were performed with the FlowJo software (FlowJo LLC, BD), gating on events showing normal forward and side scatter (FSC and SSC) parameters, on singlets and (for cell cycle analysis) upon exclusion of the sub-G_1 fraction.

4.4. Immunofluorescence Studies

Immunofluorescence studies aimed at detecting DNA damage markers in CRC-SCs were performed as reported in [26]. Briefly, upon permeabilization, the cells were blocked for 30 min in 5% (*w/v*) BSA, 5% FBS and 5% normal goat serum (NGS) in PBS. Thereafter, samples were incubated with primary antibodies directed against cleaved CASP3 (1:400; #9661) or γH2AX (S139) (1:250; #05-636) at 4 °C. After overnight incubation, slides were stained with secondary Alexa Fluor conjugates (1:500) together with 10 µM Hoechst 33342 (#H1399). Fluorescence images were visualized, analyzed and captured with a Leica DMI3000 B microscope, using a 40× objective (HCX PL Fluotar, AN 0.60) or a 100× objective (HCX PL Fluotar, AN 1.3), the Leica DFC 310FX camera, and the LAS X software (all from Leica Microsystems, Wetzlar, Germany). Analyses were carried out directly at the microscope or using the ImageJ v1.8 software (National Institute of Health, Bethesda, MD, USA; https://imagej.nih.gov/ij/). To determine γH2AX positivity, at least 61 cells per condition and per experiment were analyzed for a total of (at least) 743 cells (in five experiments). We considered three categories of positivity: (1) "focal" staining, when the cell presents > five γH2AX foci, (2) "partially-diffuse" staining, when the cell presents a diffuse γH2AX positivity covering between less than half of the nuclei, and (3) "pan-nuclear" staining, when the cell presents a diffuse γH2AX positivity covering more than half of the nuclei (see Figure 2). In Figure 3, cells were recognized as mitoses on morphological basis upon nuclear staining with Hoechst 33342. For CASP3 activation analysis, at least 261 cells per condition and per experiment were analyzed for a total of (at least) 923 cells (for three experiments). In such automated analysis, cells were considered positive for CASP3 when signal intensity calculated with ImageJ was above a predefined threshold size to account for background noise.

4.5. Immunoblotting

To detect protein levels, CRC-SCs were recollected, lysed, and subjected to western-blot as reported in [20]. Membranes were incubated with primary antibodies directed against MRE11 (1:1000; #sc-22767; RRID:AB_2145247 and #sc-135992; RRID:AB_2145244) from Santa Cruz Biotechnology (Dallas, TX, USA), PARP1 (#9542) from Cell Signaling Technology, RAD51 (1:1000; #sc-398587; RRID: AB_2756353 and #sc-377467) from Santa Cruz Biotechnology, RPA32 (#A303-874A) from Bethyl Laboratories (Montgomery, TX, USA) and pRPA32 (S4/S8) (#A300-245A) from Bethyl Laboratories, as well as with antibodies recognizing β-actin (1:2000; #A5441; RRID:AB_476744), cofilin (#3318, Cell Signaling Technology), nucleolin (1:1000; #14574; Cell Signaling Technology; RRID:AB_2798519), or β-tubulin (#T4026). After overnight incubation with primary antibodies and washes, membranes were incubated for 1 h at RT with the appropriate horseradish secondary antibodies conjugated to peroxidase: the donkey anti-rabbit IgG (#GENA934 or RRID:AB_772206), mouse anti-goat IgG (#sc-2354, Santa Cruz; RRID:AB_628490), or sheep anti-mouse IgG (#GENA931). When not specified, antibody dilutions and RRIDs can be found in [26]. Chemiluminescence imaging was carried out with the Fujifilm LAS-4000 luminescent image analyzer (GE Healthcare, Chicago, IL, USA), or the G:Box Chemi-XX9 using the GeneSys Software v1.5.6 (Synoptics, Cambridge, UK). At least two independent experiments were carried out. Representative Western-blots are shown in the figure together with the densitometry plus quantification, which were performed with ImageJ v1.8 software. Full blots are presented in Supplementary Figures S1–S4.

4.6. SYTOX Incorporation

To determine the induction of apoptosis, CRC-SCs dissociated at single cells were resuspended in 90% CSC growth medium + 10% Matrigel Basement Membrane Matrix (Matrigel, #354230, Corning), at a density of 3.5×10^5 cells in 500 μL. CRC-SCs were then seeded in 24-well plates, incubated for 24 h in standard culture condition and finally treated for 48 h with prexasertib, mirin and/or B02, alone or in combination, as reported in figure legends. After the (co)treatment, cells were incubated with the vital dye SYOX™ (at 5 μM; #S7020, Thermo Scientific) together with the DNA dye Hoechst 33342 (at 8 μM) for 10 min. Cells were immediately analyzed using the Leica DMI3000 B fluorescence microscope equipped with a 20× objective (HCX PL Fluotar AN 0.40) (see above for further details). A number of at least eight randomly selected images were analyzed with ImageJ v1.8 software (Fiji version) as follow. Briefly, nuclear fluorescence was used to discriminate single cells or small cell aggregate from spheres (threshold area: 1000 pixels), on which the green intensity (corresponding to dead cells) means were quantified and finally divided on Hoechst intensity. A minimum of 190 spheres per condition were analyzed.

4.7. Videomicroscopy

To determine the impact of prexasertib-based combinations on 3D tumorspheres, CRC-SCs were dissociated at single cells and then resuspended in 80% CSC medium + 20% Matrigel, at a density of 1×10^5 cells in a drop of 50 μL, and seeded in 24-well plates in a final volume of 500 μL. Upon 48 h incubation in standard culture conditions, cells were subjected to live videomicroscopy analysis using a Nikon LIPSI system (Nikon, Minato-ku, Tokyo, Japan) equipped with IRIS 15 photometrics camera allowing for standard culture cultivation. Images were taken every 20 min for up to 67 h, with a 20× long-range objective (S-PLAN AN 0.4) and analyzed with ImageJ v1.8 software (Fiji version). The following criteria of exclusion were adopted. Spheres were excluded from analysis when: (1) the distance from another sphere was less than 13 μm, so to avoid problems due to excessive rounds of sphere fusion; (2) underwent more than one round of fusion with other spheres, thus becoming too big to be analyzed; (3) had a length for one of the two orthogonal diameters (which have been arbitrarily chosen in the first frame of the movie) less than 13 μm; (4) had a length for the first orthogonal diameter <30 μm and the second orthogonal diameter <40 μm, (5) the fate was not clear to define; and (6) their demise occurred within the first 12 h of the video recording (and thus presumably not caused by the treatment). The following events were retained as important: (1) expulsion of apparently viable cell(s) or cell aggregate(s), as determined by morphological criteria and the capability to divide and/or grow, (2) expulsion of apparently non-viable or inert cell(s) or cell aggregate(s), as determined by morphological criteria, (3) fusion with another viable sphere, and (4) disaggregation and death of the spheres, as determined by morphological criteria. In the histograms of Figure 5B, results are expressed as percentages of abnormal spheres (comprising bot disaggregated spheres and not-growing spheres) and of spheres undergoing multiple rounds of expulsion of non-viable cells/cell aggregates. In this figure, the growth of each sphere was determined by measuring the sphere area at the beginning and at the end of the recording, and then dividing the latter on the former value. Spheres were considered as expanding when the ratio was higher than 1.3.

4.8. Statistical Procedures

All the experiments were repeated in at least three independent instances, but in case of absent sub-significant trends, the sample size was increased to more than three (see figure legends or dedicated material method sections for the exact sample size, the presence of replicates, and data/replicate exclusion criteria). We evaluated the variance equality using the F-test or the Brown-Forsythe test, when appropriate. We calculated the Shapiro-Wilk normality test for all the continuous variables. For the variables that are normally distributed and according to the number of groups compared, we used unpaired Student T test or Welch's unpaired T-test, and one-way ANOVA followed by Bonferroni post-hoc test or Brown-Forsythe and Welch one-way ANOVA followed by Dunnett's T3 post-hoc test, when appropriate. For the variables that are not normally distributed or for less than three independent experiments, we applied Mann-Whitney test or Kruskal-Wallis test followed by Dunn's post-hoc test, according to the number of groups. Data (including normalization) were calculated and visualized as reported in [26]. The following softwares were used: Microsoft Excel (Microsoft, Redmond, WA, USA), Prism (v8.3.0, GraphPad Software, San Diego, CA, USA) and SPSS (SPSS v.21, SPSS Inc-IBM, Chicago, IL, USA).

5. Conclusions

In this study we identified RAD51 and MRE11 as two targets whose inhibition increases the sensitivity of CRC-SCs to CHK1 inhibitors. We also characterized the mechanism of CSC killing by CHK1+RAD51 and CHK1+MRE11 inhibitors, showing that it involves the induction of replication stress followed by progression of replication stressed CRC-SCs through the interphase and their premature entry into mitosis, ultimately leading to caspase-dependent mitotic catastrophe. These results support the future clinical development of prexasertib-mediated regimens in colorectal cancer patients.

Supplementary Materials: The following are available online at https://www.mdpi.com/article/10.3390/cancers13081957/s1. Supplementary Figure S1. Analysis of replication stress in RES-CRC-SCs exposed to prexasertib-based regimens. Related to Figure 2. Supplementary Figure S2. Analysis of MRE11 or RAD51 levels in SENS-CRC-SCs and RES-CRC-SCs. Related to Figure 4. Supplementary Figure S3. Analysis of MRE11 or RAD51 levels in SENS-CRC-SCs and RES-CRC-SCs. Related to Figure 4. Supplementary Figure S4. Analysis of PARP1 cleavage in RES-CRC-SCs. Related to Figure 4. Supplementary Figure S5. CHK1 cooperates with MRE11 or RAD51 for RES-CRC-SC survival. Related to Figure 5. Supplementary Movies 1–12. Impact of MRE11+CHK1 inhibition and RAD51+CHK1 inhibition on CRC-SC organization, proliferation and survival. Related to Figure 5. Supplementary Table S1. Raw data of the heatmap presented in Figure 1A. Supplementary Information.

Author Contributions: L.M., S.S.A.R., and M.M. performed and analyzed experiments; A.S., S.V., C.G. and F.C. performed experiments; A.G. helped analyzing survival data; F.S. supervised and performed statistical analysis; G.M. supervised the project, designed and performed experiments, analyzed and interpreted data, and contributed to article writing; R.D.M. obtained funding, supervised the project, and contributed to article writing; I.V. obtained funding, supervised the project, designed experiments, analyzed and interpreted data and wrote the manuscript. All authors have read and agreed to the published version of the manuscript.

Funding: This work was supported by the Associazione Italiana per la Ricerca sul Cancro (AIRC, IG 2017 grant number 20417 to IV; AIRC 5 × 1000 grant number 9979 to RDM; Start-Up 2016 #18418 to AS), Ministero Italiano della Salute (RF_RF-2018-12367044 to RDM; RF_GR-2013-02357273 to AS). IV is also supported by a startup grant from the Italian Institute for Genomic Medicine (Candiolo, Turin, Italy) and Compagnia di San Paolo (Turin, Italy).

Institutional Review Board Statement: All procedures were in accordance with the ethical standards approved by Institutional Ethics Committee on human experimentation and with the Helsinki Declaration.

Data Availability Statement: The materials and data are available from the corresponding authors.

Acknowledgments: We thank Paola Di Matteo, Roberto Ricci, and Stefano Guida (Istituto Superiore di Sanità, Rome, Italy) for technical assistance. We also thank Denis Baev of the cytometry facility of the Italian Institute for Genomic Medicine (IIGM, Candiolo, Italy) for assistance in cytofluorimetric acquisitions.

Conflicts of Interest: The authors have no conflicts of interest to disclose.

References

1. Batlle, E.; Clevers, H. Cancer stem cells revisited. *Nat. Med.* **2017**, *23*, 1124–1134. [CrossRef]
2. de Sousa e Melo, F.; Kurtova, A.V.; Harnoss, J.M.; Kljavin, N.; Hoeck, J.D.; Hung, J.; Anderson, J.E.; Storm, E.E.; Modrusan, Z.; Koeppen, H.; et al. A distinct role for Lgr5(+) stem cells in primary and metastatic colon cancer. *Nature* **2017**, *543*, 676–680. [CrossRef] [PubMed]
3. Todaro, M.; Gaggianesi, M.; Catalano, V.; Benfante, A.; Iovino, F.; Biffoni, M.; Apuzzo, T.; Sperduti, I.; Volpe, S.; Cocorullo, G.; et al. CD44v6 is a marker of constitutive and reprogrammed cancer stem cells driving colon cancer metastasis. *Cell Stem. Cell* **2014**, *14*, 342–356. [CrossRef] [PubMed]
4. Schepers, A.G.; Snippert, H.J.; Stange, D.E.; van den Born, M.; van Es, J.H.; van de Wetering, M.; Clevers, H. Lineage tracing reveals Lgr5+ stem cell activity in mouse intestinal adenomas. *Science* **2012**, *337*, 730–735. [CrossRef] [PubMed]
5. O'Brien, C.A.; Pollett, A.; Gallinger, S.; Dick, J.E. A human colon cancer cell capable of initiating tumour growth in immunodeficient mice. *Nature* **2007**, *445*, 106–110. [CrossRef]
6. Ricci-Vitiani, L.; Lombardi, D.G.; Pilozzi, E.; Biffoni, M.; Todaro, M.; Peschle, C.; De Maria, R. Identification and expansion of human colon-cancer-initiating cells. *Nature* **2007**, *445*, 111–115. [CrossRef]
7. Vitale, I.; Shema, E.; Loi, S.; Galluzzi, L. Intratumoral heterogeneity in cancer progression and response to immunotherapy. *Nat. Med.* **2021**, *27*, 212–224. [CrossRef]
8. Lytle, N.K.; Barber, A.G.; Reya, T. Stem cell fate in cancer growth, progression and therapy resistance. *Nat. Rev. Cancer* **2018**, *18*, 669–680. [CrossRef]
9. Zeman, M.K.; Cimprich, K.A. Causes and consequences of replication stress. *Nat. Cell Biol.* **2014**, *16*, 2–9. [CrossRef]
10. Vitale, I.; Manic, G.; De Maria, R.; Kroemer, G.; Galluzzi, L. DNA Damage in Stem Cells. *Mol. Cell* **2017**, *66*, 306–319. [CrossRef]
11. Galluzzi, L.; Vitale, I.; Aaronson, S.A.; Abrams, J.M.; Adam, D.; Agostinis, P.; Alnemri, E.S.; Altucci, L.; Amelio, I.; Andrews, D.W.; et al. Molecular mechanisms of cell death: Recommendations of the Nomenclature Committee on Cell Death 2018. *Cell Death Differ* **2018**, *25*, 486–541. [CrossRef]
12. Chang, C.W.; Chen, Y.S.; Chou, S.H.; Han, C.L.; Chen, Y.J.; Yang, C.C.; Huang, C.Y.; Lo, J.F. Distinct subpopulations of head and neck cancer cells with different levels of intracellular reactive oxygen species exhibit diverse stemness, proliferation, and chemosensitivity. *Cancer Res.* **2014**, *74*, 6291–6305. [CrossRef]
13. Cheung, S.T.; Cheung, P.F.; Cheng, C.K.; Wong, N.C.; Fan, S.T. Granulin-epithelin precursor and ATP-dependent binding cassette(ABC)B5 regulate liver cancer cell chemoresistance. *Gastroenterology* **2011**, *140*, 344–355. [CrossRef]
14. Diehn, M.; Cho, R.W.; Lobo, N.A.; Kalisky, T.; Dorie, M.J.; Kulp, A.N.; Qian, D.; Lam, J.S.; Ailles, L.E.; Wong, M.; et al. Association of reactive oxygen species levels and radioresistance in cancer stem cells. *Nature* **2009**, *458*, 780–783. [CrossRef]
15. Milanovic, M.; Fan, D.N.Y.; Belenki, D.; Dabritz, J.H.M.; Zhao, Z.; Yu, Y.; Dorr, J.R.; Dimitrova, L.; Lenze, D.; Monteiro Barbosa, I.A.; et al. Senescence-associated reprogramming promotes cancer stemness. *Nature* **2018**, *553*, 96–100. [CrossRef]

16. Colak, S.; Zimberlin, C.D.; Fessler, E.; Hogdal, L.; Prasetyanti, P.R.; Grandela, C.M.; Letai, A.; Medema, J.P. Decreased mitochondrial priming determines chemoresistance of colon cancer stem cells. *Cell Death Differ.* **2014**, *21*, 1170–1177. [CrossRef]

17. Ahmed, S.U.; Carruthers, R.; Gilmour, L.; Yildirim, S.; Watts, C.; Chalmers, A.J. Selective Inhibition of Parallel DNA Damage Response Pathways Optimizes Radiosensitization of Glioblastoma Stem-like Cells. *Cancer Res.* **2015**, *75*, 4416–4428. [CrossRef]

18. Wang, W.J.; Wu, S.P.; Liu, J.B.; Shi, Y.S.; Huang, X.; Zhang, Q.B.; Yao, K.T. MYC regulation of CHK1 and CHK2 promotes radioresistance in a stem cell-like population of nasopharyngeal carcinoma cells. *Cancer Res.* **2013**, *73*, 1219–1231. [CrossRef]

19. Bao, S.; Wu, Q.; McLendon, R.E.; Hao, Y.; Shi, Q.; Hjelmeland, A.B.; Dewhirst, M.W.; Bigner, D.D.; Rich, J.N. Glioma stem cells promote radioresistance by preferential activation of the DNA damage response. *Nature* **2006**, *444*, 756–760. [CrossRef]

20. Manic, G.; Signore, M.; Sistigu, A.; Russo, G.; Corradi, F.; Siteni, S.; Musella, M.; Vitale, S.; De Angelis, M.L.; Pallocca, M.; et al. CHK1-targeted therapy to deplete DNA replication-stressed, p53-deficient, hyperdiploid colorectal cancer stem cells. *Gut* **2018**, *67*, 903–917. [CrossRef]

21. Carruthers, R.D.; Ahmed, S.U.; Ramachandran, S.; Strathdee, K.; Kurian, K.M.; Hedley, A.; Gomez-Roman, N.; Kalna, G.; Neilson, M.; Gilmour, L.; et al. Replication Stress Drives Constitutive Activation of the DNA Damage Response and Radioresistance in Glioblastoma Stem-like Cells. *Cancer Res.* **2018**, *78*, 5060–5071. [CrossRef]

22. Hill, S.J.; Decker, B.; Roberts, E.A.; Horowitz, N.S.; Muto, M.G.; Worley, M.J., Jr.; Feltmate, C.M.; Nucci, M.R.; Swisher, E.M.; Nguyen, H.; et al. Prediction of DNA Repair Inhibitor Response in Short-Term Patient-Derived Ovarian Cancer Organoids. *Cancer Discov.* **2018**, *8*, 1404–1421. [CrossRef] [PubMed]

23. Gallmeier, E.; Hermann, P.C.; Mueller, M.T.; Machado, J.G.; Ziesch, A.; De Toni, E.N.; Palagyi, A.; Eisen, C.; Ellwart, J.W.; Rivera, J.; et al. Inhibition of ataxia telangiectasia- and Rad3-related function abrogates the in vitro and in vivo tumorigenicity of human colon cancer cells through depletion of the CD133(+) tumor-initiating cell fraction. *Stem Cells* **2011**, *29*, 418–429. [CrossRef]

24. Carruthers, R.; Ahmed, S.U.; Strathdee, K.; Gomez-Roman, N.; Amoah-Buahin, E.; Watts, C.; Chalmers, A.J. Abrogation of radioresistance in glioblastoma stem-like cells by inhibition of ATM kinase. *Mol. Oncol.* **2015**, *9*, 192–203. [CrossRef] [PubMed]

25. Bartucci, M.; Svensson, S.; Romania, P.; Dattilo, R.; Patrizii, M.; Signore, M.; Navarra, S.; Lotti, F.; Biffoni, M.; Pilozzi, E.; et al. Therapeutic targeting of Chk1 in NSCLC stem cells during chemotherapy. *Cell Death Differ.* **2012**, *19*, 768–778. [CrossRef]

26. Manic, G.; Musella, M.; Corradi, F.; Sistigu, A.; Vitale, S.; Soliman Abdel Rehim, S.; Mattiello, L.; Malacaria, E.; Galassi, C.; Signore, M.; et al. Control of replication stress and mitosis in colorectal cancer stem cells through the interplay of PARP1, MRE11 and RAD51. *Cell Death Differ* **2021**. [CrossRef]

27. Curtin, N.J.; Szabo, C. Poly(ADP-ribose) polymerase inhibition: Past, present and future. *Nat. Rev. Drug Discov.* **2020**, *19*, 711–736. [CrossRef]

28. Sanjiv, K.; Hagenkort, A.; Calderon-Montano, J.M.; Koolmeister, T.; Reaper, P.M.; Mortusewicz, O.; Jacques, S.A.; Kuiper, R.V.; Schultz, N.; Scobie, M.; et al. Cancer-Specific Synthetic Lethality between ATR and CHK1 Kinase Activities. *Cell Rep.* **2016**, *14*, 298–309. [CrossRef]

29. Marechal, A.; Zou, L. RPA-coated single-stranded DNA as a platform for post-translational modifications in the DNA damage response. *Cell Res.* **2015**, *25*, 9–23. [CrossRef]

30. Mah, L.J.; El-Osta, A.; Karagiannis, T.C. gammaH2AX: A sensitive molecular marker of DNA damage and repair. *Leukemia* **2010**, *24*, 679–686. [CrossRef]

31. Vitale, I.; Galluzzi, L.; Castedo, M.; Kroemer, G. Mitotic catastrophe: A mechanism for avoiding genomic instability. *Nat. Rev. Mol. Cell Biol.* **2011**, *12*, 385–392. [CrossRef] [PubMed]

32. Bukhari, A.B.; Lewis, C.W.; Pearce, J.J.; Luong, D.; Chan, G.K.; Gamper, A.M. Inhibiting Wee1 and ATR kinases produces tumor-selective synthetic lethality and suppresses metastasis. *J. Clin. Investig.* **2019**, *129*, 1329–1344. [CrossRef] [PubMed]

33. Signore, M.; Buccarelli, M.; Pilozzi, E.; De Luca, G.; Cappellari, M.; Fanciulli, M.; Goeman, F.; Melucci, E.; Biffoni, M.; Ricci-Vitiani, L. UCN-01 enhances cytotoxicity of irinotecan in colorectal cancer stem-like cells by impairing DNA damage response. *Oncotarget* **2016**, *7*, 44113–44128. [CrossRef] [PubMed]

34. Al-Ejeh, F.; Pajic, M.; Shi, W.; Kalimutho, M.; Miranda, M.; Nagrial, A.M.; Chou, A.; Biankin, A.V.; Grimmond, S.M.; Australian Pancreatic Cancer Genome, I.; et al. Gemcitabine and CHK1 inhibition potentiate EGFR-directed radioimmunotherapy against pancreatic ductal adenocarcinoma. *Clin. Cancer Res.* **2014**, *20*, 3187–3197. [CrossRef] [PubMed]

35. Saldivar, J.C.; Cortez, D.; Cimprich, K.A. The essential kinase ATR: Ensuring faithful duplication of a challenging genome. *Nat. Rev. Mol. Cell Biol.* **2017**, *18*, 622–636. [CrossRef]

36. Manic, G.; Sistigu, A.; Corradi, F.; Musella, M.; De Maria, R.; Vitale, I. Replication stress response in cancer stem cells as a target for chemotherapy. *Semin. Cancer Biol.* **2018**, *53*, 31–41. [CrossRef] [PubMed]

37. Hashimoto, Y.; Ray Chaudhuri, A.; Lopes, M.; Costanzo, V. Rad51 protects nascent DNA from Mre11-dependent degradation and promotes continuous DNA synthesis. *Nat. Struct. Mol. Biol.* **2010**, *17*, 1305–1311. [CrossRef]

38. Mason, J.M.; Chan, Y.L.; Weichselbaum, R.W.; Bishop, D.K. Non-enzymatic roles of human RAD51 at stalled replication forks. *Nat. Commun.* **2019**, *10*, 4410. [CrossRef]

39. Mijic, S.; Zellweger, R.; Chappidi, N.; Berti, M.; Jacobs, K.; Mutreja, K.; Ursich, S.; Ray Chaudhuri, A.; Nussenzweig, A.; Janscak, P.; et al. Replication fork reversal triggers fork degradation in BRCA2-defective cells. *Nat. Commun.* **2017**, *8*, 859. [CrossRef]

40. Schlacher, K.; Christ, N.; Siaud, N.; Egashira, A.; Wu, H.; Jasin, M. Double-strand break repair-independent role for BRCA2 in blocking stalled replication fork degradation by MRE11. *Cell* **2011**, *145*, 529–542. [CrossRef]

41. Zellweger, R.; Dalcher, D.; Mutreja, K.; Berti, M.; Schmid, J.A.; Herrador, R.; Vindigni, A.; Lopes, M. Rad51-mediated replication fork reversal is a global response to genotoxic treatments in human cells. *J. Cell Biol.* **2015**, *208*, 563–579. [CrossRef] [PubMed]

42. Bryant, H.E.; Petermann, E.; Schultz, N.; Jemth, A.S.; Loseva, O.; Issaeva, N.; Johansson, F.; Fernandez, S.; McGlynn, P.; Helleday, T. PARP is activated at stalled forks to mediate Mre11-dependent replication restart and recombination. *EMBO J.* **2009**, *28*, 2601–2615. [CrossRef] [PubMed]

43. Lemacon, D.; Jackson, J.; Quinet, A.; Brickner, J.R.; Li, S.; Yazinski, S.; You, Z.; Ira, G.; Zou, L.; Mosammaparast, N.; et al. MRE11 and EXO1 nucleases degrade reversed forks and elicit MUS81-dependent fork rescue in BRCA2-deficient cells. *Nat. Commun.* **2017**, *8*, 860. [CrossRef] [PubMed]

44. Kolinjivadi, A.M.; Sannino, V.; De Antoni, A.; Zadorozhny, K.; Kilkenny, M.; Techer, H.; Baldi, G.; Shen, R.; Ciccia, A.; Pellegrini, L.; et al. Smarcal1-Mediated Fork Reversal Triggers Mre11-Dependent Degradation of Nascent DNA in the Absence of Brca2 and Stable Rad51 Nucleofilaments. *Mol Cell.* **2017**, *67*, 867–881. [CrossRef]

45. Taglialatela, A.; Alvarez, S.; Leuzzi, G.; Sannino, V.; Ranjha, L.; Huang, J.W.; Madubata, C.; Anand, R.; Levy, B.; Rabadan, R.; et al. Restoration of Replication Fork Stability in BRCA1- and BRCA2-Deficient Cells by Inactivation of SNF2-Family Fork Remodelers. *Mol. Cell* **2017**, *68*, 414–430. [CrossRef]

46. Neelsen, K.J.; Lopes, M. Replication fork reversal in eukaryotes: From dead end to dynamic response. *Nat. Rev. Mol. Cell Biol.* **2015**, *16*, 207–220. [CrossRef]

47. Xu, R.; Xu, Y.; Huo, W.; Lv, Z.; Yuan, J.; Ning, S.; Wang, Q.; Hou, M.; Gao, G.; Ji, J.; et al. Mitosis-specific MRN complex promotes a mitotic signaling cascade to regulate spindle dynamics and chromosome segregation. *Proc. Natl. Acad. Sci. USA* **2018**, *115*, E10079–E10088. [CrossRef]

48. Syed, A.; Tainer, J.A. The MRE11-RAD50-NBS1 Complex Conducts the Orchestration of Damage Signaling and Outcomes to Stress in DNA Replication and Repair. *Annu. Rev. Biochem.* **2018**, *87*, 263–294. [CrossRef]

49. Rozier, L.; Guo, Y.; Peterson, S.; Sato, M.; Baer, R.; Gautier, J.; Mao, Y. The MRN-CtIP pathway is required for metaphase chromosome alignment. *Mol. Cell* **2013**, *49*, 1097–1107. [CrossRef]

50. De Angelis, M.L.; Zeuner, A.; Policicchio, E.; Russo, G.; Bruselles, A.; Signore, M.; Vitale, S.; De Luca, G.; Pilozzi, E.; Boe, A.; et al. Cancer Stem Cell-Based Models of Colorectal Cancer Reveal Molecular Determinants of Therapy Resistance. *Stem. Cells Transl. Med.* **2016**, *5*, 511–523. [CrossRef] [PubMed]

Article

Pre-Operative Decitabine in Colon Cancer Patients: Analyses on WNT Target Methylation and Expression

Janneke F. Linnekamp [1,2], Raju Kandimalla [1], Evelyn Fessler [1,†], Joan H. de Jong [1,2], Hans M. Rodermond [1,2], Gregor G. W. van Bochove [1,2], Frans O. The [3], Cornelis J. A. Punt [4], Willem A. Bemelman [5], Anthony W. H. van de Ven [5,6], Pieter J. Tanis [5], Elles M. Kemper [7], Lianne Koens [8], Evelien Dekker [9], Louis Vermeulen [1,2], Hanneke W. M. van Laarhoven [4] and Jan Paul Medema [1,2,*]

1   Laboratory for Experimental Oncology and Radiobiology (LEXOR), Center for Experimental and Molecular Medicine, Amsterdam UMC, University of Amsterdam, Cancer Center Amsterdam, Meibergdreef 9, 1105 AZ Amsterdam, The Netherlands; j.f.linnekamp@amsterdamumc.nl (J.F.L.); rajbiochem@gmail.com (R.K.); fessler@genzentrum.lmu.de (E.F.); j.h.dejong@amsterdamumc.nl (J.H.d.J.); h.rodermond@amsterdamumc.nl (H.M.R.); g.g.vanbochove@amsterdamumc.nl (G.G.W.v.B.); l.vermeulen@amsterdamumc.nl (L.V.)
2   Oncode Institute, 1105 AZ Amsterdam, The Netherlands
3   Department of Gastroenterology and Hepatology, OLVG, 1105 AZ Amsterdam, The Netherlands; fransolivier.the@usz.ch
4   Department of Medical Oncology, Cancer Center Amsterdam, Amsterdam UMC, University of Amsterdam, 1105 AZ Amsterdam, The Netherlands; c.j.a.punt@umcutrecht.nl (C.J.A.P.); h.vanlaarhoven@amsterdamumc.nl (H.W.M.v.L.)
5   Department of Surgery, Amsterdam UMC, University of Amsterdam, Cancer Center Amsterdam, 1105 AZ Amsterdam, The Netherlands; w.a.bemelman@amsterdamumc.nl (W.A.B.); a.w.vandeven@amsterdamumc.nl (A.W.H.v.d.V.); p.j.tanis@amsterdamumc.nl (P.J.T.)
6   Department of Surgery, Flevo Hospital Almere, 1315 RA Almere, The Netherlands
7   Department of Pharmacology, Amsterdam UMC, University of Amsterdam, 1105 AZ Amsterdam, The Netherlands; e.m.kemper@amsterdamumc.nl
8   Department of Pathology, Amsterdam UMC, University of Amsterdam, 1105 AZ Amsterdam, The Netherlands; l.koens@amsterdamumc.nl
9   Department of Gastroenterology and Hepatology, Amsterdam UMC, University of Amsterdam, Cancer Center Amsterdam, 1105 AZ Amsterdam, The Netherlands; e.dekker@amsterdamumc.nl
*   Correspondence: j.p.medema@amsterdamumc.nl
†   Current address: Gene Center and Department of Biochemistry, Ludwig-Maximilians-Universität München, 80539 Munich, Germany.

Citation: Linnekamp, J.F.; Kandimalla, R.; Fessler, E.; de Jong, J.H.; Rodermond, H.M.; van Bochove, G.G.W.; The, F.O.; Punt, C.J.A.; Bemelman, W.A.; van de Ven, A.W.H.; et al. Pre-Operative Decitabine in Colon Cancer Patients: Analyses on WNT Target Methylation and Expression. *Cancers* **2021**, *13*, 2357. https://doi.org/10.3390/cancers13102357

Academic Editors: Marta Baiocchi, Ann Zeuner and Isabelle Van Seuningen

Received: 31 January 2021
Accepted: 8 May 2021
Published: 13 May 2021

Publisher's Note: MDPI stays neutral with regard to jurisdictional claims in published maps and institutional affiliations.

Simple Summary: Colon cancer is one of the leading causes of cancer-related death worldwide. Therefore, the development of new therapeutic strategies is of the utmost importance. Previously, we identified a subset of colon cancers that are characterised by DNA methylation and have a poor prognosis. In this study, we therefore treated ten colon cancer patients with a demethylating agent, decitabine, to investigate if reversal of methylation is feasible and can be used as a novel therapy. Unfortunately, this study revealed that while decitabine treatment is effective in vitro, it only marginally decreased global methylation in patients and had no effect on the specific regions of DNA methylation in the tumours. Future studies should therefore focus on optimisation of treatment schedules in patients with highly methylated tumours.

Abstract: DNA hypermethylation is common in colon cancer. Previously, we have shown that methylation of WNT target genes predicts poor prognosis in stage II colon cancer. The primary objective of this study was to assess whether pre-operative treatment with decitabine can decrease methylation and increase the expression of WNT target genes *APCDD1*, *AXIN2* and *DKK1* in colon cancer patients. A clinical study was conducted, investigating these potential effects of decitabine in colon cancer patients (DECO). Patients were treated two times with 25 mg/m^2 decitabine before surgery. Methylation and expression of *LINE1* and WNT target genes (primary outcome) and expression of endogenous retroviral genes (secondary outcome) were analysed in pre- and post-treatment tumour samples using pyrosequencing and rt-PCR. Ten patients were treated with decitabine and eighteen patients were used as controls. Decitabine treatment only marginally decreased *LINE1* methylation. More importantly, no differences in methylation or expression of WNT target or endogenous retroviral genes were observed. Due to the lack of an effect on primary and secondary outcomes, the study was prematurely closed. In conclusion, pre-operative treatment with decitabine is safe, but with the current dosing, the primary objective, increased WNT target gene expression, cannot be achieved.

Keywords: decitabine; colon cancer; DNA methylation; clinical translation study

1. Introduction

The genetic aberrations in colon cancer have been extensively studied and are traditionally described by "the Vogelgram", starting with loss of functional *APC*, followed by mutations in other genes including *KRAS*, *TP53* and *SMAD4* [1]. In addition to genetic events, epigenetic alterations are frequently found and have been shown to be essential for the initiation and progression of colon cancer [2]. DNA methylation is associated with changes in the chromatin structure and results in altered gene expression without permanently changing the DNA sequence itself [3]. In various types of tumours, genome-wide hypomethylation mainly occurs in repetitive sequences and can lead to genomic instability [4,5]. In contrast, DNA hypermethylation occurs in CpG islands in promotor regions of specific genes, resulting in transcriptional silencing (e.g., tumour suppressor genes), methylation of *CDKN2A* in many cancers being an example [3,6,7]. Besides being an important step in tumourigenesis, DNA hypermethylation has also been suggested to cause resistance to systemic therapy [8,9].

In colon cancer, relevant tumour suppressor genes are epigenetically silenced by DNA hypermethylation. For example, silencing of *MLH1*, a DNA mismatch repair gene, results in microsatellite instable (MSI) tumours. Methylation of *MLH1* as well as other genes is encompassed in the CpG island Methylator Phenotype (CIMP). This phenotype is characterised by global hypermethylation and, in proximal tumours, is associated with worse prognosis [10,11]. Methylation of several other genes with biological, predictive or prognostic relevance has also been reported [12,13]. Previously, we have shown that methylation of the WNT target genes *APCDD1*, *AXIN2* and *DKK1* predicts poor prognosis in stage II colon cancer patients [14,15]. These genes can be methylated in both CIMP high, low and negative samples and are all negative regulators of the WNT pathway by negative feedback [16–18]. Therefore, inactivation of these genes by methylation can lead to activation of the WNT pathway. Importantly, even in *APC* mutant CRC, some level of WNT pathway regulation is still observed and inactivation of WNT pathway inhibitors is therefore thought to further tune the pathway.

DNA hypermethylation is facilitated by a group of enzymes called DNA methyltransferases (DNMTs) [19]. Azacitidine and decitabine are the best-known examples of DNMT inhibitors and are FDA-approved for myelodysplastic syndrome (MDS) and acute myeloid leukaemia [20,21]. In preclinical studies, we showed re-expression of WNT target genes in xenografted tumours after treatment of mice with the demethylating agent azacitidine [14]. Moreover, a subsequent decrease in tumour growth was observed [14]. These findings suggest that DNA methylation could be a therapeutic target in colon cancer and inducing re-expression could potentially lead to improved patient outcomes, especially in tumours characterised by extensive WNT target gene methylation.

The relationship between the clinical efficacy and the underlying molecular mechanisms of demethylating agents remains unclear, especially whether clinical response is a direct result of global demethylation [22,23]. The discrepancy between changes in methylation and clinical effect in several studies suggests that other factors in addition to methylation, such as immune regulation, are involved in patient response. One recent hypothesis is that endogenous retroviruses (ERV), integral parts of the human genome and silenced by methylation, are re-activated upon demethylation by DNMT inhibiting agents.

This results in an interferon-like immune response in tumour cells, which finally leads to cell death [23,24]. Whether this indeed explains the therapeutic effect in patients needs to be further investigated. Facilitating this immune recognition and response could therefore be a promising new strategy.

The aim of this study was to examine the effect of decitabine in colon cancer. A translational clinical study was conducted, investigating the effect of pre-operative decitabine on the methylation and expression of WNT target genes *APCDD1*, *AXIN2* and *DKK1* and global methylation in colon cancer patients.

2. Materials and Methods

2.1. Patient Recruitment and Inclusion Criteria

The DECO study (NCT01882660) was conducted from February 2014 until December 2017. The study was approved by the Medical Ethical Committee of the Academic Medical Center (AMC), Amsterdam. Patients were approached in the outpatient clinic from the AMC and Onze Lieve Vrouwe Gasthuis (OLVG) in Amsterdam and Flevo Ziekenhuis in Almere, all in the Netherlands. Initial clinical staging was performed based on CT scan. Diagnosis was based on endoscopical view, CT scan and/or biopsies, and indication for tumour resection was determined by a multidisciplinary panel. Eligible participants included both male and female patients of 18 years or older, with colon cancer, who had an indication for primary tumour resection. Other inclusion criteria for decitabine treatment included: Karnofsky Performance Score > 70, adequate bone marrow function and adequate hepatic and renal function. Finally, written informed consent had to be signed. Exclusion criteria included known hypersensitivity to decitabine or its additives or if surgery was not planned according to time frame of the study. Moreover, patients who received other systemic or local treatment of the primary tumour in the waiting time until surgery and administration of any experimental drug within 60 days prior to the first dose of decitabine were also excluded.

Pre-treatment samples were taken during endoscopy. For a detailed description of tumour samples, see Section 2.2. Ten $\pm$ two days before surgery, patients were treated with decitabine (kindly donated by Janssen-Cilag, The Netherlands) as two one-hour infusions at a dose of 25 mg/m^2 on two consecutive days. On the day of surgery, directly after resection, a second (post-treatment) sample was taken from the resected primary tumour. Furthermore, pathological staging was performed. Predefined primary endpoint was re-expression of WNT target genes (*APCDD1*, *ASCL2*, *AXIN2* and *DKK1*) measured by quantitative real-time PCR (rt-PCR) in both pre-treatment samples taken during endoscopy and compared with post-treatment samples taken directly after resection. Secondary endpoints included global (*LINE1*) and WNT target gene methylation (*APCDD1*, *ASCL2*, *AXIN2* and *DKK1*) and proliferation assessed by immunohistochemistry in the described pre- and post-treatment tumour samples. During the study, we performed a separate validation study on prediction of prognosis of WNT target gene methylation and showed no additional value of *ASCL2* in analyses [15]. Therefore, results of expression and methylation of *ASCL2* were not included in this study.

To investigate if tumour material from the same patient collected with different procedures (endoscopy vs. resection) was comparable considering methylation, expression and proliferation, a non-treated control cohort was included. Inclusion criteria were patients with colon cancer, older than 18 years, with a Karnofsky performance score > 70. Moreover, if an extra endoscopy procedure was performed, written informed consent was obtained.

2.2. Patient Samples

For pre-treatment samples, biopsies from endoscopy were used. Post-treatment samples were collected from resection specimens. In the decitabine-treated cohort, initially, only freshly frozen material was used. In order to obtain fresh-frozen pre-treatment biopsies, an extra endoscopy was performed before surgery. Due to the invasiveness of this procedure, we experienced a low inclusion rate. As a result, the protocol was amended after five patients were included. For the next five patients, FFPE material from a previously performed diagnostic endoscopy was used and compared with FFPE material from surgery.

For the control cohort, six patients with fresh-frozen material were enrolled before the amendment; however, three were excluded. Exclusion reasons were: neo-adjuvant treatment ($n = 1$), only tumour samples from endoscopy were freshly frozen ($n = 1$) and quality of material was insufficient ($n = 1$). We completed the control cohort with twelve colon cancer patients for whom FFPE material was stored. In total, 18 patients were enrolled for the control group, of which 15 could be evaluated for methylation and expression.

For fresh-frozen samples, tissues were immediately snap-frozen using liquid nitrogen and stored at -80 °C. FFPE samples were incubated in 4% buffered formaldehyde for a maximum of 24 h and then transferred to 70% ethanol. Thereafter, samples were dehydrated through 80%, 90%, 96% and 100% of ethanol and finally in 1-butanol and paraffin. For all tumours, multiple (2–5) pre- and post-treatment biopsies were obtained and tumour percentage was determined by HE staining. Two biopsies from each sampling were used for the final analyses. For *LINE1* methylation in treated patients, also a technical replicate was performed and results were averaged for final outcome. For all treated patients, MSI/MSS status, CIMP status and mutation of *TP53*, *KRAS* and *BRAF* were determined. Since the numbers of patients were low, no subgroup analysis could be performed.

Pre- and post-treatment blood samples were collected for haematological toxicity, and Common Terminology Criteria for Adverse Events were used to monitor other toxicities. Pre-treatment blood samples were taken at the time of diagnosis as standard of care and did not require an extra sample. Post-treatment blood samples were taken at the day of surgery.

2.3. DNA/RNA Isolation

Genomic DNA (gDNA) and RNA from the fresh-frozen patient samples were extracted with the AllPrep DNA/RNA/miRNA Universal Kit (Qiagen, Hilden, Germany) according to the manufacturer's instructions. For RNA, RNA integrity number values were determined using the Agilent 2100 bioanalyzer (see Table S1). FFPE tissue was cut into 10 μm sections and deparaffinised. gDNA was isolated using a Nucleospin DNA FFPE xs kit (Machery-Nagel, Düren, Germany) following the manufacturer's instructions.

2.4. Bisulfite Conversion and Pyrosequencing

Bisulfite conversion was performed with 600–800 ng of gDNA using EZ DNA Methylation-Gold Kit (Zymo research, Irvine, CA, USA) according to the manufacturer's protocol. For the PCR of the bisulfite converted DNA (bcDNA), PyroMark PCR kit (Qiagen, Hilden, Germany) was used. In short, 20 ng of bcDNA was mixed with kit reagents and a subsequent amplification was performed on a thermocycler. Annealing temperatures were adjusted for different primers: for *LINE1* and *AXIN2*, 56 °C was used; for *APCDD1*, 58 °C; and for *DKK1*, 52 °C. Next, pyrosequencing was performed using 12 ng of bcDNA. PyroMark Assay Design Software 2.0 (Qiagen, Hilden, Germany) was used for primer design and PCR and sequencing primers are listed in Table S2. For the WNT target genes, the exact location within the gene and CpG sites tested have been described before [15]. LINE1 sequence used was derived from Woloszynska-Read et al. [25]. The sequence analysed was 206–352 (Genbank accession number X52235.1) and contained three CpG sites. For validation, primers were also tested on DNA isolated from FFPE material and compared to DNA obtained from freshly frozen tissue before analysing patient material. This material originated from a previously conducted study, in which xenografts obtained from multiple colon cancer cell lines were used [26]. Results for the validation are shown in Figure S1 and show perfect correlation in methylation levels between the two distinct sample preparations.

2.5. Quantitative Real-Time PCR

Complementary DNA (cDNA) was synthesised from 1 μg of RNA using Superscript III reverse transcriptase (ThermoFisher, Waltham, MA, USA). For quantitative real-time PCR, 5 ng of cDNA was used in a total reaction volume of 5 μl containing 2.5 μL of SYBR green and 0.5 μM forward and reverse primer (see Table S3). Reaction was performed in a Lightcycler LC480 II (Roche).

2.6. Ki67 Staining

FFPE samples were used for Ki67 staining. Sections of a thickness of 4 μm were prepared and deparaffinised using xylene and rehydrated through ethanol. Antigen retrieval was achieved using 10 mM sodium citrate buffer (pH = 6) (Vector Laboratories, Burlingame, CA, USA) for 20 min at 98 °C. Samples were blocked using Dako REAL Peroxidase-Blocking Solution (Agilent technologies, Santa Clara, CA, USA) for 5 min at room temperature. Ki67 antibody (Sigma, SAB5500134, Saint Louis, MI, USA) was diluted 1:1000, in normal antibody diluent (Klinipath, ABB999, Duiven, The Netherlands), and incubated overnight at 4 °C. After washing with PBS, poly HRP-anti Rabbit IgG (Bright vision, DPVR-55HRP, Immunologic, Duiven, The Netherlands) was added for 30 min at room temperature and finally stained using Bright DAB solution (3,3′ diaminobenzidine, Immunologic, Duiven, The Netherlands). Counterstaining with haemotoxylin (Klinipath, 4085–9002, Duiven, The Netherlands) was incubated for 1 min. After dehydration, slides were mounted using Pertex (HistoLab, Västra Frölunda, Sweden). For material from one patient (patient 9), no staining could be performed due to low quality of the material. For quantification of stainings, haemotoxylin colour was separated from DAB using the plugin "color deconvolution" in ImageJ. Positive nuclei were calculated as a percentage of total nuclei.

2.7. CIMP Analysis

For CIMP analyses, a panel of eight genes was used containing *CACNA1G*, *CDKN2A*, *CRABP1*, *IGF2*, *MLH1*, *NEUROG1*, *RUNX3* and *SOCS1*. Furthermore, *ALU* was used to normalise for the amount of input bcDNA. Methods, primers and probes have been described previously [26,27]. Percentage of methylated reference (PMR) > 10 was considered as positive. Tumours where 1–5 out of 8 CIMP markers had a PMR > 10 were defined as CIMP low. Tumours that had ≥6 out of 8 markers with a PMR > 10 were defined as CIMP high. Tumours were considered CIMP negative if none of the markers had a PMR > 10.

2.8. MSI/MSS

Microsatellite stability was tested during standard of clinical care in the pathology department of our institute using immunohistochemistry for MLH1, MSH2, MSH6 and PMS2. If no results were available, a PCR-based MSI Analysis System, version 1.2 (Promega, Leiden, The Netherlands), was used. In this assay, the markers NR-21, BAT-26, BAT-25, NR-24, MONO-27, Penta C and Penta D were used. Assays were performed according to the manufacturers' instructions. Samples were considered as microsatellite instable if no staining was present in one of the four immunohistochemical stainings or more than 2 out of 5 markers of the PCR based analyses were instable.

2.9. Mutational Status

For mutational status, we used tumour samples obtained from resection. *KRAS* exon 2 and 3 and *TP53* exon 1–11 were amplified by PCR, using 20 ng of gDNA (*KRAS*) or cDNA (*TP53*), 12.5 μL Reddymix (ThermoFisher scientific), 1 μL forward primer and reverse primer 10 μM and 8.5 μL H₂O in a total volume of 25 μL. For *KRAS*, thermocycler program was as follows: 5 min 95 °C, 40 cycles of 30 s 95 °C, 30 s 50 °C, 1 min 30 s 72 °C, followed by 5 min 72 °C. We used the same protocol for *TP53* only with an annealing temperature of 60 °C. Then, 0.1 μL of the PCR product was sequenced by Big Dye Terminator 1.1 and subsequently analysed by direct Sanger sequencing. Primers are listed in Table S4. *BRAF* mutation was tested via quantitative rt-PCR with a wild type and *BRAF* V600E specific primer (Table S4). Reaction was performed using SYBR green by Lightcycler 480. Ct value from *BRAF* mutant was subtracted with Ct value of *BRAF* wild type. Samples with differences of < 4 Ct values were considered as *BRAF* mutant.

2.10. Statistical Analyses

The planned maximum sample size for the decitabine treatment group as well as the control cohort was 44 with a 10% loss due to insufficient quality of material. We aimed to include twenty patients with high methylation of WNT target genes and twenty with lowly methylated WNT target genes. The group size was determined based on the incidence of methylation and the expected effect size. We expected a quarter of the tumours to be highly methylated for at least one gene based on results from a previous study [15]. The first interim analysis was performed after ten patients were included and treated with decitabine. For statistical analyses, GraphPad Prism 7 was used. To study the biological effect of decitabine in the clinical samples, we used a paired *t*-test to evaluate significant differences between pre- and post-treatment samples. For comparing the results of the Ki67 staining, a paired *t*-test was used. For all statistical comparisons, the level of significance was set at $p < 0.05$.

3. Results

3.1. Patient Characteristics

To determine whether decitabine could decrease methylation and thereby re-express WNT target genes, a clinical trial (DECO) was conducted between February 2014 and December 2017. A total of ten colon cancer patients, nine male and one female, were enrolled and pre-operatively treated with decitabine, after which we performed an interim analysis that is reported here. Baseline characteristics and flow chart for inclusion are presented in Figure 1.

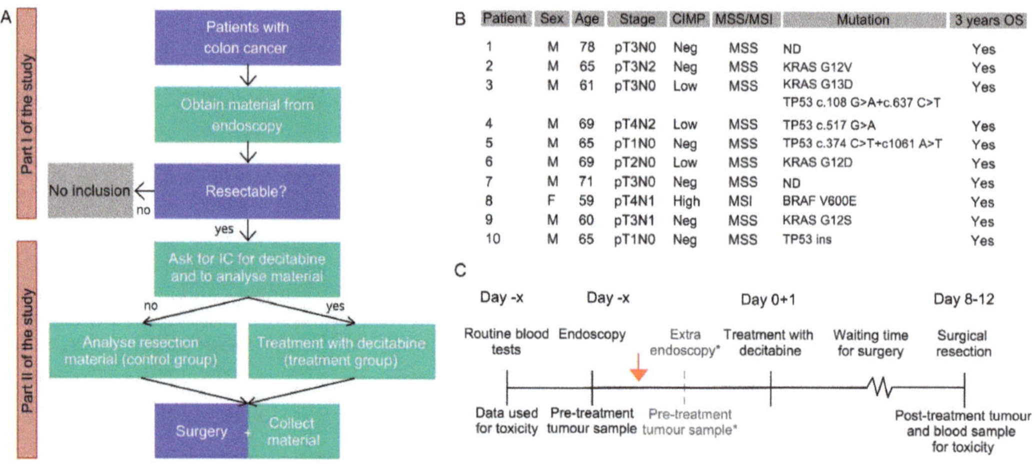

B Patient	Sex	Age	Stage	CIMP	MSS/MSI	Mutation	3 years OS
1	M	78	pT3N0	Neg	MSS	ND	Yes
2	M	65	pT3N2	Neg	MSS	KRAS G12V	Yes
3	M	61	pT3N0	Low	MSS	KRAS G13D TP53 c.108 G>A+c.637 C>T	Yes
4	M	69	pT4N2	Low	MSS	TP53 c.517 G>A	Yes
5	M	65	pT1N0	Neg	MSS	TP53 c.374 C>T+c1061 A>T	Yes
6	M	69	pT2N0	Low	MSS	KRAS G12D	Yes
7	M	71	pT3N0	Neg	MSS	ND	Yes
8	F	59	pT4N1	High	MSI	BRAF V600E	Yes
9	M	60	pT3N1	Neg	MSS	KRAS G12S	Yes
10	M	65	pT1N0	Neg	MSS	TP53 ins	Yes

Figure 1. (**A**) Flow chart of DECO study. In blue squares, daily routine steps in clinical care are shown; in green squares, additional steps for DECO study are shown; (**B**) Baseline characteristics of decitabine-treated patients included in DECO study. Indicated stage is pathological staging after resection was done. Initial staging (for inclusion) was performed based on CT scan; (**C**) Timeline for DECO study. Red arrow shows the moment of inclusion. Blood tests before inclusion were part of standard of care. * For the patients with fresh frozen material, an extra endoscopy was performed to obtain freshly frozen biopsies as pre-treatment sample. For FFPE patients, pre-treatment samples were obtained from the diagnostic endoscopy performed for clinical purposes. IC = informed consent, ND = not detected, OS = overall survival.

The median age was 65 years (range 59–78) and all patients were evaluable for toxicity, and, from all patients, material was available for methylation analyses. Furthermore, a non-decitabine-treated control cohort (n = 18) was enrolled, of which 15 patients were eligible for analyses. Although clinical outcome was not an endpoint of this study, three-year overall survival was documented and all patients reached this endpoint. In the follow-up after four years, two events occurred.

3.2. Effect of Pre-Operative Treatment with Decitabine on Methylation in Colon Cancer Patients

Before performing analyses on our primary endpoint, we verified if biopsies had a comparable percentage of methylation to tumour samples from resection using a control cohort. In this cohort, a total of 18 patients were included, of which 15 patients were available for analyses. To determine levels of global methylation, *LINE1* methylation was used as a surrogate marker [28]. In this cohort, the average of *LINE1* methylation for biopsies was 69.0 ± 6.1% and resection material was 69.0 ± 4.3% (n = 15). Paired analysis revealed that biopsies and resection material could be directly compared, showing no statistical differences in methylation (p = 0.9718) (Figure 2A). Next, *LINE1* methylation was assessed in decitabine-treated patients (n = 10). The average *LINE1* methylation from the ten patients before treatment, as analysed on the biopsy material, was 71.2 ± 6.4%, while after treatment, the average was numerically lower (67.2 ± 6.5%). Paired analysis of the patients indicated that all but one patient showed a decrease in methylation and that this was significant when analysing the group (p = 0.0075) (Figure 2A). Nevertheless, this decrease was relatively small for all patients tested, indicating that decitabine could modulate *LINE1* methylation, but with the dosing used, the impact was minimal. To determine the impact of decitabine on WNT target gene methylation, CpG methylation of *APCDD1*, *AXIN2* and *DKK1* was measured in the first five patients from which fresh-frozen samples were available. Importantly, analysis of five patient sample pairs showed similar WNT target CpG methylation in pre- and post-treatment samples (Figure 2B) and no clear decrease could be detected. However, firm conclusions cannot be drawn for these data due to low patient numbers and the fact that these patients did not display high WNT target methylation at the start of treatment.

3.3. Effect of Pre-Operative Treatment with Decitabine on Gene Expression and Proliferation in Colon Cancer Patients

Despite the fact that only a small difference in LINE1 methylation and no clear impact on WNT target methylation could be detected, differences in gene expression or cell biological features, such as cell proliferation, could potentially be orchestrated without overt changes in methylation. To this end, WNT target gene and LINE1 expression was first analysed with quantitative rt-PCR in the fresh-frozen samples (n = 5). This revealed that both LINE1 and WNT target gene expression were not significantly different between pre- and post-treatment tumour samples (Figure 2C).

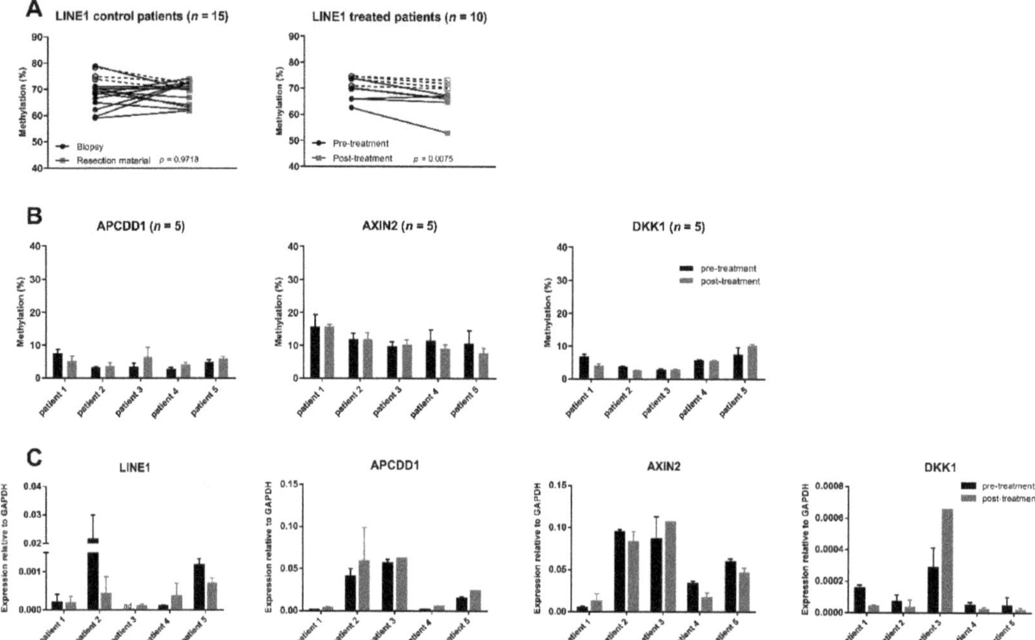

Figure 2. (**A**) Methylation of *LINE1* in the control group (n = 15) and in the treated group (n = 10) (before and after treatment with 25 mg/m^2 decitabine two times) measured by pyrosequencing. In both cohorts, FF tumour samples and FFPE samples were included. Open symbols and dotted lines represent FF samples, and closed symbols and lines represent FFPE samples. In the treated cohort, two technical replicates per time point were averaged and two biological replicates (two different samples from the same tumour) were used. For patient 6 to patient 10, no biological replicate was available for the pre-treatment sample. For statistical analyses, for pre-treatment and post-treatment samples, the average of all measurements was used. A paired *t*-test revealed no significant difference in the control cohort (p = 0.9718). For the treated cohort, a significant (p = 0.0075) difference was shown; (**B**) Methylation of WNT target genes before and after treatment with decitabine in colon cancer patients measured by pyrosequencing (n = 5); (**C**) Expression of *LINE1* and WNT target genes after treatment with decitabine measured by quantitative rt-PCR in fresh-frozen samples (n = 5). Values are the average of two samples (both for pre- and post-treatment samples), except for patient 3, where only one post-treatment sample was available.

A reduction in proliferation after treatment with decitabine has been reported [29] and could also significantly impact the efficacy of demethylation as this is suggested to require cell cycle progression. Therefore, immunohistochemical staining for Ki67 was used to determine the expression on protein level (Figure 3). A large variation in Ki67 positive cells between patients as well as tumour region was observed, which is in line with earlier results [30]. However, no consistent difference between pre-treatment samples and post-treatment samples was detected (Figure 3B) (p = 0.7618). Although this may relate to the relatively small group size, we conclude that a strong impact on proliferation was not detected. This likely aligns with the lack of impact of a short course decitabine treatment on tumour growth.

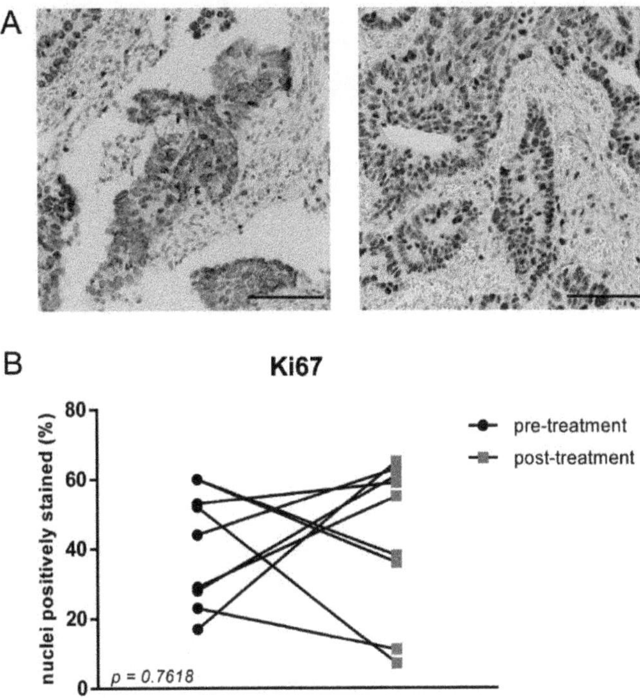

Figure 3. (**A**) Representable Ki67 staining of biopsy and resection material of the tumour from one patient. The scale bar represents 100 μm; (**B**) Percentage of Ki67 positive cells compared to total cells from nine treated patients. A representative area of the tumour block was used for quantification. No significant difference between pre- and post-treatment samples (*p* = 0.7618) was observed using a paired *t*-test.

3.4. Decitabine Does Not Induce Expression of Endogenous Retrovirus ERVL and Interferon Associated Genes in Colon Cancer Patients

Endogenous retroviruses (ERVs) are heavily encoded in our human genome and effectively silenced by CpG methylation. Recent evidence suggests that decitabine can result in effective demethylation of these silencing CpG islands and result in reactivation of ERVs [23,24]. The cellular response towards ERV reactivation is rapid induction of interferon and interferon-related gene expression mounting an anti-viral and, as a result, anti-tumour immune response [23]. Importantly, as recent studies have also suggested that the inhibition of immune checkpoints in a neo-adjuvant setting is effective in colorectal cancer [31], we wondered whether decitabine could activate ERVs and hence provide an anti-tumour response. Therefore, the impact of decitabine on the gene expression of interferon-related genes and ERV *ERVL* was assessed in our patient tumour samples (*n* = 5; DECO patient 1–5). However, although the patient numbers were limited, neither the reactivation of *ERVL* nor the activation of the interferon response was evident (Figure 4), suggesting that the levels of decitabine used in this study do not lead to ERV activation.

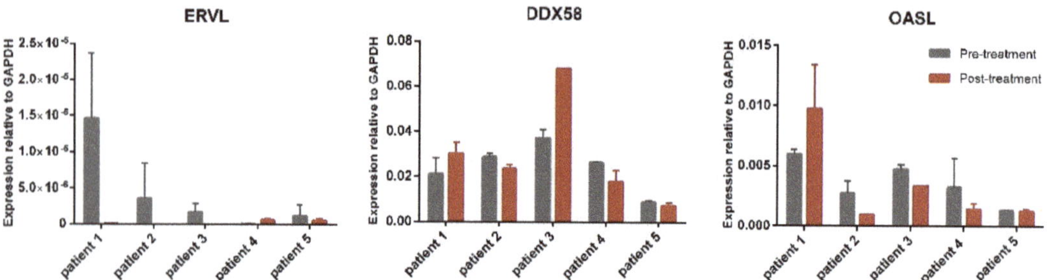

Figure 4. Expression of *ERVL* and interferon genes *DDX58* and *OASL* in pre- and post-treatment samples (*n* = 2 per patients except from patient 3) from patients treated with decitabine. Only fresh-frozen samples were used.

3.5. Toxicity

Grade 1 adverse events are summarised in Table S5 and pre- and post-treatment laboratory tests of the patients in Table S6. In addition, decitabine administration had no effect on the timing of surgery. This indicates that decitabine can be used safely pre-operatively at the concentrations and timing used in this study.

3.6. Study Closure

Due to low patient inclusion, we amended the protocol after two years to use FFPE material to avoid the need for an extra endoscopy for patients. After including ten patients (not pre-specified) for the decitabine arm, the current analysis was performed. This revealed that decitabine treatment with the employed scheme did not result in demethylation and/or subsequent upregulation of WNT target genes partially controlled by methylation. The effect of decitabine on *LINE1* methylation was significant, yet too limited to be impactful when analysing the expression of *LINE1*, while we also did not observe an impact on WNT target methylation or expression or on *ERV* expression. Initially, we anticipated to include forty evaluable patients for decitabine treatment, aiming to change WNT target expression and methylation. However, with the results of the first ten patients, a different conclusion after forty patients was unlikely and we closed the study for further patient inclusion to avoid unnecessary impacts on patients.

4. Discussion

For early-stage colon cancer, surgery remains the cornerstone of treatment. However, in the case of stage III or high-risk stage II disease, adjuvant treatment with cytotoxic drugs improves patient outcomes. Although an overall survival benefit for cytotoxic treatment in these groups has been clearly documented, the proportion of patients with increased survival because of adjuvant therapy remains low. In clinical stage I-III colon cancer, neo-adjuvant therapy is currently not standard therapy, but a recent study shows that neo-adjuvant FOLFOX is safe, with no increase in perioperative morbidity [32]. This not only paves the way for studies assessing the long-term benefit of this strategy but also for the use of neo-adjuvant treatment for expeditious evaluation of response and thus for the development of new therapeutic options in colon cancer. CpG hypermethylation is an important event in tumourigenesis and its reversible nature makes it an attractive target for therapy. In addition to its role in tumour progression, CpG hypermethylation is associated with poor prognosis [14,33,34], underlining the relevance of studying the effect of demethylating agents in colon cancer.

In this study, the biological effect of the demethylating drug, decitabine, was studied in colon cancer patients. A translational clinical study was conducted in which patients were treated with decitabine prior to surgery. The impact on *LINE1* methylation using this treatment was significant but, compared to earlier studies, very small. In agreement, the observed decrease did not lead to an increase in *LINE1* gene expression, nor did we observe a change in methylation of WNT target genes or WNT target gene or ERV expression. The minor effect on methylation was further corroborated by the observation that tumour proliferation did not change in the treated patients. In conclusion, this suggests that dosing decitabine twice at the concentration used, and after 8–12 days, is not sufficient to obtain impactful changes in tumour cells. It should be mentioned that the study was, due to pre-mature closure, underpowered. Furthermore, none of the patients analysed for WNT target gene methylation showed high levels of methylation for these genes and this could hamper the effect of decitabine. However, this level of methylation was not unexpected, as a previous study showed positive WNT target gene methylation in 26% of the tumours [15]. Nevertheless, the low methylation levels were also not affected by decitabine treatment, nor was the expression of the WNT target genes. In line with the limited effect on demethylation, a re-activation of endogenous retrovirus *ERVL* as well as the linked interferon response was not observed in tumour material from a limited number of colon cancer patients treated with decitabine.

There are several explanations for the limited effect of decitabine in the patients in our study. Firstly, to prevent toxicity and, more importantly, an impact on the timing or success of the surgery, a relatively low dose of decitabine at only two injections was used, which could have resulted in a relatively low effective concentration in the patients. However, data from previous clinical studies suggest that the dose and timing that was used could be appropriate for demethylation in patients with solid tumours or myelodysplastic syndrome [35,36]. Nevertheless, repetitive administration and longer time intervals have been reported to optimise the effect [35,37–39]. Another explanation for our findings could be that decitabine is less effective in tumour tissue than in PBMCs, which are commonly used to monitor the effect on DNA methylation in patients [35,40,41]. Studies that compared the effect of methylation in PBMCs with tumour samples are limited and show conflicting results [35,42–45]. It is therefore difficult to extrapolate results from PBMCs to tumour samples as PBMCs are, by virtue of their location, more accessible for decitabine. PBMCs were not collected during our study as our focus was on the resected tumour tissue. However, for additional information about treatment schedules, collection of PBMCs might be useful in future studies. Moreover, collection of circulating tumour DNA could be insightful to evaluate response on methylation and prevent invasive biopsies [42,45–47]. Finally, a lower proliferation rate in tumour cells in patients compared to xenografts or in vitro cultures could also impair the effect of decitabine. Decitabine is only active during cell proliferation and demethylation is progressive with each cell division. Although the proliferation rate in these patient samples measured by Ki67 staining was relatively high, this still could be lower than in vitro. Nevertheless, no correlation between Ki67 staining in pre-treatment material and effect on *LINE* methylation was observed (data not shown).

The findings of a demethylating agent in colon cancer patients in this study are in line with several previously reported clinical studies in solid tumours [48]. Thus far, four studies on demethylating agents in colon cancer have been conducted. In a clinical trial using a combination of decitabine with panitumumab, a 10% response rate was shown, but no effect on *MAGE* re-expression was observed [9]. In another study, capecitabine and oxaliplatin were combined with azacitidine in twenty-six colon cancer patients [46]. In this study, no objective response was observed, neither in CIMP high nor in CIMP low patients. In 60% of patients, methylation of vimentin was decreased; however, this effect was limited and did not outperform technical variation of methylation testing. More recently, guadecitabine was combined with irinotecan to treat metastatic colorectal cancer patients [37]. No consistent *LINE1* demethylation was detected in tumour biopsies or circulating tumour DNA after 8 days. However, a reduction was seen after 15 days, although no correlation with clinical response could be observed [37]. This was in line with results from a study with 47 colon cancer patients, where demethylation was shown in post-treatment samples but was unrelated to response or overall survival [49]. Overall, the results of trials with demethylating agents in colon cancer are disappointing, although responses in individual patients are seen, emphasizing the importance of biomarkers to predict response or to find synergy with other drugs, e.g., immune checkpoint inhibitors.

Despite our findings, the study setup in which patients received neo-adjuvant treatment for a short period and tissue pre- and post-treatment was analysed is of interest for future drug studies. Our method facilitated the measurement of treatment response in colon cancer patients on primary tumour tissue on individual basis in a short time frame. Recent data on neo-adjuvant immunotherapy in colorectal cancer patients support this approach, and a more extensive analysis of the role of neo-adjuvant therapy in colon cancer is warranted. This study setup also allows for a quick evaluation of hypotheses and drugs that emanate from preclinical work. If an effect is detected, this strategy would allow for a rapid dissemination and identification of biomarkers to select patients for certain treatments. Thereby, this setup could potentially be used to personalise adjuvant treatment in colon cancer.

5. Conclusions

No decrease in WNT target gene methylation was observed after short-term pre-operative treatment with decitabine in a limited amount of tumour tissue from five colon cancer patients. Future methylation studies should focus on optimisation of treatment regimens in patients with highly methylated tumours and perform parallel collection of PBMCs with tumour material.

Supplementary Materials: The following are available online at https://www.mdpi.com/article/10.3390/cancers13102357/s1, Figure S1: Correlation between methylation of WNT target genes in DNA from fresh frozen (FF) and Formalin Fixed Paraffin Embedded (FFPE) samples from the same xenograft. All genes showed a high correlation, indicated with the correlation efficient r2 which was significant, Table S1: RIN values of samples used for quantitative real-time PCR. Table S2: Sequencing primers used for pyrosequencing, Table S3: Primers used for quantitative real-time PCR, Table S4: Primers used for mutation analyses, Table S5: Adverse events, Table S6: Laboratory test from pre- and posttreatment.

Author Contributions: Conceptualisation, J.F.L., F.O.T., C.J.A.P., A.W.H.v.d.V., P.J.T., E.M.K., E.D., L.V., H.W.M.v.L. and J.P.M.; Formal analysis, J.F.L., R.K. and J.P.M.; Funding acquisition, H.W.M.v.L. and J.P.M.; Investigation, J.F.L., R.K., J.H.d.J., H.M.R. and G.G.W.v.B.; Methodology, R.K., E.F. and J.P.M.; Project administration, J.F.L.; Resources, J.F.L., E.F., J.H.d.J., H.M.R., G.G.W.v.B., F.O.T., W.A.B., A.W.H.v.d.V., P.J.T., E.M.K., L.K. and E.D.; Supervision, E.D., L.V., H.W.M.v.L. and J.P.M.; Validation, L.K.; Visualisation, J.F.L.; Writing—original draft, J.F.L., H.W.M.v.L. and J.P.M.; Writing—review and editing, E.F., F.O.T., C.J.A.P., W.A.B., A.W.H.v.d.V., P.J.T., E.M.K., L.K., E.D. and L.V. All authors have read and agreed to the published version of the manuscript.

Funding: This research was funded by the KWF grant from the Dutch Cancer Society, grant number KWF 2012-542, and the Dutch Gastrointestinal and Liver disorder Foundation, grant number MLDS FP13-07, awarded to Jan Paul Medema. L.V. is a New York Stem Cell Foundation—Robertson Investigator.

Institutional Review Board Statement: The study was conducted according to the guidelines of the Declaration of Helsinki and approved by the Ethics Committee of the Academic Medical Center Amsterdam (2013_060, NL44048.018.13, Eudract number 2013-001060-38, 28 May 2013).

Informed Consent Statement: Informed consent was obtained from all subjects involved in the study.

Data Availability Statement: No new data were created or analyzed in this study. Data sharing is not applicable to this article.

Acknowledgments: We would like to thank M.W. Mundt for his help with patient inclusion. We would also like to thank A. Karpf for the help in sequence alignment. Decitabine was kindly donated by Janssen-Cilag.

Conflicts of Interest: The authors declare no conflict of interest. The funders had no role in the design of the study; in the collection, analyses, or interpretation of data; in the writing of the manuscript, or in the decision to publish the results.

References

1. Fearon, E.R.; Vogelstein, B. A genetic model for colorectal tumorigenesis. *Cell* **1990**, *61*, 759–767. [CrossRef]
2. Okugawa, Y.; Grady, W.M.; Goel, A. Epigenetic Alterations in Colorectal Cancer: Emerging Biomarkers. *Gastroenterology* **2015**, *149*, 1204–1225.e12. [CrossRef]
3. Herman, J.G.; Baylin, S.B. Gene silencing in cancer in association with promoter hypermethylation. *N. Engl. J. Med.* **2003**, *349*, 2042–2054. [CrossRef]
4. Feinberg, A.P.; Vogelstein, B. Hypomethylation distinguishes genes of some human cancers from their normal counterparts. *Nature* **1983**, *301*, 89–92. [CrossRef]
5. Gama-Sosa, M.A.; Slagel, V.A.; Trewyn, R.W.; Oxenhandler, R.; Kuo, K.C.; Gehrke, C.W.; Ehrlich, M. The 5-methylcytosine content of DNA from human tumors. *Nucleic Acids Res.* **1983**, *11*, 6883–6894. [CrossRef]
6. Jones, P.A. Functions of DNA methylation: Islands, start sites, gene bodies and beyond. *Nat. Rev. Genet.* **2012**, *13*, 484–492. [CrossRef]
7. Herman, J.G.; Merlo, A.; Mao, L.I.; Lapidus, R.G.; Issa, J.P.J.; Davidson, N.E.; Sidransky, D.; Baylin, S.B. Inactivation of the CDKN2/p16/MTS1 gene is frequently associated with aberrant DNA methylation in all common human cancers. *Cancer Res.* **1995**, *55*, 4525–4530.
8. Glasspool, R.M.; Teodoridis, J.M.; Brown, R. Epigenetics as a mechanism driving polygenic clinical drug resistance. *Br. J. Cancer* **2006**, *94*, 1087–1092. [CrossRef]
9. Garrido-Laguna, I.; McGregor, K.A.; Wade, M.; Weis, J.; Gilcrease, W.; Burr, L.; Soldi, R.; Jakubowski, L.; Davidson, C.; Morrell, G.; et al. A phase I/II study of decitabine in combination with panitumumab in patients with wild-type (wt) KRAS metastatic colorectal cancer. *Investig. New Drugs* **2013**, *31*, 1257–1264. [CrossRef]
10. Ahn, J.B.; Chung, W.B.; Maeda, O.; Shin, S.J.; Kim, H.S.; Chung, H.C.; Kim, N.K.; Issa, J.-P.J. DNA methylation predicts recurrence from resected stage III proximal colon cancer. *Cancer* **2011**, *117*, 1847–1854. [CrossRef]
11. Juo, Y.Y.; Johnston, F.M.; Zhang, D.Y.; Juo, H.H.; Wang, H.; Pappou, E.P.; Yu, T.; Easwaran, H.; Baylin, S.; van Engeland, M.; et al. Prognostic value of CpG island methylator phenotype among colorectal cancer patients: A systematic review and meta-analysis. *Ann. Oncol.* **2014**, *25*, 2314–2327. [CrossRef] [PubMed]
12. Lam, K.; Pan, K.; Linnekamp, J.F.; Medema, J.P.; Kandimalla, R. DNA methylation based biomarkers in colorectal cancer: A systematic review. *Biochim. Biophys. Acta* **2016**, *1866*, 106–120. [CrossRef] [PubMed]
13. Draht, M.X.G.; Goudkade, D.; Koch, A.; Grabsch, H.I.; Weijenberg, M.P.; Van Engeland, M.; Melotte, V.; Smits, K.M. Prognostic DNA methylation markers for sporadic colorectal cancer: A systematic review. *Clin. Epigenetics* **2018**, *10*, 35. [CrossRef]
14. de Sousa, E.M.F.; Colak, S.; Buikhuisen, J.; Koster, J.; Cameron, K.; de Jong, J.H.; Tuynman, J.B.; Prasetyanti, P.R.; Fessler, E.; van den Bergh, S.P.; et al. Methylation of cancer-stem-cell-associated Wnt target genes predicts poor prognosis in colorectal cancer patients. *Cell Stem Cell* **2011**, *9*, 476–485.
15. Kandimalla, R.; Linnekamp, J.F.; Van Hooff, S.; Castells, A.; Llor, X.; Andreu, M.; Jover, R.; Goel, A.; Medema, J.P. Methylation of WNT target genes AXIN2 and DKK1 as robust biomarkers for recurrence prediction in stage II colon cancer. *Oncogenesis* **2017**, *6*, e308. [CrossRef]
16. Shimomura, Y.; Agalliu, D.; Vonica, A.; Luria, V.; Wajid, M.; Baumer, A.; Belli, S.; Petukhova, L.; Schinzel, A.; Brivanlou, A.H.; et al. APCDD1 is a novel Wnt inhibitor mutated in hereditary hypotrichosis simplex. *Nature* **2010**, *464*, 1043–1047. [CrossRef]
17. Behrens, J.; Jerchow, B.-A.; Würtele, M.; Grimm, J.; Asbrand, C.; Wirtz, R.; Kühl, M.; Wedlich, D.; Birchmeier, W. Functional interaction of an axin homolog, conductin, with beta-catenin, APC, and GSK3beta. *Science* **1998**, *280*, 596–599. [CrossRef]
18. Zorn, A.M. Wnt signalling: Antagonistic Dickkopfs. *Curr. Biol.* **2001**, *11*, R592–R595. [CrossRef]
19. Okano, M.; Bell, D.W.; Haber, D.A.; Li, E. DNA methyltransferases Dnmt3a and Dnmt3b are essential for de novo methylation and mammalian development. *Cell* **1999**, *99*, 247–257. [CrossRef]
20. Gore, S.D.; Jones, C.; Kirkpatrick, P. Decitabine. *Nat. Rev. Drug Discov.* **2006**, *5*, 891–892. [CrossRef]
21. Issa, J.P.; Kantarjian, H.M.; Kirkpatrick, P. Azacitidine. *Nat. Rev. Drug Discov.* **2005**, *4*, 275–276. [CrossRef] [PubMed]
22. Issa, J.P.; Kantarjian, H.M. Targeting DNA methylation. *Clin. Cancer Res.* **2009**, *15*, 3938–3946. [CrossRef]
23. Roulois, D.; Yau, H.L.; Singhania, R.; Wang, Y.; Danesh, A.; Shen, S.Y.; Han, H.; Liang, G.; Jones, P.A.; Pugh, T.J.; et al. DNA-Demethylating Agents Target Colorectal Cancer Cells by Inducing Viral Mimicry by Endogenous Transcripts. *Cell* **2015**, *162*, 961–973. [CrossRef] [PubMed]
24. Chiappinelli, K.B.; Strissel, P.L.; Desrichard, A.; Li, H.; Henke, C.; Akman, B.; Hein, A.; Rote, N.S.; Cope, L.M.; Snyder, A.; et al. Inhibiting DNA Methylation Causes an Interferon Response in Cancer via dsRNA Including Endogenous Retroviruses. *Cell* **2015**, *162*, 974–986. [CrossRef]
25. Woloszynska-Read, A.; Mhawech-Fauceglia, P.; Yu, J.; Odunsi, K.; Karpf, A.R. Intertumor and intratumor NY-ESO-1 expression heterogeneity is associated with promoter-specific and global DNA methylation status in ovarian cancer. *Clin. Cancer Res.* **2008**, *14*, 3283–3290. [CrossRef] [PubMed]
26. Linnekamp, J.F.; Hooff, S.R.V.; Prasetyanti, P.R.; Kandimalla, R.; Buikhuisen, J.Y.; Fessler, E.; Ramesh, P.; Lee, K.; Bochove, G.G.W.; de Jong, J.H.; et al. Consensus molecular subtypes of colorectal cancer are recapitulated in in vitro and in vivo models. *Cell Death Differ.* **2018**, *25*, 616–633. [CrossRef] [PubMed]

27. Ogino, S.; Kawasaki, T.; Kirkner, G.J.; Kraft, P.; Loda, M.; Fuchs, C.S. Evaluation of markers for CpG island methylator phenotype (CIMP) in colorectal cancer by a large population-based sample. *J. Mol. Diagn.* **2007**, *9*, 305–314. [CrossRef] [PubMed]

28. Yang, A.S.; Estecio, M.R.; Doshi, K.; Kondo, Y.; Tajara, E.H.; Issa, J.P. A simple method for estimating global DNA methylation using bisulfite PCR of repetitive DNA elements. *Nucleic Acids Res.* **2004**, *32*, e38. [CrossRef] [PubMed]

29. Stresemann, C.; Brueckner, B.; Musch, T.; Stopper, H.; Lyko, F. Functional diversity of DNA methyltransferase inhibitors in human cancer cell lines. *Cancer Res.* **2006**, *66*, 2794–2800. [CrossRef] [PubMed]

30. Prasetyanti, P.R.; van Hooff, S.R.; van Herwaarden, T.; de Vries, N.; Kalloe, K.; Rodermond, H.; van Leersum, R.; de Jong, J.H.; Franitza, M.; Nürnberg, P.; et al. Capturing colorectal cancer inter-tumor heterogeneity in patient-derived xenograft (PDX) models. *Int. J. Cancer* **2019**, *144*, 366–371. [CrossRef]

31. Chalabi, M.; Fanchi, L.F.; Dijkstra, K.K.; Berg, J.G.V.D.; Aalbers, A.G.; Sikorska, K.; Lopez-Yurda, M.; Grootscholten, C.; Beets, G.L.; Snaebjornsson, P.; et al. Neoadjuvant immunotherapy leads to pathological responses in MMR-proficient and MMR-deficient early-stage colon cancers. *Nat. Med.* **2020**, *26*, 566–576. [CrossRef]

32. Seymour, M.T.; Morton, D.; International FOxTROT Trial Investigators. FOxTROT: An international randomised controlled trial in 1052 patients (pts) evaluating neoadjuvant chemotherapy (NAC) for colon cancer. *J. Clin. Oncol.* **2019**, *37* (Suppl. 15), 3504. [CrossRef]

33. Hiranuma, C.; Kawakami, K.; Oyama, K.; Ota, N.; Omura, K.; Watanabe, G. Hypermethylation of the MYOD1 gene is a novel prognostic factor in patients with colorectal cancer. *Int. J. Mol. Med.* **2004**, *13*, 413–417. [CrossRef]

34. Maeda, K.; Kawakami, K.; Ishida, Y.; Ishiguro, K.; Omura, K.; Watanabe, G. Hypermethylation of the CDKN2A gene in colorectal cancer is associated with shorter survival. *Oncol. Rep.* **2003**, *10*, 935–938. [CrossRef]

35. Appleton, K.; Mackay, H.J.; Judson, I.; Plumb, J.A.; McCormick, C.; Strathdee, G.; Lee, C.; Barrett, S.; Reade, S.; Jadayel, D.; et al. Phase I and pharmacodynamic trial of the DNA methyltransferase inhibitor decitabine and carboplatin in solid tumors. *J. Clin. Oncol.* **2007**, *25*, 4603–4609. [CrossRef]

36. Kantarjian, H.; Oki, Y.; Garcia-Manero, G.; Huang, X.; O'Brien, S.; Cortes, J.; Faderl, S.; Bueso-Ramos, C.; Ravandi, F.; Estrov, Z.; et al. Results of a randomized study of 3 schedules of low-dose decitabine in higher-risk myelodysplastic syndrome and chronic myelomonocytic leukemia. *Blood* **2007**, *109*, 52–57. [CrossRef]

37. Lee, V.; Wang, J.S.; Zahurak, M.L.; Gootjes, E.C.; Verheul, H.M.; Parkinson, R.M.; Kerner, Z.; Sharma, A.; Rosner, G.L.; De Jesus-Acosta, A.; et al. A Phase I Trial of a Guadecitabine (SGI-110) and Irinotecan in Metastatic Colorectal Cancer Patients Previously Exposed to Irinotecan. *Clin. Cancer Res.* **2018**, *24*, 6160–6167. [CrossRef]

38. Samlowski, W.E.; Leachman, S.A.; Wade, M.; Cassidy, P.; Porter-Gill, P.; Busby, L.; Wheeler, R.; Boucher, K.; Fitzpatrick, F.; Jones, D.A.; et al. Evaluation of a 7-day continuous intravenous infusion of decitabine: Inhibition of promoter-specific and global genomic DNA methylation. *J. Clin. Oncol.* **2005**, *23*, 3897–3905. [CrossRef]

39. Glasspool, R.M.; Brown, R.; Gore, M.E.; Rustin, G.J.S.; McNeish, I.A.; Wilson, R.H.; Pledge, S.; Paul, J.; MacKean, M.; Hall, G.D.; et al. A randomised, phase II trial of the DNA-hypomethylating agent 5-aza-2'-deoxycytidine (decitabine) in combination with carboplatin vs carboplatin alone in patients with recurrent, partially platinum-sensitive ovarian cancer. *Br. J. Cancer* **2014**, *110*, 1923–1929. [CrossRef]

40. Candelaria, M.; Gallardo-Rincón, D.; Arce, C.; Cetina, L.; Aguilar-Ponce, J.; Arrieta, Ó.; González-Fierro, A.; Chávez-Blanco, A.; de la Cruz-Hernández, E.; Camargo, M.; et al. A phase II study of epigenetic therapy with hydralazine and magnesium valproate to overcome chemotherapy resistance in refractory solid tumors. *Ann. Oncol.* **2007**, *18*, 1529–1538. [CrossRef]

41. Falchook, G.S.; Fu, S.; Naing, A.; Hong, D.S.; Hu, W.; Moulder, S.; Wheler, J.J.; Sood, A.K.; Bustinza-Linares, E.; Parkhurst, K.L.; et al. Methylation and histone deacetylase inhibition in combination with platinum treatment in patients with advanced malignancies. *Investig. New Drugs* **2013**, *31*, 1192–1200. [CrossRef]

42. Matei, D.; Fang, F.; Shen, C.; Schilder, J.; Arnold, A.; Zeng, Y.; Berry, W.A.; Huang, T.; Nephew, K.P. Epigenetic resensitization to platinum in ovarian cancer. *Cancer Res.* **2012**, *72*, 2197–2205. [CrossRef] [PubMed]

43. Zambrano, P.; Segura-Pacheco, B.; Perez-Cardenas, E.; Cetina, L.; Revilla-Vazquez, A.; Taja-Chayeb, L.; Chavez-Blanco, A.; Angeles, E.; Cabrera, G.; Sandoval, K.; et al. A phase I study of hydralazine to demethylate and reactivate the expression of tumor suppressor genes. *BMC Cancer* **2005**, *5*, 44. [CrossRef] [PubMed]

44. Aparicio, A.; Eads, C.A.; Leong, L.A.; Laird, P.W.; Newman, E.M.; Synold, T.W.; Baker, S.D.; Zhao, M.; Weber, J.S. Phase I trial of continuous infusion 5-aza-2'-deoxycytidine. *Cancer Chemother. Pharmacol.* **2003**, *51*, 231–239. [CrossRef] [PubMed]

45. Fang, F.; Balch, C.; Schilder, J.; Breen, T.; Zhang, S.; Shen, C.; Li, L.; Kulesavage, C.; Bs, A.J.S.; Nephew, K.P.; et al. A phase 1 and pharmacodynamic study of decitabine in combination with carboplatin in patients with recurrent, platinum-resistant, epithelial ovarian cancer. *Cancer* **2010**, *116*, 4043–4053. [CrossRef]

46. Overman, M.J.; Morris, V.; Moinova, H.; Manyam, G.; Ensor, J.; Lee, M.S.; Eng, C.; Kee, B.; Fogelman, D.; Shroff, R.T.; et al. Phase I/II study of azacitidine and capecitabine/oxaliplatin (CAPOX) in refractory CIMP-high metastatic colorectal cancer: Evaluation of circulating methylated vimentin. *Oncotarget* **2016**, *7*, 67495–67506. [CrossRef]

47. Juergens, R.A.; Wrangle, J.; Vendetti, F.P.; Murphy, S.C.; Zhao, M.; Coleman, B.; Sebree, R.; Rodgers, K.; Hooker, C.M.; Franco, N.; et al. Combination epigenetic therapy has efficacy in patients with refractory advanced non-small cell lung cancer. *Cancer Discov.* **2011**, *1*, 598–607. [CrossRef]

48. Linnekamp, J.F.; Butter, R.; Spijker, R.; Medema, J.P.; van Laarhoven, H.W.M. Clinical and biological effects of demethylating agents on solid tumours—A systematic review. *Cancer Treat. Rev.* **2017**, *54*, 10–23. [CrossRef] [PubMed]
49. Azad, N.S.; El-Khoueiry, A.; Yin, J.; Oberg, A.L.; Flynn, P.; Adkins, D.; Sharma, A.; Weisenberger, D.J.; Brown, T.; Medvari, P.; et al. Combination epigenetic therapy in metastatic colorectal cancer (mCRC) with subcutaneous 5-azacitidine and entinostat: A phase 2 consortium/stand up 2 cancer study. *Oncotarget* **2017**, *8*, 35326–35338. [CrossRef] [PubMed]

Review

Colorectal Cancer and Immunity: From the Wet Lab to Individuals

Elodie Pramil [1,2], Clémentine Dillard [1,2] and Alexandre E. Escargueil [1,*]

1 Sorbonne Université, INSERM U938, Centre de Recherche Saint-Antoine, F-75012 Paris, France; elodie.pramil@hotmail.fr (E.P.); clementine.dillard@inserm.fr (C.D.)
2 Alliance Pour la Recherche en Cancérologie—APREC, Tenon Hospital, F-75012 Paris, France
* Correspondence: alexandre.escargueil@inserm.fr; Tel.: +33-(0)1-49-28-46-44

Simple Summary: Tackling the current dilemma of colorectal cancer resistance to immunotherapy is puzzling and requires novel therapeutic strategies to emerge. However, characterizing the intricate interactions between cancer and immune cells remains difficult because of the complexity and heterogeneity of both compartments. Developing rationales is intellectually feasible but testing them can be experimentally challenging and requires the development of innovative procedures and protocols. In this review, we delineated useful in vitro and in vivo models used for research in the field of immunotherapy that are or could be applied to colorectal cancer management and lead to major breakthroughs in the coming years.

Abstract: Immunotherapy is a very promising field of research and application for treating cancers, in particular for those that are resistant to chemotherapeutics. Immunotherapy aims at enhancing immune cell activation to increase tumor cells recognition and killing. However, some specific cancer types, such as colorectal cancer (CRC), are less responsive than others to the current immunotherapies. Intrinsic resistance can be mediated by the development of an immuno-suppressive environment in CRC. The mutational status of cancer cells also plays a role in this process. CRC can indeed be distinguished in two main subtypes. Microsatellite instable (MSI) tumors show a hyper-mutable phenotype caused by the deficiency of the DNA mismatch repair machinery (MMR) while microsatellite stable (MSS) tumors show a comparatively more "stable" mutational phenotype. Several studies demonstrated that MSI CRC generally display good prognoses for patients and immunotherapy is considered as a therapeutic option for this type of tumors. On the contrary, MSS metastatic CRC usually presents a worse prognosis and is not responsive to immunotherapy. According to this, developing new and innovative models for studying CRC response towards immune targeted therapies has become essential in the last years. Herein, we review the in vitro and in vivo models used for research in the field of immunotherapy applied to colorectal cancer.

Keywords: colorectal cancer; immunotherapy; methods

Citation: Pramil, E.; Dillard, C.; Escargueil, A.E. Colorectal Cancer and Immunity: From the Wet Lab to Individuals. *Cancers* **2021**, *13*, 1713. https://doi.org/10.3390/cancers13071713

Academic Editors: Marta Baiocchi and Ann Zeuner

Received: 1 March 2021
Accepted: 30 March 2021
Published: 4 April 2021

Publisher's Note: MDPI stays neutral with regard to jurisdictional claims in published maps and institutional affiliations.

1. Introduction

Colorectal cancer (CRC) is the third most commonly diagnosed malignancy worldwide, and the second leading cause of cancer related-deaths among men and women with 1.8 million estimated cases and more than 800,000 deaths annually [1]. As the disease mostly progresses indolently at the initial stages, becomes symptomatic late, and is often diagnosed at an advanced stage (about 35% of patients presenting with a metastatic cancer). This issue is of importance because the prognosis for CRC patients is strongly dependent on the stage of the tumor at diagnosis [2]. Following the Tumor Node Metastasis (TNM) staging, the 5-year survival rates following surgical removal of tumors for localized (stage I), regional (stages II and III) and metastatic (stage IV) diseases reach in the USA 90, 72, and 14%, respectively [3]. For stage I and most stage II CRCs, the standard of care is surgery alone [2]. For high-risk stage II and stage III CRCs, surgical removal is followed by adjuvant 5-fluoruracil (5-FU) or capecitabine-based chemotherapy [4,5]. For metastatic disease, surgical removal of the primary and/or distant lesions is followed by therapies using a set of chemotherapies and targeted agents [2]. However, as mentioned before, the prognosis of patients with metastatic CRC (mCRC) remains poor, with a median overall survival (OS) of about 30 months [6]. In addition to the TNM classification, recent advances have led to the development of immune-based classifications of colorectal tumors. Indeed, numerous reports demonstrated that an enhanced T-lymphocytic infiltration in tumor tissues is associated with an improved prognosis [7,8]. However, the composition of the tumor microenvironment (TME) varies substantially between colorectal tumors [2,9]. Thus, in an effort to translate these findings to the clinic, an international consortium developed the "Immunoscore" [10]. This scoring system is based on the histological quantification and localization of cytotoxic and memory T-cells in the center of the tumor and invasive margin. Importantly, time to recurrence was significantly improved in patients with stage I–III colon cancer presenting a high "Immunoscore" [11]. These observations thus supported the role of this scoring system in providing a reliable estimate of the risk of recurrence in patients with colon cancer and its additional prognostic value when combined with conventional TNM-staging [9]. They also underline the impact of immune cell infiltration on CRC outcome, thus opening new therapeutic opportunities.

In addition to the characterization of immune cells infiltrates in CRC, significant research has also helped in the last years to better understand the complex interplay between cancer and immune cells. This knowledge has led to the emergence of novel immunotherapies including the development of immune checkpoint inhibitors (anti-CTLA-4, anti-PD-1, and anti-PD-L1 monoclonal antibodies). These molecules have dramatically changed the therapeutic situation for several types of cancer [12]. However, for mCRC, only few objective responses have been observed in unselected colorectal cancer patients. Long-lasting responses were only restricted to 4 to 5% of patients who presented tumors harboring microsatellite instability (MSI-H) and/or mismatch repair deficiency (dMMR) [12–15]. For this small subset of patients, the therapeutic scenario was nonetheless significantly changed thanks to the introduction of immune checkpoint inhibitors. In 2020, pembrolizumab (anti PD-1) was approved by the U.S. Food and Drug Administration (FDA) for the first-line treatment of patients with unresectable or metastatic dMMR CRC. The success of immunotherapies for treating metastatic dMMR CRC also paved the way for clinical research aiming at introducing immunotherapy in the adjuvant setting used for treating patients with localized MSI/dMMR CRC [14].

This success for this specific subtype of CRC is, however, not surprising in terms of biological understanding. Indeed, MSI tumors are known for being highly intruded by tumor-infiltrating lymphocytes (TILs) such as CD8+ cytotoxic lymphocytes, Th1-activated cells that produce IFNγ, and CD45 RO+ T memory cells [8,16,17]. This phenomenon is explained by the hypermutated phenotype of these tumors, leading to high mutational burden (TMB) with highly immunogenic neoantigens as a consequence of a large number of deletions, insertions, and frameshift mutations accumulated during cancer cell replication [12,14,15]. The accumulation of tumor-associated neoantigens indeed favors the identification of cancer cells by the host immune system [18,19]. This hypothesis was recently confirmed in a controlled murine syngeneic model of CRC [20]. By genetically inactivating DNA mismatch repair in an otherwise MMR proficient (pMMR) cell line, the authors clearly demonstrated that MMR loss caused a tumor hyper-mutated status associated with an increased load of tumor neoantigens. In turn, those triggered long-lasting immune surveillance that could be further enhanced by immune modulators [20]. Importantly, MSI/dMMR tumors are often associated with an upregulation of checkpoint inhibitors that exhaust intra-tumor cytotoxic T lymphocytes and consequently protect MSI/dMMR cancer cells from their hostile immune microenvironment [21,22]. Together, these made metastatic MSI/dMMR tumors a valuable candidate for immune checkpoint inhibitors (ICI).

Unfortunately, this type of ICI responsive tumors (e.g. MSI/dMMR) represents only 15 to 20% of total CRC and about 4% of stage IV CRC [12,23,24]. Therefore, a vast majority of mCRC is cold refractory to this therapeutic strategy. As with most cancers, CRC is a genetically heterogeneous disease. However, heterogeneity also emerges from the composition of the surrounding tissue and cells, commonly called the tumor microenvironment. This includes epithelial cells, blood and lymphatic vessels, stromal and infiltrating immune cells, as well as extracellular components (e.g., chemokines, cytokines, and extracellular matrix) [9]. This general context and the subsequent crosstalks established between TME and tumor cells are key features to determine the effect of infiltrating immune cells on clinical outcome. According to these observations, and based on both tumor and infiltrating stroma gene expression profiles, a consensus molecular subtype (CMS) classification has been set up in the last years [25]. According to it, four major groups were distinguished: CMS1 (approximately 14% of cases) are hypermutated tumors, mostly with MSI-H features and showing a robust immune cells infiltration; CMS2 (approximately 37% of cases) are canonical CRC tumors characterized by the activation of the Wnt and Myc pathways; CMS3 (approximately 13% of the cases) are tumors frequently mutated in *KRAS* and displaying a deregulated cancer cell metabolism; CMS4 (approximately 23% of the cases) are mesenchymal tumors characterized by transforming growth factor beta (TGF-β) pathway activation, enhanced angiogenesis, stromal activation, and inflammatory infiltrate [9,12,25]. The 13% of missing samples corresponded to tumors with mixed features (13%) [25]. Interestingly, in this classification, CMS1 and CMS4 were considered as "hot" tumors with an intense immune infiltration, whereas CMS2 and CMS3 were defined as "cold" tumors with a lack of immune activation [12]. However, CMS4 tumors, despite their immune infiltration, displayed the worse overall and relapse-free survival [25]. This is explained by the specific immune infiltrate seen in these tumors mostly composed of T regulatory cells (Treg), myeloid-derived suppressor cells (MDSCs), and monocyte-derived cells. This inflamed immune-tolerant TME is characterized by marked upregulation of immunosuppressive factors, such as TGF-β, Vascular endothelial growth factor (VEGF), and CXCL12 [26]. On the contrary, CMS1 CRC has a diffuse immune infiltrate with notable CD8+ TILs. As discussed above, those MSI/dMMR tumors also upregulate immune checkpoint molecules (CTLA-4, PD-1, PD-L1) [21,22,26]. Though the use of ICIs can activate an effective antitumor immune response for CMS1 but not for CMS4 CRC subtypes [12], nonetheless, strategies aiming at targeting the TGF-β and/or VEGF pathways might prove useful for CMS4 CRC [27,28]. In contrast to CMS1 and CMS4 tumors, CMS2 and CMS3 tumors were defined as "immune desert" cancers [12]. Different mechanisms can be responsible for this phenomenon, including lack of major histocompatibility complex (MHC) class I molecules and/or upregulation of nonclassical human leukocyte antigens (HLA) [9,12]. Interestingly, targetable oncogenic-driven cancer cell pathways have been identified as potential sources of the above immune evasion processes. For example, MEK inhibition has been shown to rescue low class I MHC expression and augment anti-tumor T-cell immunity [29]. However, acting on the sole tumor compartment is likely not to be sufficient to overcome resistance in "immune desert" tumors. To that end, combinatorial strategies aiming at increasing immune cells infiltrate and/or neoantigens generation or release have been proposed to synergize with immunotherapies [15]. For example, current chemotherapies, like oxaliplatin and 5-FU can improve TME immune-competency by inducing immunogenic cell death and/or depleting MDSCs [30–32]. VEGF-targeted therapy can also improve TME immune competency by reducing the proportion and number of Tregs in CRC murine tumors as well as in the peripheral blood of patients with mCRCs [33]. On the other hand, cetuximab which targets the extracellular domain of epidermal growth factor receptor (EGFR) might promote activation of the immune response in CRC patients in addition to its direct action on cancer cells [34]. In the last years, several clinical trials have been launched to evaluate strategies combining chemotherapies and targeted therapies to extend the efficacy of immunotherapy to pMMR CRC [15]. However, to date, those approaches and rationales have not been successfully transformed in terms of clinical benefits [15]. Further insights into the molecular mechanisms underlying the immune competence and/or immunotherapy resistance are therefore urgently needed for developing predictive biomarkers and/or improving pharmacological combination strategies for mCRC resistant to immunotherapy. To this end, substantial help is awaited from translational research with the aim of turning all cold CRC into hot responsive immunogenic tumors.

Herein, we review the in vitro and in vivo models used for research in the field of immunotherapy applied to CRC. We discuss their useful meaning and propose to define the most accurate approaches to expand our knowledge on immunological-based therapies in pMMR colorectal cancer with a special emphasis on models allowing a better characterization of the resistance mechanisms, as well as the identification of predictive biomarkers and the assessment of novel combinatorial therapeutic strategies.

2. In Vitro Models for Immunotherapy Studies

2.1. 2-Dimensional Methods

Tumors, including CRC, are not composed of homogeneous cell populations but contain a multitude of cells with different characteristics [35]. This heterogeneity leads to different treatment responses within the tumor itself as well as between patients and has to be considered early while developing new therapies. Cancer cell lines are a widely used tool for pre-clinical in vitro research. Their major advantage is their simple manipulation. Due to the heterogeneity of CRC, there is a multitude of derived cell lines available to date with different molecular patterns. Hence, the selection of the most relevant model is a crucial step during the development of new pharmaceutics. Numerous multi-omics studies have dealt with the analysis and classification of CRC. In particular, following the CMS classification [25], Berg et al. analyzed 34 CRC cell lines and classified them among the 4 CMS groups mentioned above, bringing new resources for CRC model selection (Table 1) [36]. As discussed, MSI/dMMR tumors are preferentially immunogenic, heavily infiltrated by lymphocytes and good responders to immunotherapies [37]. Accordingly, in vitro studies on CMS1 CRC cell lines gave interesting responses to immunotherapies. However, to better understand the heterogeneous behaviors seen in tumors, experiments need to be extended to other cell lines. More specifically, cells corresponding to the major types of mCRC should carefully be characterized to respond to the huge therapeutic challenge we are facing now [38]. The initial choice of the cell line is therefore an important criterion. Testing and comparing responses in cells originating from different CMS clusters will therefore prove useful for assessing the underlying molecular mechanisms of resistance to immunotherapies. We thereafter discuss different experimental models based on the use of commercially available cancer cell lines for the study of immunotherapies in vitro.

Table 1. Classification of 34 colorectal cancer (CRC) cell lines into consensus molecular subtype (CMS) subgroups. Human colorectal cancer cell lines are assigned into the best fitting CMS group according to Berg et al. [36].

CMS	Type	Colorectal Cell Lines
CMS1	MSI, Immune	Co115, DLD-1, HCT15, KM12, LoVo, SW48, Colo205, HCC2998
CMS2	Canonical	EB, FRI, IS3, LS1034, NCI-H508, SW116, SW1463, SW403, V9P
CMS3	Metabolic	CL-34, LS174T, CL-40, HT29, SW948, WiDr
CMS4	Mesenchymal	HCT116, RKO, TC71, CaCo2, CL-11, Colo678, IS1, SW480, SW837

2.1.1. Cancer Cells: Secretome Assessment

During culturing, cells naturally release proteins, soluble factors, exosomes, or microvesicles capable to act on cell interaction, proliferation, death, metabolism, or drug resistance [39]. In the context of immunotherapy, transferring a conditioned medium (CM) from one cell culture to another is a simple experiment to address the effect of the cancer cells' secretome on immune cells' phenotype (Figure 1) [40]. In the recent years, numerous studies have focused on the establishment of therapeutic strategies for converting tumor-associated macrophages (TAMs) displaying an immunosuppressive M2 phenotype into a pro-inflammatory M1 phenotype [41–43]. In order to generate TAMs in vitro, the use of tumor-based CM has proven to be an effective approach. Several research groups have indeed demonstrated that TAMs, differentiated in tumor-based CM, display the same genetic, phenotypic, and functional characteristics as the tumor-associated macrophages derived from patients [44,45]. Benner et al. used conditioned media derived from two breast cancer cell lines which was complemented with a cocktail of cytokines (IL-4, IL-10, M-CSF) and incubated it with healthy donor monocytes to successfully generate tumor-associated macrophages (TAMs). Those TAMs which differentiated in M2 macrophages showed an increased co-expression of the CD163/CD206 TAM surface markers as well as several functional TAM markers. Importantly, those TAMs also secreted factors in vitro able to promote tumor cells survival and growth [46]. In a similar approach using tumor-based CM, Dong et al. also produced TAMs and demonstrated that the immunocomplex formed between lactoferrin and anti-lactoferrin was capable of converting M2-TAMs towards the M1 phenotype [47]. These studies demonstrate the potential of producing in vitro TAMs for studying new immunological opportunities. For CRC, this strategy also was successful since CRC-based CM prepared from 4 distinctive cell lines was able to activate and induce differentiation of the human monocytic cell line THP-I towards a TAM-associated phenotype displaying immunosuppressive properties [48].

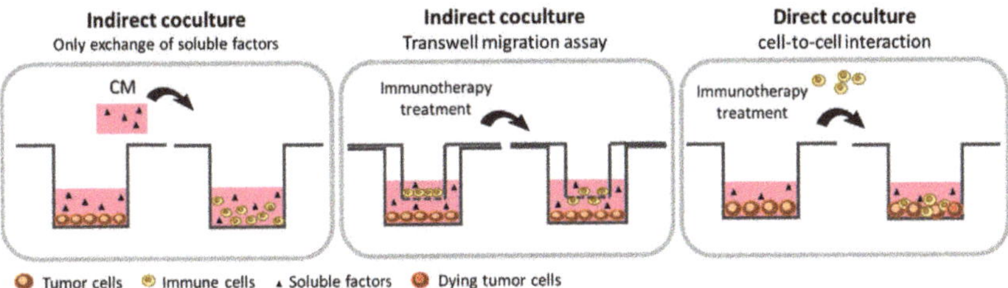

Figure 1. Schematic representation of the 2-dimensional (2D) co-culture methods that are suitable for in vitro study of immunotherapies. The in vitro 2D co-cultures using commercialized cell lines are a first approach to evaluate the activation, migration, or cytotoxic potential of immune cells following an immunomodulatory treatment because of its simplicity to set up, its low cost, and its reproducibility. Indirect co-culture consists of the transfer of conditioned medium from one cell to another. This allows the effects of soluble factors on immune cells biology to be studied. The indirect co-culture method using the Transwell assay allows the study of the migratory capacity of immune cells in the presence of tumor-derived conditioned media. Finally, the direct co-culture assay permits cell-to-cell interactions, thus allowing studies on immune cells activation and cytotoxic activity towards tumor cells.

In addition to TAMs, T lymphocytes are another immune cell population that is highly targeted by immunotherapies. Adil et al. studied the effects of cancer cell based-CM on Peripheral Blood Mononuclear Cell (PBMCs) originating from healthy donors. They showed an anti-proliferative effect of both MCF7 and HeLa conditioned media. However, CM prepared from the leukemic K562 cell line demonstrated a pro-proliferative effect on PBMC associated with an increased expression of Treg markers and of the CD4+/Helios+ sub-population. These results correlate with the induction of immunosuppressive functions of PBMC promoted by CM [49]. Similarly, it was shown that CM prepared from the RENCA mouse kidney cancer cell line converted CD4+CD25- T lymphocytes into CD4+CD25+ Treg cells [50]. Together, these methodological approaches underline the important crosstalk existing between immune and tumor cells and the influence that secreted soluble factors can exert on the fate of immune cells. This, combined with the development of new omics technologies, can help future studies aiming at identifying new targetable immunologic molecules involved in cell–cell trans-communication. Such approaches have yet been useful for CRC by demonstrating the capacity of the Treg supernatant to enhance chemoresistance [51]. Recently, conditioned media were prepared from rectal cancer and non-cancer control biopsies and 19 oversecreted inflammatory proteins were identified in the rectal cancer secretome [52]. By comparing CRC-CM-induced effects on immune cells, and conversely by comparing immune cell-CM-induced effects on colorectal cancer cells, new interventional opportunities can therefore be identified and secretome components targeted to eventually be blocked [53]. Alternatively, targetable cellular pathways involved in the secretion of specific soluble factors having an immunogenic potential might directly be identified and pharmacologically evaluated through new screening approaches in CRC tumor cells selected among those classified as CMS2-4.

2.1.2. Co-Culture with Paracrine Interaction: The Transwell Technology

Immunotherapies are ineffective for most pMMR CRC cancer because of the poor number of infiltrating immune cells in the tumor microenvironment [54]. The development of immunotherapies able to promote migration and recruitment of immune cells within the tumor microenvironment is thus essential [8,55,56]. The Transwell technology can be used in vitro to study the impact of immunotherapeutic molecules to act on the ability of cancer cells to attract immune cells. The Transwell consists of an upper insert containing a permeable membrane allowing the exchange of soluble factors and/or cell migration (Figure 1). Several Transwell pore sizes are indeed commercially available. Membrane with a pore size of 0.4 μm exclusively permits measuring the exchange of soluble factors like cytokines between the two compartments. In contrast, a larger pore size allows cells to migrate through the membrane. For the study of human-derived immune cell migration, a 3 to 5 μm pore size membrane is usually sufficient [57–59]. In Transwell coculture, immune cells (PBMCs or isolated subpopulations of immune cells) are seeded onto the upper layer of the insert, while the tested molecules, CM, or attached cells are deposited in the lower chamber. Tumor cells can then be stimulated or treated (chemotherapeutics or radiotherapy) before positioning the Transwell upper layer to trigger the secretion of soluble factors. After diffusion throughout the well, those factors can reach immune cells with their subsequent activation, proliferation, cytokines production, and/or migration easily monitored (Figure 1). Hence, activation can be studied by flow cytometry through the expression of specific surface markers. Proliferation can be assessed by counting cells in the upper chamber. Cytokine secretion can be quantified by ELISA assay and/or cytokine array [60]. Finally, immune cells migration can be evaluated either after fixation, coloration, and counting with a microscope [61,62] or by flow cytometry, after Transwell centrifugation and cell harvesting, with antibodies directed against specific surface antigens. The percentage of each cell subpopulation can thus be precisely determined [63,64]. Transwell is commonly used to test chemotaxis of immune or cancer cells. For example, Harlin et al. demonstrated that chemokines (CCL2, CCL3, CCL4, CCL5, CXCL9, and CXCL10) are able to induce migration of CD8+ T-cells from the upper to the lower chamber of the Transwell. They also demonstrated that, in contrast to the culture medium alone, the presence of M537-CM melanoma cells in the lower compartment stimulates the recruitment of CD8+ T-cells [63]. Similarly, Hennel et al. used the Transwell technology to study the ability of breast cancer cells dying after radiotherapy to release factors capable to stimulate monocyte migration. To do so, they seeded in the upper insert THP-1 macrophages while supernatants from mock-irradiated and irradiated breast cancer cells were put in the lower chamber. This approach allowed them to demonstrate that radiation-induced necrosis of HCC1937 cells is particularly efficient for stimulating THP-1 cell migration and identifying apyrase-sensitive nucleotides as molecules responsible for attracting monocytes [65]. Transwell assays were also used to study the migratory capacity of mast cells in CRC [66]. This immune cell type is one of the earliest to be recruited during CRC tumorigenesis. In this work, the authors plated human CD34+-derived mast cells in the upper chamber of the Transwell while CM prepared from either HT29 or Caco2 CRC cells were positioned in the lower chamber. Their results demonstrated a significant increase of mast cell migration in both conditions. However, mast cells' chemo-attraction originated from two distinct mechanisms according to the CRC cell line used. The stem cell factor (SCF) seems to be involved in the Caco-2-CM while CCL15 chemokine is responsible for the mast cell migration in HT29-CM [66].

These works highlight the effectiveness of using the Transwell assay to deal with the migratory functions of immune cells and could be an asset for the understanding of the migration mechanisms following treatment by immunotherapies.

2.1.3. Co-Culture with Direct Cell-to-Cell Interaction

The use of direct co-culturing conditions allows physical contact between tumor and immune cells. This approach should be considered as an alternative method for studying in vitro anti-cancer immunotherapies [67]. It indeed allows the role of cell–cell physical contact in addition to the action of secreted soluble factors to be evaluated. It also the direct cytotoxic activities of immune cells towards cancer cells to be evaluated (Figure 1). Co-cultures can be performed with either PBMC from healthy donors, mice, patients' peripheral blood, or immune cells collected and isolated from colorectal carcinoma specimens. Prior to establish the co-cultures, immune cells can be sorted out in order to isolate specific immune cell subpopulations (LT, LB, NK, DC, Mono) or differentiated in vitro (macrophages) [68,69]. However, an important issue is the need to precisely establish the ratio that is used in the co-culture experiment between tumor cells and immune cells. Moreover, defining the precise cytokines/antibodies cocktail required for activating immune cells can be particularly tricky. Anti-CD3/CD28 beads, IL-12, and M-CSF or GM-CSF are usually used for activating T-cells, NK-cells, and macrophages, respectively [70–72]. In addition, defining the incubation time of the co-culture prior to performing analyses is an important issue and is likely to depend on the tumor cell type used. In that sense, a detailed protocol used for performing co-cultures of tumor cells with T-cells was described by Melief et al. [73]. In addition, Minute et al. performed co-cultures combining modified MC38 CRC murine cells (MC38EGFROVA) with either cytotoxic T lymphocytes (CTL) or activated NK-cells. Interestingly, prior to establishing co-cultures, tumor cells were here pre-treated with IFN-γ in order to increase their expression of MHC-class I. Human gp100 peptide was also added to load MHC-class I. The first co-culture was then established with CD8+ splenocytes which were preactivated in vivo in mice. Activated splenocytes and preconditioned MC38 were then co-cultured for 3 days at a 10:1 ratio, in the presence of IL-2 and human gp100. The second co-culture model was established with NK-cells which were also preactivated in vivo in mice. In that case, NK-cells were co-cultured for 3 days with MC38 cells which were not preconditioned, at a 5:1 ratio and in presence of IL-2. At the end of the co-cultures, the authors highlighted the presence of two alarmins in the extracellular compartment, HMGB1, which was released in the culture medium and the calreticulin which was exposed on the cell surface. This study presented evidence that T- and NK-cells induce features of immunogenic cell death (ICD) on tumor cells and that in vitro co-culture can trigger immune response against tumor cells. Importantly, in the same study, the authors demonstrated the same capacity of CTL and NK-cells to induce ICD on the human CRC cell line HT29. In the presence of a bispecific antibody EpCAM-CD3ε, HMGB1 and calreticulin were indeed exposed on the HT29 cells when co-cultured with CTL or NK-cells (preactivated by IL-2 and IL-15). However, the ratio used here (1:1) was strikingly different from those used for the murine cells [74]. This underlines the difficulties of setting up general protocols and the sometimes difficult interpretation in terms of biological relevance. In a similar context, others performed co-cultures of HT29 CRC cells with CD8+ T-cells isolated from either healthy donors, CRC patients' peripheral blood, or from tumor immune infiltrates. The CD8+ T-cells were then stimulated for 2 h in the presence of anti-CD3/CD28 beads and then directly and indirectly (Transwell) co-cultured with HT29 cells (4:1 ratio) for 48 h in the presence of anti-CD3/CD28 beads. In this study, the authors evaluated the effect of the indirect pharmacological inhibition of Notch by the γ-secretase inhibitor DAPT on the anti-tumor immunity. Interestingly, they evidenced an increased production of pro-inflammatory cytokines (IFN-γ, IL-1β, IL-6, TNF-α) in the supernatant of both direct and indirect co-cultures when HT29 cells were co-cultured with either peripheral or infiltrated tumor CD8+ T-cells in presence of DAPT. However, a significant increase in CD8+ T-cell-induced HT29 cell death was only observed when co-cultures were made with CD8+ T-cells which were purified from colorectal carcinoma, not CD8+ T-cells isolated from peripheral blood. This phenomenon was also reported only when cells were co-cultured in conditions permitting direct contacts between cells. Remarkably, in the absence of Notch inhibition, this study also demonstrated a greater cytotoxic effect on HT29 cells of peripheral CD8+ T-cells isolated from healthy donors compared to peripheral CD8+ T-cells isolated from CRC patients. Thus, this study evidenced the immunosuppressive potential of Notch signaling in CRC and demonstrated the in vitro ability of Notch inhibition to stimulate and restore anti-tumor immunity [75]. Co-cultures with direct cell-to-cell interaction were also used for deciphering the mechanism of action of the TIGIT immune checkpoint (present on T-cells' surface) and its role in the impairment of metabolism and function of CD8+ T-cells [76]. In this study, CD8+ T-cells were isolated from PBMCs of healthy donors and co-cultured with the SGC7901 gastric cancer cell line at a 5:1 ratio. Intriguingly, the results initially showed that tumor cells were capable to inhibit T-cell metabolism and this effect could be reversed by addition of glucose in the culture medium. The authors also demonstrated that co-culturing CD8+ T-cells with tumor cells enhanced the expression of TIGIT on their cellular surface. Interestingly, the blockade of TIGIT antigen with an anti-TIGIT antibody increased CD8+ T-cell metabolism, glucose consumption, as well as lactate production. It also restored T-cell effector functions by reversing gastric cancer cell-mediated inhibition of IFNγ production. This study thus highlights the potential of targeting TIGIT immune checkpoint to restore immune T-cells' anti-tumor functions [76].

Finally, assessing the interaction of M1 macrophages with tumor cells can also be performed through direct co-culturing experiments [77]. To that end, co-cultures were carried out with RAW264.7 murine macrophage cells, which were previously polarized as M1 macrophage by LPS and INFγ treatment and the 4T1 breast cancer murine cell line. The latter cells were also primarily labeled with the CFSE fluorescent probe. Through a very detailed protocol, the authors described in their manuscript a clever method permitting visualizing 4T1 cells' engulfment by macrophages which could be used for further immunotherapy studies aiming at targeting this specific subtype of immune cells [77].

Together, these data show that co-cultures involving direct cell-to-cell contact are becoming a widely used method for evaluating immunotherapies' efficacy and the underlying molecular mechanisms associated with it. As discussed earlier, it has the advantage of being easy to set up and to give quick results. However, the experimental conditions should be carefully defined for not misinterpreting or over-interpreting the data obtained. Moreover, one of the major drawbacks of these experimental models is the lack of predictive value in terms of tumor heterogeneity, complexity, and 3D organization.

2.2. 3-Dimensional Methods

The major limitation for the development of new therapies acting on the TME, including immunotherapies, is the lack of consistent in vitro models. The traditional two-dimensional (2D) cancer models are still in use to study molecular and cellular features of tumors and sensitivity to treatments. However, many drawbacks have been identified for those models [78,79]. 2D culture methods indeed poorly represent patient's tumor complexity, thus limiting their reliability. This is particularly true for solid tumors for which 3-dimensional organization is an important characteristic affecting their biological properties and survival capacities. 2D cell line cultures are indeed unable to fully reproduce tumor features such as microenvironment, immune system interaction, stromal compartment, and heterogeneity of cancer cells [80,81]. These limitations led, in the past years, to tremendous efforts for developing and designing new models capable of reproducing 3D tumor structuring, thus giving a rational intermediate between in vitro 2D culture and in vivo animal models.

2.2.1. Spheroids

The easiest method to switch from 2D to 3D models is culturing spheroids, also called multicellular tumor spheroids (MCTS). These 3D structures can be easily obtained by incubating cell lines in low attachment plate. This specific surface favors cell-to-cell interactions, thus promoting spontaneous homotypic aggregation [82,83]. An alternative hanging drop method can also be used to generate spherical cell growth. Here, a drop of cell suspension is placed on a dish lid and then inverted onto the bottom chamber. Cells in the drop then aggregate and form spheroids (Figure 2) [84–86]. Importantly, it should be noted that not all cell lines are capable of assembling themselves to a 3D spheroid structure [87]. However, even if spheroids do not contain all the cell types and soluble factors that are present in the tumor microenvironment, they allow by their 3D organization mimicking cell-to-cell interactions, hypoxic conditions and low nutrient concentrations that otherwise characterize tumors. Moreover, it establishes a phenotypic heterogeneity which is not, or is poorly, observed in 2D cultures. Many studies have indeed highlighted a metabolic (oxygen consumption and lactate production) and proliferative gradient between the core and the periphery of the spheroids [88–90]. A necrotic part is also observed in the center of the spheroid as it is in vivo for solid tumors [85,91]. Another point which is important to consider is the sensitivity of cellular models to anti-cancer drugs. Indeed, many drugs show anti-cancer activity in 2D cellular models but their observed effects often do not predict activity in patients [79,92]. The 3D culture of spheroids, by more closely mimicking tumor complexity, has thus unraveled resistance mechanisms not found in a 2D cellular context [79,93–95]. Moreover, cell lines are able to form spheroids of varying density and it is known that dense spheroids show higher chemoresistance [93]. Therefore, 3D spheroids allow, early in their development, for the assessment of the anti-cancer activity of compounds on cancer cell populations displaying phenotypically distinctive traits and exposed to either high or low concentrations of tested molecules [96].

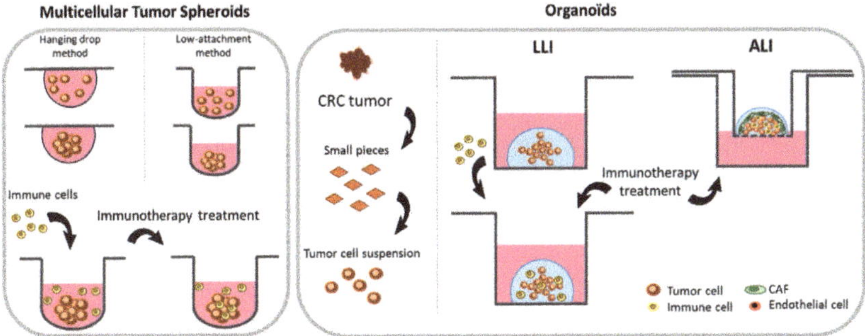

Figure 2. Schematic representation of 3-dimensional (3D) cultures with cell lines (spheroids) or patient tumor tissues (organoids) helpful for studying immunotherapies. The growth of tumor cell lines in 3D allows the formation of spheroids characterized by a necrotic core and a proliferative and metabolic gradient mimicking the 3D structure of a tumor. The spheroid allows the easy assessment of immune cells infiltration and the evaluation of strategies with pro-immunogenic potential. The main limitation of this model is the lack of heterogeneity related to the use of cell lines. 3D models made from small pieces of tumor tissues, also called organoids, have shown their ability to mimic tumor heterogeneity in terms of cellular components, TME, or tumor histology. Co-cultures of 3D-cells isolated from tumor tissues with immune cells in the liquid–liquid interface (LLI) method allow immune cell infiltration, activation, and their associated anti-tumor effect to be studied in a context closely reproducing tumor complexity, heterogeneity, and histology. On the other hand, the air–liquid interface (ALI) culture method has been developed to preserve the micro-environmental cellular components to further improve studies on immunotherapies in a context as close as possible than those observed in clinical solid tumors.

Spheroids also became a powerful tool for studying immunotherapies. Immune cells are indeed able to infiltrate them and to exert their biological effects [97–99]. Moreover, 3D-co-culturing of tumor cells with immune cells and fibroblasts demonstrated the accumulation of cytokines, chemokines, extracellular matrix components, and metalloproteins in the TME [97]. Interestingly, Rebelo et al. recently studied the increased infiltration capacity of THP-1 macrophages or donor blood-derived macrophages into heterotypic spheroids composed of the NCI-H157 lung cancer cell line and cancer-associated fibroblasts (CAF). The authors also reported the polarization of the infiltrating macrophages towards the M2 phenotype [97]. Similar experiments were conducted with spheroids formed by the LS174T CRC cell line co-cultured with immortalized fibroblasts (MRC-5). In this model, the authors showed the capability of leukocytes and monocytes to efficiently infiltrate heterotypic spheroids while T-cell infiltrate was limited [98]. Such co-cultured heterotypic spheroids are also a useful tool to evaluate therapeutic strategies. In line with this, Alonso–Nocelo et al. performed 3D co-cultures of A549 lung cancer cells with Jurkat T-cells (ratio 1:1) and evaluated T-cell infiltration into A549 cell line spheroids [99]. Interestingly, they reported the establishment of an inflammatory and immunosuppressive environment mediated by the 3D structuring of cancer cell spheroids which was increased when Jurkat T-cells were co-cultured. This makes such models of particular interest for studying anti-tumor immunity and evaluating anti-tumor activity of drugs [99] This new paradigm can be exemplified by the recent work published by Courau et al. [100]. Spheroids were grown with either HT29 or DLD-1 CRC cell lines and PBMCs added to the culture medium (ratio 1:10). This study first revealed the capacity of T- and NK-cells to infiltrate the spheroid and to be activated in the presence of IL-15, thus leading to spheroid destruction. Second, and maybe more importantly, this work identified the potential of using immunomodulatory antibodies targeting NKG2D ligands, a central activator of the NK cytotoxic response. The authors indeed observed an increased NK-cell infiltration and expression of the CD137 activation marker at their surface as well as a decreased expression of the CD16 receptor. Together, those results highlighted a NK-mediated anti-tumor response against CRC spheroids [100].

In addition to their usefulness and relevance for cancer research, 3D culturing approaches also permit animal uses to be reduced and should be considered as an alternative to them. Their ability to closely mimic crosstalks between immune, stromal (mostly fibroblasts), and tumor cells offers a good template to initially assess strategies aiming at targeting the TME [97]. Obviously, this method, which is easily available and of relatively low cost, also has some limitations. In particular, it does not permit the tumor structure and heterogeneity as well as its complex microenvironment to be completely represented. In addition, because they are often transformed or genetically modified, the cell lines used to form spheroids lack predictive power. However, to work around these limits, new methodologies such as organoids are now using patient-derived tumor slices to preserve the heterogeneous nature of tumors.

2.2.2. Organoids

The organoid model is a 3-dimensional technology allowing the growth of a small-scale tissue in vitro, leading to its structure mimicking the in vivo parent organ [101,102]. Organoid culture is a promising approach to study the efficacy of immunotherapies in a context close to the patient's physiology. The culture of the CRC organoid has long been studied and is now well characterized [103]. Particularly, it has been evidenced that organoids, developed from patient-derived colorectal tumor slices, allow the preservation of the tumor's genetic heterogeneity, and, from a histological point of view, cells in the organoid are able to self-organize and to reproduce the morphological architecture of the original tumor. Hence, depending on the localization of the original surgical resection, organoids develop specific organizations reproducing the organ-like tissue [104–106].

In practice, organoids are established from small pieces of tumors isolated from surgical resections or biopsies. These fragments are crushed with an enzyme mix, filtered, and included into a Matrigel to subsequently be cultured in culture medium. A complete protocol has been designed by van de Wetering et al. [104]. With such approaches, development of co-cultures with immune cells can also be considered for immunotherapy development. The so-called liquid–liquid interface (LLI) method consists in separately culturing first organoids and immune cells before to establish co-cultures. The immune cells used in these studies can therefore be isolated either from healthy donors' blood, and patients' blood or tumors. Moreover, specific immune cell subpopulation (LT, NK, DC) can be cell sorted from the entire pool of PBMCs prior to co-culturing in order to study the response and effect of specific subpopulations on the organoids [68,107]. To initiate co-cultures, two distinctive approaches are usually considered. The first one consists of the digestion of organoids, addition of immune cells to the tumor suspension, and the regrowth of the organoid with immune cells. The second approach consists of the addition of immune cells directly to the culture medium without prior digestion of the organoids (Figure 2) [108].

The importance of organoids in cancer research can be exemplified by the work of Gonzales–Exposito et al. [109]. Co-cultures were performed with patient-derived CRC organoids and CD8+ T-cells isolated from PBMC of healthy donors (added 24 h after organoid formation). Suspensions were then treated with cibisatamab, a bispecific antibody recognizing CD3+ T-cells as well as the carcinoembryonic antigen (CEA) which is overexpressed by CRC cells. In this setting, the authors showed that tumors strongly expressing the CEA antigen responded to cibisatamab treatment while those expressing low CEA levels did not. Importantly, this study demonstrated the ability to redirect T-cells' response against tumor cells and the potential of using organoids co-cultured with allogeneic immune cells [109]. To avoid non-specific allogeneic response of immune cells against "non-self" organoids, proper controls should, however, be used in this setting to ensure correct interpretation. To work around this problem, the co-culture of dMMR CRC organoids with PBMC isolated from the same patient can also be performed [110]. In the presence of IL-2- and CD28-coated antibodies, as well as by targeting PD-1, the authors succeeded in enriching the tumor-reactive T lymphocytes fraction and showed that reactive T-cells which were generated and were capable of effectively killing organoids [110]. This study also demonstrated that organoids express antigens permitting the recruitment, proliferation, and activation of T-cells.

To go on with the improvement of organoids, several research groups developed methods for culturing organoids in vitro in conditions favoring the maintenance of the whole microenvironment cell components. The so-called "air–liquid interface" (ALI) model consists of culturing minced tumor biopsy fragments that contain the entire cell populations actually present in the tumor (endogenous immune cells, fibroblasts, endothelial cells, epithelial cells, tumor cells) [111,112]. To do so, cells are embedded in a Matrigel solution and placed in an insert that was pre-coated with Matrigel. The insert is then positioned in a well containing the appropriate culture medium (Figure 2) [68]. This original approach helps mimic the intestinal membrane consisting of a monolayer of polarized epithelial cells with an apical surface towards the lumen and a basal surface towards the lamina propria [113]. An important issue for studying immune responses is the particular composition of TILs within the tumor microenvironment as it can generate an immunosuppressive environment promoting tumor progression. Importantly, it was reported that organoids prepared in such conditions can maintain the expression of the CD45 surface marker on leukocytes for 8 days. In the same study, the authors however noted a loss of CD3+ T-lymphocytes [113]. More recently, an alternative ALI culture method was developed in order to preserve TILs and the original tumor T-cell receptors for up to 30 days [114]. Therefore, co-culturing ALI-prepared organoids with infiltrating leukocytes now allows short-term studies for assessing the response to immunotherapies in the in vitro 3D model still harboring their original immune microenvironment. Importantly, after treatment, the cells that are present in the organoid can be harvested and analyzed (qRT-PCR, imaging, or flow cytometry) to study the efficacy of the therapy and to understand the underlying molecular mechanisms. By using this model, Neal et al. prepared organoids from murine tumors which were established in syngeneic models. Then, the authors treated the organoids for 7 days with immune checkpoint inhibitors (targeting PD-1 and PD-L1) and demonstrated by flow cytometry an increased number of CD8+ cytotoxic T-cells among the whole set of CD3+ T-cells as well as an increased cellular death evidenced by Annexin V/7-AAD double labeling. Finally, by performing RT-qPCR, the authors identified genes involved in this stimulation [114].

The ability to co-culture organoids with either heterologous or autologous PBMC as well as tumor resident immune cells is paving the way for studies aiming at characterizing in vitro the effect of the immune components on heterogeneous tumor models closely related to what can be found in vivo. These methodological developments open new perspectives in terms of testing drugs on immune cells as well as on their recruitment and/or activation. Moreover, the easy access to the tumor-like structure facilitates the understanding of the underlying molecular mechanisms and the identification of novel therapeutic opportunities. Importantly, organoids can be cryopreserved, allowing the development of tumoroid biobanks. Such models are of particular interest notably for screening molecules [104]. Finally, the development of the two LLI and ALI approaches offers two distinctive methodological options depending on the purpose of the experiments. If the objectives of the study are identifying molecules capable of attracting and/or activating peripheral immune cells or a subset of them to the tumor, the LLI approach seems to be the most appropriate. On the contrary, the ALI approach should be used for testing in vitro drugs and screening molecules capable of acting on the intrinsic immune cells and/or the immunosuppressive microenvironment. As a patient-derived xenograft, ALI organoids can also be thought of as a tool for personalized medicine. Organoids can therefore be considered as an excellent pre-clinical model bringing patients into basic cancer research and facilitating the transfer of knowledge into the clinical practice.

3. In Vivo CRC Models for Immunotherapy Studies

Even if organoids brought new options for drug discovery, in vitro models usually remain an initial step in the development of novel immunotherapies. Validation in animal models is indeed required to gain access to the whole parameters involved in the anti-tumor response such as pharmacokinetics, metabolism, immunity, or organ toxicity. It is therefore important to use animal models before translating new findings to humans. For most anti-cancer therapies, immunodeficient mice xenografted with human cancer cell lines are used. However, due to the lack of immune system, those models cannot apply to immunotherapy and therefore specific models should be considered.

3.1. Syngeneic Models

The in vivo model the most commonly used by research groups working on immunotherapy is the syngeneic mouse model. This model consists of engraft murine cell lines previously grown in vitro into immunocompetent BALB/c or C57BL/6 mice. These murine models have an effective immune system. Hence, their treatment with immunotherapies allows treatment efficacy in terms of immune system activation and/or cytotoxic activity against tumors to be investigated (Figure 3) [29,67,115]. The major advantage of working with engrafted mouse cell lines is that the effect of the molecule can be easily determined by measuring the size of tumors. At the end of the procedure, tumors can be harvested, and several biological parameters possibly influenced by the treatment monitored. Immune cell infiltration can thus be evaluated by flow cytometry or immunohistochemistry. Expression of intra-tumor cytokines can be evaluated by cytokine array. Western blotting or RT-qPCR can also be performed to assess post-translational modifications or protein/genes expression levels. The whole set of data thus generated then improves our understanding of the mechanisms involved in either sensitivity or resistance to therapeutic interventions (Figure 3) [116–120]. Another major advantage of using syngeneic models is that engrafted cells can easily be manipulated and genetically modified in vitro prior to inoculation. Specific genes can be turned off to assess their impact on therapy and thus better ascribe the mechanism of action [121–123]. Genes encoding luciferase can also be introduced in the genome of the tumor cells to monitor their outcome in living organisms [124,125].

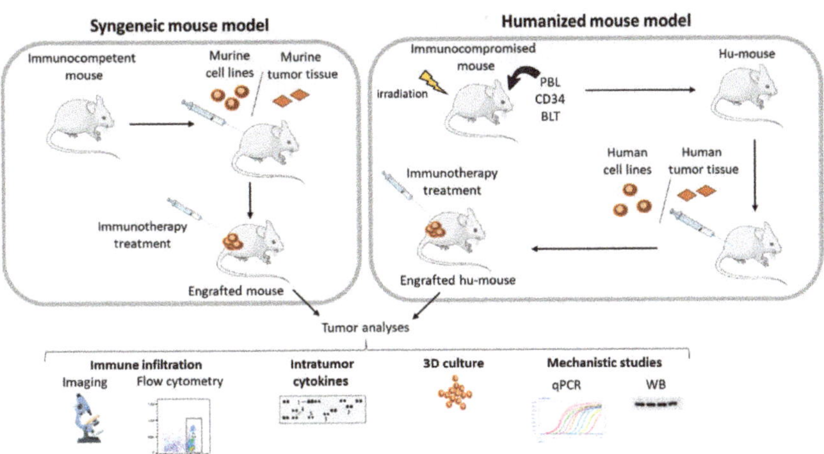

Figure 3. Mouse models used for immunotherapy research. The use of in vivo animal models is crucial for studying anti-tumor molecules acting on the TME. For immunotherapies, the easiest and most efficient model is the so-called syngeneic mouse model. A murine cancer cell line is injected into an immunocompetent mouse and the anti-cancer activity of the molecule of interest assessed through tumor growth inhibition, immune cell infiltration, and activation. Humanized mice models allow the efficacy of immunotherapies used in clinic on mice expressing human immune cells to be studied. Mice are humanized by injecting PBMC (PBL model), hematopoietic stem cells (CD34 model) or hematopoietic stem cells with grafting of human fetal liver and thymus (BLT model). These humanized mice are then grafted with either human tumor cell lines or human tumors samples (patient-derived xenograft). As for syngeneic models, tumors can be harvested and analyzed ex vivo for measuring immune infiltrate, cytokine release, and performed mechanistic studies. PBMC: Peripheral Blood Mononuclear Cell, PBL: Peripheral Blood Leukocyte, BLT: Bone marrow, Liver, Thymus.

For CRC, the most commonly used syngeneic mouse models are the two CT26 and MC38 murine cell lines inoculated in immunocompetent BALB/c and C57BL/6 mouse, respectively [126]. The CT26 cell line is derived from a colon tumor formed in BALB/c mice exposed to N-nitroso-N-methylurethane, while MC38 cells were isolated from a colon tumor formed in a C57BL/6 mouse exposed to 1,2-dimethylhydrazine dihydrochloride (DMH). In human CRC, the majority of tumors present mutations in APC, KRAS, and TP53 genes. However, CT26 cells only have identified mutations in KRAS (G12D, V8M) but not in either APC or TP53 genes. Similarly, MC38 cells are mutated in the TP53 gene (G242V, S2581) but not in either KRAS or APC genes [127]. CT26 cells are pMMR and express CMH class I with a robust binding capability, but not CMH class II antigen presentation molecules. This model is often considered as the most immunogenic syngeneic mouse model and is described as a good responsive model for immunotherapeutic research [127,128]. In particular, CT26 cells show a high mutational load in contrast to most pMMR human tumor. Moreover, a large NK-cell infiltrate is observed in CT26 syngeneic models, a characteristic which does not mirror human colon tumors known to be poorly infiltrated by NK-cells [127,129]. On the other hand, the MC38 cell line is defined as an MSI model of CRC and present the highest mutational load among the ten most commonly used syngeneic mouse models evaluated to date [127]. Several groups used these models to evaluate CRC responses to classical immunotherapeutic agents. In a detailed study made on commonly used syngeneic models for different cancer location, CT26 tumor growth was significantly inhibited by both anti-CTLA4 and anti-PD-1 antibodies. In contrast, no anti-tumor activity was detected against the MC38 cells [130]. However, in another study, a complete inhibition of the MC38 tumor growth was observed after treatment with an anti-PD-L1 antibody [123].

In addition to the direct evaluation of immunotherapies, several research groups are interested in studying drug combinations capable of enhancing immune-directed molecule activity. For CRC, a strong interest is focused on combining oxaliplatin with molecules in order to potentiate its action. Oxaliplatin is indeed an effective chemotherapy used for treating CRC patients in clinics but its effect is often limited by the development of resistance [131,132]. Interestingly, oxaliplatin is also known to strongly induce ICD which favors the establishment of an immune-favorable microenvironment, thus facilitating the recruitment and activation of immune cells [133,134]. In a recent work, to overcome oxaliplatin resistance, this platinated compound was combined with an inhibitor of the ataxia telangiectasia and Rad3-related protein (ATR) kinase. The results first showed a strong in vitro synergistic effect in six different human colorectal cancer cell lines and their oxaliplatin-resistant counterparts. Importantly, this combination was also evaluated in the MC38 syngeneic mouse model. In this setting, a synergistic effect was also demonstrated in terms of tumor growth. In addition, the use of the immunocompetent model permitted further insights in the in vivo activity of this combination to be gained. Indeed, the authors clearly demonstrated that the combination of oxaliplatin with VE-822 (an ATR inhibitor) promoted an anti-tumor T-cell response which was characterized by an increased number of MC38-targeting IFNγ-producing CD8+ T-cells in mice that received the combined treatment compared to those treated with oxaliplatin alone [135]. Another interesting study investigated the effects of combining oxaliplatin with the anti-PD-L1 immune checkpoint inhibitor in the CT26 syngeneic mouse model. In this work, the authors successfully demonstrated the interest of combining these two classes of molecules to inhibit tumor growth. Maybe more importantly, this work clearly showed that combining oxaliplatin with an anti-PD-L1 monoclonal antibody led to an increased tumor infiltration of CD8+ T-cells, especially when the anti-PD-L1 molecule was injected before oxaliplatin. This study thus highlights the interest of these murine models to evaluate novel drug combinations and to optimize drug administration scheduling [136]. In agreement, the CT26 syngeneic mouse model was also used to evaluate the combination of MEK inhibitors with anti-PD-L1 monoclonal antibodies [29]. As discussed above, CT26 cells are cancer cells bearing activating mutations in the Ras pathway. However, beside the direct effect of MEK inhibition on cancer cells, the authors showed that G-38963 (which is similar to the MEK inhibitor Cobimetinib) can increase the number of effector-phenotype antigen-specific CD8+ T-cells within the tumor and act on the tumor-infiltrating CD8+ T-cells' biology and survival. Importantly, the authors also reported that combining MEK inhibition with an anti-PD-L1 monoclonal antibody resulted in a synergistic and durable tumor response in mice [29]. However, in clinics, the effect of combining atezolizumab, a humanized IgG1 monoclonal antibody selectively targeting PD-L1, with Cobimetinib did not show any significant improvement of the overall survival of heavily pretreated pMMR mCRC patients compared with regorafenib or atezolizumab alone [137]. The discrepancy between preclinical and clinical data might be partly explained by the heterogeneous nature of mCRC at the advanced stages of the disease. This is particularly true for third-line-treated CRC patients who present otherwise chemo-refractory metastatic tumors. Moreover, the IMblaze370 clinical trial was not initially designed to assess the activity of the combination in different subgroups of patients. In addition, the CT26 cell line is known to be mutated in KRAS as well as for presenting MAPK1 and MET loci amplification [128]. This specific feature might therefore also impact their global response to MEK inhibitors. Finally, and as discussed above, the known immune responsiveness of the mouse CT26 syngeneic model is likely not to mirror the immune phenotype found in advanced pMMR tumors. These results underline the difficulties to translate preclinical data obtained on single cell type homogeneous tumors displaying genetic and immune-specific features to advanced clinical settings with inter- and intra-patient heterogeneous diseases. However, even if translation to humans might be tricky in the absence of biomarker assessments, the use of syngeneic mouse models remains a useful tool for deciphering clinical observations. This can be exemplified by the recent work demonstrating how liver metastasis might impact immunotherapy efficacy in patients with cancer [138]. In this work, the authors demonstrated that mice bearing subcutaneous syngeneic MC38 CRC tumors efficiently responded to anti-PD-L1 therapy while the same tumors failed to respond to this therapy in the presence of liver metastases. This phenomenon was related to a systemic loss of antigen-specific T-cells in mice bearing liver metastasis due an altered liver immune microenvironment favoring T-cell apoptosis. This model mirrors the systemic T-cell loss and decreased immunotherapy efficacy observed in patients with liver metastases. Importantly, this study also demonstrated that radiotherapy might reshape the liver immune microenvironment and abolish immunotherapy resistance. Together, these results suggest that liver metastases could serve as a potential biomarker for predicting immunotherapy response. Finally, the development of novel syngeneic mouse models, capable of better recapitulating the genetic origin of human CRC, might also prove useful for dealing with "immune desert" CRC. By crossing mice, mutated in four genes involved in colorectal cancer (APC, KRAS, TGFBR2, PTR53), Tauriello et al. established mutant mice developing pMMR metastatic intestinal tumors. From these tumors, the authors prepared organoids and engrafted them into immunocompetent mice. Interestingly, only a limited effect of anti-PD-L1 therapy could be observed on these MSS tumors. In contrast, TGF-β inhibition induced a significant reduction of the tumor mass thanks to the induction of a strong anti-tumor cytotoxic T-cell response. Moreover, TGF-β inhibition also prevented the formation of distant metastases and improved the response to anti-PD-L1 monoclonal antibody. Hence, the authors identified the TGF-β signaling as an interesting target for developing new strategies aimed at treating pMMR mCRC tumors that are otherwise resistant to immunotherapies [28].

Another important impact of mouse models in the field of immunotherapies is the easy access researchers have to immunodeficient as well as immunocompetent mice from the same origin. Indeed, by comparing the effects of molecules in mice with the same genetic background but in which the immune system is active or compromised, it is possible to evaluate the impact of the immune system has on the response to novel therapeutics or strategies. This approach was used for example to demonstrate that caloric restriction or hydroxycitrate improved the therapeutic outcome in CT26 CRC treated by chemotherapeutics in a T-cell-dependent fashion [139]. This strategy was also employed to explore the role of T-cells in the response to a novel immunotherapy targeting the phagocytic CD47 immune checkpoint [140]. There, the authors first demonstrated the anti-tumor activity of an anti-CD47 antibody in the treatment of A20 B lymphoma and MC38 CRC cells inoculated into wild-type BALB/c and C57BL/6 mice, respectively. However, when they looked at the efficacy of the treatment in immunocompromised nude mice, no effect could be observed on tumor growth, thus underlying the need for a T-cell-competent immune system in this process. Further experiments allowed the authors to demonstrate that the anti-cancer activity of the anti-CD47 antibody actually relied on DC cross-priming of CD8+ T-cells [140]. Finally, this approach can favor the emergence of new drug combinations which can be exemplified by the development of the prostaglandin E2 receptor 4 inhibitor, called TP-16 [141]. This molecule reduces the immunosuppressive myeloid cell functions. In this work, the authors first treated CT26 and MC38 CRC syngeneic tumor models with TP-16 and showed a significant reduction of the tumor mass. However, this effect was completely lost when the CRC cell lines were engrafted in immunocompromised nude mice. Again, these results stressed on the important role of having an intact immune system for observing an activity of this new molecule. Using in vitro approaches, the authors demonstrated that TP-16 reverses the immunosuppressive functions of PGE2 leading to an increased proportion of M1 macrophages, and a decreased fraction of M2 macrophages leading to a diminution of myeloid-derived suppressor cells thus favoring T-cell proliferation. These observations led the authors to combine TP-16 with an anti-PD-1 antibody in an immunocompetent CT26 syngeneic mouse model. Importantly, the combination showed a much more potent anti-tumor activity than either drug alone. The authors also reported that the combined effect of both compounds led to a more immune-favorable tumor microenvironment [141].

Together, these studies highlight the need of using syngeneic murine models for determining not only the efficacy of immunotherapies but also their underlying mechanism of action as well as the immune cells populations involved in. For colorectal cancer, syngeneic mouse models have yet proven to be a powerful tool for evaluating immunotherapies as well as strategies capable of turning cold tumors into hot responsive ones.

3.2. Humanized Mouse Model

One of the major limitations of syngeneic murine models is by definition the presence of a murine immune system which does not fully recapitulate the human one. Phenotypic and functional differences indeed exist between the two immune systems, thus leading to potential failures when findings are translated to the clinic [142]. Moreover, monoclonal antibodies used for targeting ICI in mice or in humans should be duplicated in order to target proteins in each species.

To circumvent this problem, humanized mice (hu-mice) were developed. They are models of immunocompromised mice displaying a reconstituted human immune system. The use of such models is a good alternative for testing the efficacy of immunotherapies on mice bearing human cell xenografts (Figure 3) [143,144]. Different strains of immuno-compromised mice with deficiencies in specific immune cells populations are available to date: (i) Nude mice (Foxn1 mutated) have no T-cells, (ii) scid mice (severe combined immunodeficient) have no T- or B-cells, (iii) NOD-scid mice (non-obese diabetic severe combined immunodeficiency) have reduced NK-cell and myeloid cell functions, express human SIRP-α, and have no C5 complement, (iv) NSG (NOD/SCID/IL2R^{null}) and NOG (NOD/SCID/IL2R$^{partial\ deficiency}$) mice have no T-, B-, or NK-cells and show a reduced myeloid cell functions and no complement, (v) NRG mice (NOD/RAG1^{null}/IL2R^{null}) have no T-, B-, or NK-cells as well as no complement or myeloid cells, and (vi) BRG mice (BALB/c/RAG2^{null}/IL2R^{null}) have no T-, B-, or NK-cells, and no complement but, when compared to the NSG model, display more functional myeloid cells [145,146].

Mice humanization can be performed following different procedures. PMBCs (Peripheral Blood Leukocyte or PBL model) or hematopoietic stem cells (HSC) isolated from cord blood, bone marrow, mobilized peripheral human blood, or fetal liver (CD34 model) can be injected in immunocompromised mice. Alternatively, transplanting human tissues (thymus + liver) in immunodeficient mice (BLT model) can be done (Figure 3). The PBL model is probably the easiest to establish since it relies on a simple intra-splenic, intraperitoneal, or intravenous injection of human PBMC and it allows an efficient engraftment of T-cells [147]. However, the engraftment of myeloid cells is generally low and mice survival is relatively short because of the occurrence of graft-versus-host disease (GvHD) [148,149]. On the other hand, the CD34 model implies intra-femoral or intravenous injection of HSC and allows the development of multiple hematopoietic lineages and insures primary immune responses. However, the education of T-cells which occurs in the murine thymus is achieved by the murine major histocompatibility complex (MHC), thus preventing the development of human MHC T-cells [147]. To overcome this problem, it is now possible to graft fragments of the human fetal liver and thymus under the kidney capsule of the mouse in addition to intravenously injecting human CD34+ HSC (BLT method). Hence, the education of T-cells is taking place in human thymus and it allows for the development of multiple hematopoietic lineages and functional human MHC T-cells [147]. Nonetheless, this technique requires complex surgery and depends on the availability of human fetal tissue. Moreover, these mice are often subject to lethality because of xenogeneic GvHD [150,151].

Despite these limitations, those models can help to better understand what might happen in the human context. This can be illustrated by the work of Wang et al. in which irradiated NSG mice were humanized by injecting HSC isolated from fetal liver [152]. The authors first confirmed the humanization by labeling human CD45+ cells in the blood of the mice. Then, they implanted a human breast cancer cell line (MDA-MB-231) into the mammary fat pad and assessed the treatment efficacy of pembrolizumab, an anti-PD-1 monoclonal antibody used in clinics, on those humanized mice. Interestingly, while a significant reduction of the tumor mass was observed, the authors could not evidence any increase of CD45+, CD3+/CD4+, and CD3+/CD8+ infiltrating cells in the whole tumor. However, they demonstrated that CD8+ T-cells were relocalized from the tumor burden to the center after treatment with pembrolizumab. By depleting the human-derived CD8+ T-cells with an anti-human CD8+ antibody, the authors confirmed in hu-mice the absolute requirement of a competent CD8+ T-cell population for observing a cytotoxic effect of pembrolizumab. The anti-PD-1 monoclonal antibody indeed showed no activity on humanized mice when they were depleted of their hu-CD8+ T-cells [152]. Hu-mice are also a powerful tool to evaluate the activity of immune cells expressing a chimeric antigen receptor. In that sense, we can cite the work of Klichinsky et al. who generated chimeric antigen receptor macrophages (CAR-M) and tested their anti-cancer activity in the Hu-NSGS (NOD/SCID/IL2R^{null}/hIL3/hGMCSF/hSF) mice model obtained after intravenous injection of human HSC [153]. The authors engrafted the human ovarian cancer SKOV3 cell line subcutaneously and then started intra-tumor injections of either CAR-M or unmodified macrophages. After 5 days of treatment, tumors were harvested, and RNA sequencing performed. The results clearly demonstrated the establishment of a pro-inflammatory tumor microenvironment promoting anti-tumor immunity when the hu-mice were treated with CAR-M [153].

Today, human cancer cell lines are still mostly used to engraft hu-mice. However, those cellular models are known to not strictly represent the tumor complexity and heterogeneity found in human cancers. Therefore, to better model the human pathology, humanized mice can be engrafted with patient-derived tumor tissue, the so-called patient-derived xenografts (PDX) (Figure 3) [154]. For CRC, PDX have been widely studied and have demonstrated high engraftment success rates [155,156]. PDX are isolated from patient tumor samples and transplanted subcutaneously in anaesthetized mice. The major advantage of PDXs is that they preserve the characteristics of the original tumor, both in terms of the gene expression profile and tumor heterogeneity [157,158]. Humanized mice bearing PDXs are therefore becoming a powerful tool for analyzing tumor and immune cell interaction and evaluating the efficacy of immunotherapies [159]. Recently, this approach was used to evaluate the response of CRC-PDX to Nivolumab, an anti-PD-1 monoclonal antibody used in clinics [160]. In this study, the authors first established their Hu-BRGS mice model by injecting into the facial vein or, when unsuccessful, into the liver CD34+ human HSC (isolated from cord blood) in irradiated BRGS (BALB/c/RAG2^{null}/IL2R^{null}/SIRPα^{NOD}) mice. CRC-PDX were then established in nude mice in order to favor their initial growth and thereafter implanted in the Hu-CB-BRGS mice. When tumors volumes reached 150–300 mm^3, treatments with Nivolumab were started. Strikingly, the authors demonstrated a significant tumor growth inhibition when the humanized mice were transplanted with CRC-PDX derived from MSI CRC patients while no efficacy could be observed in non-humanized immunodeficient BRGS mice. Moreover, the researchers could evidence, after treatment with Nivolumab, an increased infiltration of human TIL, T-cells, and CD8+ T-cells in PDX as well as an increased secretion of IFN-γ, thus demonstrating an anti-tumor immune response. Remarkably, when CRC-PDX derived from MSS CRC patients was used to engraft Hu-mice, the authors only showed a transient and partial anti-tumor effect of Nivolumab. This limited anti-cancer activity was accompanied by an absence of increased human TIL infiltration in the tumor [160]. This study thus recapitulated the observations made in the clinic, namely that dMMR CRC usually respond to ICI while pMMR CRC do not. This technological breakthrough opens new opportunities to better predict in patient-derived samples the clinical benefit of immunotherapies, but also to assess new strategies to overcome ICI resistance in CMS2-4 mCRC.

4. Conclusions

Our understanding of the impact of the immune system on the progression and prognostic of CRC led to tremendous efforts to better define tumor immune phenotypes. However, targeting most mCRC with immunotherapies remains to date challenging. Nonetheless, the development of novel experimental models led, in the last years, to the identification of some of the sources responsible for CRC non-responsiveness to immune-targeted therapies, thus identifying novel pharmacological opportunities to turn "cold" tumors into "hot" ones.

The better characterization of the intricate relationships existing between immune and tumor cells should, however, not limit our focus on these two unique compartments; the whole TME should indeed be considered. This assumption is particularly true for CRC if we consider that immunotherapies are, to date, mostly developed to treat metastatic diseases. However, nowadays, the characterization of metastasized CRC is still relatively poor, notably in terms of understanding of distant tissue TME [9]. In particular, the liver, which is one of the preferential location for CRC metastases, displays an immunosuppressive TME which likely facilitates CRC cells settlement but also shows specificities when compared to the original site of the tumor [9,161]. Accordingly, it has been reported that primary CRC tumors and CRC liver metastases diverged in terms of immune phenotypes [162]. Therefore, future successes in the field of CRC immunity are permitted due to the development of technologies allowing the evaluation of new innovative therapeutic strategies in this specific setting. In that sense, liver CRC metastasis-derived PDX established in humanized mice is likely to prove itself as a powerful tool to improve our understanding. However, the use of humanized mice remains complex, in particular for screening novel molecules or assessing innovative combinatorial strategies. Thus, emerging in vitro approaches should be designed to help circumventing the complexity of the in vivo models. In that sense, the development of multi-organoids-on-a-chip fluidic devices will prove useful for modeling primary and distant diseases as well as screening new drug modalities [163]. Development and implementation of novel microfluidics technologies might indeed provide solutions for managing small heterogeneous samples coming from patient-derived tumors or biopsy materials and for studying cancer cells' behavior in a closed physiological context [164]. Interestingly, such a metastasis-on-a-chip fluidic system has yet to be set up to study early stages of human CRC metastasis [165]. A 3D microfluidic platform has also been developed for evaluating the migration of interferon-α-conditioned dendritic cells toward SW620 CRC cells [166]. However, while substantial progress has been made in developing chip modeling organ physiology, efforts for incorporating immune components in these micro-physiological systems have only been expanded recently [167]. In the future, such devices might be further improved by adding immune cells as well as tumoroids derived from either the primary or distant tissues originating from one single patient. Such a system will therefore permit simultaneous immunity to be assessed on both locations and to determine whether therapeutic strategies equally apply to both CRC tumors' organ implantation sites.

Author Contributions: Conceptualization, E.P., C.D. and A.E.E.; writing—original draft preparation, E.P.; writing—review and editing, E.P. and A.E.E.; project administration, E.P. and A.E.E.; funding acquisition, A.E.E. All authors have read and agreed to the published version of the manuscript.

Funding: C.D. was supported by a fellowship from APREC ("Alliance Pour la Recherche En Cancérologie", France). This work was supported by APREC.

Institutional Review Board Statement: Not applicable.

Informed Consent Statement: Not applicable.

Data Availability Statement: Not applicable.

Conflicts of Interest: The authors declare no conflict of interest.

References

1. Bray, F.; Ferlay, J.; Soerjomataram, I.; Siegel, R.L.; Torre, L.A.; Jemal, A. Global Cancer Statistics 2018: GLOBOCAN Estimates of Incidence and Mortality Worldwide for 36 Cancers in 185 Countries. *CA Cancer J. Clin.* **2018**, *68*, 394–424. [CrossRef] [PubMed]
2. Koi, M.; Carethers, J.M. The Colorectal Cancer Immune Microenvironment and Approach to Immunotherapies. *Future Oncol.* **2017**, *13*, 1633–1647. [CrossRef] [PubMed]
3. Siegel, R.L.; Miller, K.D.; Fuchs, H.E.; Jemal, A. Cancer Statistics, 2021. *CA Cancer J. Clin.* **2021**, *71*, 7–33. [CrossRef] [PubMed]

4. Sobrero, A.; Lonardi, S.; Rosati, G.; Di Bartolomeo, M.; Ronzoni, M.; Pella, N.; Scartozzi, M.; Banzi, M.; Zampino, M.G.; Pasini, F.; et al. FOLFOX or CAPOX in Stage II to III Colon Cancer: Efficacy Results of the Italian Three or Six Colon Adjuvant Trial. *J. Clin. Oncol.* **2018**, *36*, 1478–1485. [CrossRef] [PubMed]

5. André, T.; Meyerhardt, J.; Iveson, T.; Sobrero, A.; Yoshino, T.; Souglakos, I.; Grothey, A.; Niedzwiecki, D.; Saunders, M.; Labianca, R.; et al. Effect of Duration of Adjuvant Chemotherapy for Patients with Stage III Colon Cancer (IDEA Collaboration): Final Results from a Prospective, Pooled Analysis of Six Randomised, Phase 3 Trials. *Lancet Oncol.* **2020**, *21*, 1620–1629. [CrossRef]

6. Van Cutsem, E.; Cervantes, A.; Adam, R.; Sobrero, A.; Van Krieken, J.H.; Aderka, D.; Aranda Aguilar, E.; Bardelli, A.; Benson, A.; Bodoky, G.; et al. ESMO Consensus Guidelines for the Management of Patients with Metastatic Colorectal Cancer. *Ann. Oncol.* **2016**, *27*, 1386–1422. [CrossRef]

7. Pagès, F.; Berger, A.; Camus, M.; Sanchez-Cabo, F.; Costes, A.; Molidor, R.; Mlecnik, B.; Kirilovsky, A.; Nilsson, M.; Damotte, D.; et al. Effector Memory T Cells, Early Metastasis, and Survival in Colorectal Cancer. *N. Engl. J. Med.* **2005**, *353*, 2654–2666. [CrossRef]

8. Galon, J.; Costes, A.; Sanchez-Cabo, F.; Kirilovsky, A.; Mlecnik, B.; Lagorce-Pagès, C.; Tosolini, M.; Camus, M.; Berger, A.; Wind, P.; et al. Type, Density, and Location of Immune Cells within Human Colorectal Tumors Predict Clinical Outcome. *Science* **2006**, *313*, 1960–1964. [CrossRef]

9. Roelands, J.; Kuppen, P.J.K.; Vermeulen, L.; Maccalli, C.; Decock, J.; Wang, E.; Marincola, F.M.; Bedognetti, D.; Hendrickx, W. Immunogenomic Classification of Colorectal Cancer and Therapeutic Implications. *Int. J. Mol. Sci.* **2017**, *18*, 2229. [CrossRef]

10. Galon, J.; Pagès, F.; Marincola, F.M.; Angell, H.K.; Thurin, M.; Lugli, A.; Zlobec, I.; Berger, A.; Bifulco, C.; Botti, G.; et al. Cancer Classification Using the Immunoscore: A Worldwide Task Force. *J. Transl. Med.* **2012**, *10*, 205. [CrossRef]

11. Pagès, F.; Mlecnik, B.; Marliot, F.; Bindea, G.; Ou, F.-S.; Bifulco, C.; Lugli, A.; Zlobec, I.; Rau, T.T.; Berger, M.D.; et al. International Validation of the Consensus Immunoscore for the Classification of Colon Cancer: A Prognostic and Accuracy Study. *Lancet* **2018**, *391*, 2128–2139. [CrossRef]

12. Ciardiello, D.; Vitiello, P.P.; Cardone, C.; Martini, G.; Troiani, T.; Martinelli, E.; Ciardiello, F. Immunotherapy of Colorectal Cancer: Challenges for Therapeutic Efficacy. *Cancer Treat. Rev.* **2019**, *76*, 22–32. [CrossRef] [PubMed]

13. Le, D.T.; Uram, J.N.; Wang, H.; Bartlett, B.R.; Kemberling, H.; Eyring, A.D.; Skora, A.D.; Luber, B.S.; Azad, N.S.; Laheru, D.; et al. PD-1 Blockade in Tumors with Mismatch-Repair Deficiency. *N. Engl. J. Med.* **2015**, *372*, 2509–2520. [CrossRef] [PubMed]

14. Cohen, R.; Rousseau, B.; Vidal, J.; Colle, R.; Diaz, L.A.; André, T. Immune Checkpoint Inhibition in Colorectal Cancer: Microsatellite Instability and Beyond. *Target. Oncol.* **2020**, *15*, 11–24. [CrossRef] [PubMed]

15. Marmorino, F.; Boccaccino, A.; Germani, M.M.; Falcone, A.; Cremolini, C. Immune Checkpoint Inhibitors in PMMR Metastatic Colorectal Cancer: A Tough Challenge. *Cancers* **2020**, *12*, 2317. [CrossRef] [PubMed]

16. Jass, J.R. Classification of Colorectal Cancer Based on Correlation of Clinical, Morphological and Molecular Features. *Histopathology* **2007**, *50*, 113–130. [CrossRef]

17. Bindea, G.; Mlecnik, B.; Tosolini, M.; Kirilovsky, A.; Waldner, M.; Obenauf, A.C.; Angell, H.; Fredriksen, T.; Lafontaine, L.; Berger, A.; et al. Spatiotemporal Dynamics of Intratumoral Immune Cells Reveal the Immune Landscape in Human Cancer. *Immunity* **2013**, *39*, 782–795. [CrossRef]

18. McGranahan, N.; Furness, A.J.S.; Rosenthal, R.; Ramskov, S.; Lyngaa, R.; Saini, S.K.; Jamal-Hanjani, M.; Wilson, G.A.; Birkbak, N.J.; Hiley, C.T.; et al. Clonal Neoantigens Elicit T Cell Immunoreactivity and Sensitivity to Immune Checkpoint Blockade. *Science* **2016**, *351*, 1463–1469. [CrossRef]

19. Mouw, K.W.; Goldberg, M.S.; Konstantinopoulos, P.A.; D'Andrea, A.D. DNA Damage and Repair Biomarkers of Immunotherapy Response. *Cancer Discov.* **2017**, *7*, 675–693. [CrossRef]

20. Germano, G.; Lamba, S.; Rospo, G.; Barault, L.; Magrì, A.; Maione, F.; Russo, M.; Crisafulli, G.; Bartolini, A.; Lerda, G.; et al. Inactivation of DNA Repair Triggers Neoantigen Generation and Impairs Tumour Growth. *Nature* **2017**, *552*, 116–120. [CrossRef]

21. Llosa, N.J.; Cruise, M.; Tam, A.; Wicks, E.C.; Hechenbleikner, E.M.; Taube, J.M.; Blosser, R.L.; Fan, H.; Wang, H.; Luber, B.S.; et al. The Vigorous Immune Microenvironment of Microsatellite Instable Colon Cancer Is Balanced by Multiple Counter-Inhibitory Checkpoints. *Cancer Discov.* **2015**, *5*, 43–51. [CrossRef] [PubMed]

22. Becht, E.; de Reyniès, A.; Giraldo, N.A.; Pilati, C.; Buttard, B.; Lacroix, L.; Selves, J.; Sautès-Fridman, C.; Laurent-Puig, P.; Fridman, W.H. Immune and Stromal Classification of Colorectal Cancer Is Associated with Molecular Subtypes and Relevant for Precision Immunotherapy. *Clin. Cancer Res.* **2016**, *22*, 4057–4066. [CrossRef] [PubMed]

23. Malesci, A.; Laghi, L.; Bianchi, P.; Delconte, G.; Randolph, A.; Torri, V.; Carnaghi, C.; Doci, R.; Rosati, R.; Montorsi, M.; et al. Reduced Likelihood of Metastases in Patients with Microsatellite-Unstable Colorectal Cancer. *Clin. Cancer Res.* **2007**, *13*, 3831–3839. [CrossRef] [PubMed]

24. Boland, C.R.; Goel, A. Microsatellite Instability in Colorectal Cancer. *Gastroenterology* **2010**, *138*, 2073–2087.e3. [CrossRef]

25. Guinney, J.; Dienstmann, R.; Wang, X.; de Reyniès, A.; Schlicker, A.; Soneson, C.; Marisa, L.; Roepman, P.; Nyamundanda, G.; Angelino, P.; et al. The Consensus Molecular Subtypes of Colorectal Cancer. *Nat. Med.* **2015**, *21*, 1350–1356. [CrossRef] [PubMed]

26. Angelova, M.; Charoentong, P.; Hackl, H.; Fischer, M.L.; Snajder, R.; Krogsdam, A.M.; Waldner, M.J.; Bindea, G.; Mlecnik, B.; Galon, J.; et al. Characterization of the Immunophenotypes and Antigenomes of Colorectal Cancers Reveals Distinct Tumor Escape Mechanisms and Novel Targets for Immunotherapy. *Genome Biol.* **2015**, *16*, 64. [CrossRef]

27. Courau, T.; Nehar-Belaid, D.; Florez, L.; Levacher, B.; Vazquez, T.; Brimaud, F.; Bellier, B.; Klatzmann, D. TGF-β and VEGF Cooperatively Control the Immunotolerant Tumor Environment and the Efficacy of Cancer Immunotherapies. *JCI Insight* **2016**, *1*, e85974. [CrossRef] [PubMed]

28. Tauriello, D.V.F.; Palomo-Ponce, S.; Stork, D.; Berenguer-Llergo, A.; Badia-Ramentol, J.; Iglesias, M.; Sevillano, M.; Ibiza, S.; Cañellas, A.; Hernando-Momblona, X.; et al. TGFβ Drives Immune Evasion in Genetically Reconstituted Colon Cancer Metastasis. *Nature* **2018**, *554*, 538–543. [CrossRef]

29. Ebert, P.J.R.; Cheung, J.; Yang, Y.; McNamara, E.; Hong, R.; Moskalenko, M.; Gould, S.E.; Maecker, H.; Irving, B.A.; Kim, J.M.; et al. MAP Kinase Inhibition Promotes T Cell and Anti-Tumor Activity in Combination with PD-L1 Checkpoint Blockade. *Immunity* **2016**, *44*, 609–621. [CrossRef]

30. Kanterman, J.; Sade-Feldman, M.; Biton, M.; Ish-Shalom, E.; Lasry, A.; Goldshtein, A.; Hubert, A.; Baniyash, M. Adverse Immunoregulatory Effects of 5FU and CPT11 Chemotherapy on Myeloid-Derived Suppressor Cells and Colorectal Cancer Outcomes. *Cancer Res.* **2014**, *74*, 6022–6035. [CrossRef]

31. Emens, L.A.; Middleton, G. The Interplay of Immunotherapy and Chemotherapy: Harnessing Potential Synergies. *Cancer Immunol. Res.* **2015**, *3*, 436–443. [CrossRef]

32. Galluzzi, L.; Buqué, A.; Kepp, O.; Zitvogel, L.; Kroemer, G. Immunological Effects of Conventional Chemotherapy and Targeted Anticancer Agents. *Cancer Cell* **2015**, *28*, 690–714. [CrossRef]

33. Terme, M.; Pernot, S.; Marcheteau, E.; Sandoval, F.; Benhamouda, N.; Colussi, O.; Dubreuil, O.; Carpentier, A.F.; Tartour, E.; Taieb, J. VEGFA-VEGFR Pathway Blockade Inhibits Tumor-Induced Regulatory T-Cell Proliferation in Colorectal Cancer. *Cancer Res.* **2013**, *73*, 539–549. [CrossRef] [PubMed]

34. Wang, L.; Wei, Y.; Fang, W.; Lu, C.; Chen, J.; Cui, G.; Diao, H. Cetuximab Enhanced the Cytotoxic Activity of Immune Cells during Treatment of Colorectal Cancer. *Cell. Physiol. Biochem.* **2017**, *44*, 1038–1050. [CrossRef] [PubMed]

35. Raimondi, C.; Nicolazzo, C.; Gradilone, A.; Giannini, G.; De Falco, E.; Chimenti, I.; Varriale, E.; Hauch, S.; Plappert, L.; Cortesi, E.; et al. Circulating Tumor Cells: Exploring Intratumor Heterogeneity of Colorectal Cancer. *Cancer Biol. Ther.* **2014**, *15*, 496–503. [CrossRef] [PubMed]

36. Berg, K.C.G.; Eide, P.W.; Eilertsen, I.A.; Johannessen, B.; Bruun, J.; Danielsen, S.A.; Bjørnslett, M.; Meza-Zepeda, L.A.; Eknæs, M.; Lind, G.E.; et al. Multi-Omics of 34 Colorectal Cancer Cell Lines—A Resource for Biomedical Studies. *Mol. Cancer* **2017**, *16*, 116. [CrossRef]

37. Sahin, I.H.; Akce, M.; Alese, O.; Shaib, W.; Lesinski, G.B.; El-Rayes, B.; Wu, C. Immune Checkpoint Inhibitors for the Treatment of MSI-H/MMR-D Colorectal Cancer and a Perspective on Resistance Mechanisms. *Br. J. Cancer* **2019**, *121*, 809–818. [CrossRef] [PubMed]

38. Gupta, R.; Sinha, S.; Paul, R.N. The Impact of Microsatellite Stability Status in Colorectal Cancer. *Curr. Probl. Cancer* **2018**, *42*, 548–559. [CrossRef]

39. Brandi, J.; Manfredi, M.; Speziali, G.; Gosetti, F.; Marengo, E.; Cecconi, D. Proteomic Approaches to Decipher Cancer Cell Secretome. *Semin. Cell Dev. Biol.* **2018**, *78*, 93–101. [CrossRef]

40. Mohebbi, B.; Ashtibaghaei, K.; Hashemi, M.; Hashemi, M.; Asadzadeh Aghdaei, H.; Zali, M.R. Conditioned Medium from Cultured Colorectal Cancer Cells Affects Peripheral Blood Mononuclear Cells Inflammatory Phenotype in Vitro. *Iran. J. Med. Sci.* **2019**, *44*, 334–341. [CrossRef]

41. Lee, C.; Jeong, H.; Bae, Y.; Shin, K.; Kang, S.; Kim, H.; Oh, J.; Bae, H. Targeting of M2-like Tumor-Associated Macrophages with a Melittin-Based pro-Apoptotic Peptide. *J. Immunother. Cancer* **2019**, *7*, 147. [CrossRef] [PubMed]

42. Di Mitri, D.; Mirenda, M.; Vasilevska, J.; Calcinotto, A.; Delaleu, N.; Revandkar, A.; Gil, V.; Boysen, G.; Losa, M.; Mosole, S.; et al. Re-Education of Tumor-Associated Macrophages by CXCR2 Blockade Drives Senescence and Tumor Inhibition in Advanced Prostate Cancer. *Cell Rep.* **2019**, *28*, 2156–2168.e5. [CrossRef] [PubMed]

43. Andón, F.T.; Digifico, E.; Maeda, A.; Erreni, M.; Mantovani, A.; Alonso, M.J.; Allavena, P. Targeting Tumor Associated Macrophages: The New Challenge for Nanomedicine. *Semin. Immunol.* **2017**, *34*, 103–113. [CrossRef]

44. Grugan, K.D.; McCabe, F.L.; Kinder, M.; Greenplate, A.R.; Harman, B.C.; Ekert, J.E.; van Rooijen, N.; Anderson, G.M.; Nemeth, J.A.; Strohl, W.R.; et al. Tumor-Associated Macrophages Promote Invasion While Retaining Fc-Dependent Anti-Tumor Function. *J. Immunol.* **2012**, *189*, 5457–5466. [CrossRef] [PubMed]

45. Solinas, G.; Schiarea, S.; Liguori, M.; Fabbri, M.; Pesce, S.; Zammataro, L.; Pasqualini, F.; Nebuloni, M.; Chiabrando, C.; Mantovani, A.; et al. Tumor-Conditioned Macrophages Secrete Migration-Stimulating Factor: A New Marker for M2-Polarization, Influencing Tumor Cell Motility. *J. Immunol.* **2010**, *185*, 642–652. [CrossRef]

46. Benner, B.; Scarberry, L.; Suarez-Kelly, L.P.; Duggan, M.C.; Campbell, A.R.; Smith, E.; Lapurga, G.; Jiang, K.; Butchar, J.P.; Tridandapani, S.; et al. Generation of Monocyte-Derived Tumor-Associated Macrophages Using Tumor-Conditioned Media Provides a Novel Method to Study Tumor-Associated Macrophages in Vitro. *J. Immunother. Cancer* **2019**, *7*, 140. [CrossRef]

47. Dong, H.; Yang, Y.; Gao, C.; Sun, H.; Wang, H.; Hong, C.; Wang, J.; Gong, F.; Gao, X. Lactoferrin-Containing Immunocomplex Mediates Antitumor Effects by Resetting Tumor-Associated Macrophages to M1 Phenotype. *J. Immunother. Cancer* **2020**, *8*. [CrossRef]

48. Sawa-Wejksza, K.; Dudek, A.; Lemieszek, M.; Kaławaj, K.; Kandefer-Szerszeń, M. Colon Cancer-Derived Conditioned Medium Induces Differentiation of THP-1 Monocytes into a Mixed Population of M1/M2 Cells. *Tumour Biol.* **2018**, *40*, 1010428318797880. [CrossRef]

49. Adil, A.A.M.; Vallinayagam, L.; Chitra, K.; Jamal, S.; Pandurangan, A.K.; Ahmed, N. Increased Expression of TGF-β and IFN-γ in Peripheral Blood Mononuclear Cells (PBMCs) Cultured in Conditioned Medium (CM) of K562 Cell Culture. *J. Environ. Pathol. Toxicol. Oncol.* **2019**, *38*, 173–183. [CrossRef]

50. Teng, L.; Liu, L.; Su, Y.; Yuan, X.; Li, J.; Fu, Q.; Chen, S.; Wang, C. Suppression of Alloimmunity in Mice by Regulatory T Cells Converted with Conditioned Media. *J. Surg. Res.* **2011**, *171*, 797–806. [CrossRef]

51. Wang, D.; Yang, L.; Yu, W.; Wu, Q.; Lian, J.; Li, F.; Liu, S.; Li, A.; He, Z.; Liu, J.; et al. Colorectal Cancer Cell-Derived CCL20 Recruits Regulatory T Cells to Promote Chemoresistance via FOXO1/CEBPB/NF-KB Signaling. *J. Immunother. Cancer* **2019**, *7*, 215. [CrossRef]

52. Heeran, A.B.; Dunne, M.R.; Morrissey, M.E.; Buckley, C.E.; Clarke, N.; Cannon, A.; Donlon, N.E.; Nugent, T.S.; Durand, M.; Dunne, C.; et al. The Protein Secretome Is Altered in Rectal Cancer Tissue Compared to Normal Rectal Tissue, and Alterations in the Secretome Induce Enhanced Innate Immune Responses. *Cancers* **2021**, *13*, 571. [CrossRef]

53. Madden, E.C.; Gorman, A.M.; Logue, S.E.; Samali, A. Tumour Cell Secretome in Chemoresistance and Tumour Recurrence. *Trends Cancer* **2020**, *6*, 489–505. [CrossRef] [PubMed]

54. Kather, J.N.; Halama, N.; Jaeger, D. Genomics and Emerging Biomarkers for Immunotherapy of Colorectal Cancer. *Semin. Cancer Biol.* **2018**, *52*, 189–197. [CrossRef] [PubMed]

55. Slaney, C.Y.; Kershaw, M.H.; Darcy, P.K. Trafficking of T Cells into Tumors. *Cancer Res.* **2014**, *74*, 7168–7174. [CrossRef] [PubMed]

56. Kmiecik, J.; Poli, A.; Brons, N.H.C.; Waha, A.; Eide, G.E.; Enger, P.Ø.; Zimmer, J.; Chekenya, M. Elevated CD3+ and CD8+ Tumor-Infiltrating Immune Cells Correlate with Prolonged Survival in Glioblastoma Patients despite Integrated Immunosuppressive Mechanisms in the Tumor Microenvironment and at the Systemic Level. *J. Neuroimmunol.* **2013**, *264*, 71–83. [CrossRef] [PubMed]

57. Tyrer, P.C.; Bean, E.G.; Ruth Foxwell, A.; Pavli, P. Effects of Bacterial Products on Enterocyte-Macrophage Interactions in Vitro. *Biochem. Biophys. Res. Commun.* **2011**, *413*, 336–341. [CrossRef]

58. Nagata, M.; Yamamoto, H.; Tabe, K.; Sakamoto, Y. Eosinophil Transmigration across VCAM-1-Expressing Endothelial Cells Is Upregulated by Antigen-Stimulated Mononuclear Cells. *Int. Arch. Allergy Immunol.* **2001**, *125* (Suppl. S1), 7–11. [CrossRef]

59. Wang, Y.; Zhang, H.; He, H.; Ai, K.; Yu, W.; Xiao, X.; Qin, Y.; Zhang, L.; Xiong, H.; Zhou, G. LRCH1 Suppresses Migration of CD4+ T Cells and Refers to Disease Activity in Ulcerative Colitis. *Int. J. Med. Sci.* **2020**, *17*, 599–608. [CrossRef]

60. Huang, R.-P. An Array of Possibilities in Cancer Research Using Cytokine Antibody Arrays. *Expert Rev. Proteom.* **2007**, *4*, 299–308. [CrossRef]

61. Zhang, Y.; Du, W.; Chen, Z.; Xiang, C. Upregulation of PD-L1 by SPP1 Mediates Macrophage Polarization and Facilitates Immune Escape in Lung Adenocarcinoma. *Exp. Cell Res.* **2017**, *359*, 449–457. [CrossRef]

62. Wang, C.-J.; Zhu, C.-C.; Xu, J.; Wang, M.; Zhao, W.-Y.; Liu, Q.; Zhao, G.; Zhang, Z.-Z. The LncRNA UCA1 Promotes Proliferation, Migration, Immune Escape and Inhibits Apoptosis in Gastric Cancer by Sponging Anti-Tumor MiRNAs. *Mol. Cancer* **2019**, *18*, 115. [CrossRef] [PubMed]

63. Harlin, H.; Meng, Y.; Peterson, A.C.; Zha, Y.; Tretiakova, M.; Slingluff, C.; McKee, M.; Gajewski, T.F. Chemokine Expression in Melanoma Metastases Associated with CD8+ T-Cell Recruitment. *Cancer Res.* **2009**, *69*, 3077–3085. [CrossRef] [PubMed]

64. Wang, X.-Q.; Zhou, W.-J.; Luo, X.-Z.; Tao, Y.; Li, D.-J. Synergistic Effect of Regulatory T Cells and Proinflammatory Cytokines in Angiogenesis in the Endometriotic Milieu. *Hum. Reprod.* **2017**, *32*, 1304–1317. [CrossRef]

65. Hennel, R.; Brix, N.; Seidl, K.; Ernst, A.; Scheithauer, H.; Belka, C.; Lauber, K. Release of Monocyte Migration Signals by Breast Cancer Cell Lines after Ablative and Fractionated γ-Irradiation. *Radiat. Oncol.* **2014**, *9*, 85. [CrossRef]

66. Yu, Y.; Blokhuis, B.; Derks, Y.; Kumari, S.; Garssen, J.; Redegeld, F. Human Mast Cells Promote Colon Cancer Growth via Bidirectional Crosstalk: Studies in 2D and 3D Coculture Models. *Oncoimmunology* **2018**, *7*, e1504729. [CrossRef] [PubMed]

67. Liu, C.; Liu, R.; Wang, B.; Lian, J.; Yao, Y.; Sun, H.; Zhang, C.; Fang, L.; Guan, X.; Shi, J.; et al. Blocking IL-17A Enhances Tumor Response to Anti-PD-1 Immunotherapy in Microsatellite Stable Colorectal Cancer. *J. Immunother. Cancer* **2021**, *9*. [CrossRef]

68. Grützkau, A.; Radbruch, A. Small but Mighty: How the MACS-Technology Based on Nanosized Superparamagnetic Particles Has Helped to Analyze the Immune System within the Last 20 Years. *Cytom. A* **2010**, *77*, 643–647. [CrossRef] [PubMed]

69. Rios, F.J.; Touyz, R.M.; Montezano, A.C. Isolation and Differentiation of Human Macrophages. *Methods Mol. Biol.* **2017**, *1527*, 311–320. [CrossRef]

70. Trickett, A.; Kwan, Y.L. T Cell Stimulation and Expansion Using Anti-CD3/CD28 Beads. *J. Immunol. Methods* **2003**, *275*, 251–255. [CrossRef]

71. Tomchuck, S.L.; Leung, W.H.; Dallas, M.H. Enhanced Cytotoxic Function of Natural Killer and CD3+CD56+ Cells in Cord Blood after Culture. *Biol. Blood Marrow Transplant.* **2015**, *21*, 39–49. [CrossRef]

72. Vogel, D.Y.S.; Glim, J.E.; Stavenuiter, A.W.D.; Breur, M.; Heijnen, P.; Amor, S.; Dijkstra, C.D.; Beelen, R.H.J. Human Macrophage Polarization in Vitro: Maturation and Activation Methods Compared. *Immunobiology* **2014**, *219*, 695–703. [CrossRef]

73. Melief, J.; Wickström, S.; Kiessling, R.; Pico de Coaña, Y. Assessment of Antitumor T-Cell Responses by Flow Cytometry After Coculture of Tumor Cells with Autologous Tumor-Infiltrating Lymphocytes. *Methods Mol. Biol.* **2019**, *1913*, 133–140. [CrossRef]

74. Minute, L.; Teijeira, A.; Sanchez-Paulete, A.R.; Ochoa, M.C.; Alvarez, M.; Otano, I.; Etxeberrria, I.; Bolaños, E.; Azpilikueta, A.; Garasa, S.; et al. Cellular Cytotoxicity Is a Form of Immunogenic Cell Death. *J. Immunother. Cancer* **2020**, *8*. [CrossRef]

75. Yu, W.; Wang, Y.; Guo, P. Notch Signaling Pathway Dampens Tumor-Infiltrating CD8+ T Cells Activity in Patients with Colorectal Carcinoma. *Biomed. Pharmacother. Biomed. Pharmacother.* **2018**, *97*, 535–542. [CrossRef]

76. He, W.; Zhang, H.; Han, F.; Chen, X.; Lin, R.; Wang, W.; Qiu, H.; Zhuang, Z.; Liao, Q.; Zhang, W.; et al. CD155T/TIGIT Signaling Regulates CD8+ T-Cell Metabolism and Promotes Tumor Progression in Human Gastric Cancer. *Cancer Res.* **2017**, *77*, 6375–6388. [CrossRef]

77. Zaidi, N.E.; Shazali, N.A.H.; Chor, A.L.T.; Osman, M.A.; Ibrahim, K.; Jaoi-Edward, M.; Afizan Nik Abd Rahman, N.M. Time-Lapse 2D Imaging of Phagocytic Activity in M1 Macrophage-4T1 Mouse Mammary Carcinoma Cells in Co-Cultures. *J. Vis. Exp. JoVE* **2019**, *154*, e60281. [CrossRef] [PubMed]

78. Souza, A.G.; Silva, I.B.B.; Campos-Fernandez, E.; Barcelos, L.S.; Souza, J.B.; Marangoni, K.; Goulart, L.R.; Alonso-Goulart, V. Comparative Assay of 2D and 3D Cell Culture Models: Proliferation, Gene Expression and Anticancer Drug Response. *Curr. Pharm. Des.* **2018**, *24*, 1689–1694. [CrossRef] [PubMed]

79. Tung, Y.-C.; Hsiao, A.Y.; Allen, S.G.; Torisawa, Y.; Ho, M.; Takayama, S. High-Throughput 3D Spheroid Culture and Drug Testing Using a 384 Hanging Drop Array. *Analyst* **2011**, *136*, 473–478. [CrossRef] [PubMed]

80. Breslin, S.; O'Driscoll, L. Three-Dimensional Cell Culture: The Missing Link in Drug Discovery. *Drug Discov. Today* **2013**, *18*, 240–249. [CrossRef] [PubMed]

81. Silva-Almeida, C.; Ewart, M.-A.; Wilde, C. 3D Gastrointestinal Models and Organoids to Study Metabolism in Human Colon Cancer. *Semin. Cell Dev. Biol.* **2020**, *98*, 98–104. [CrossRef]

82. Vinci, M.; Gowan, S.; Boxall, F.; Patterson, L.; Zimmermann, M.; Court, W.; Lomas, C.; Mendiola, M.; Hardisson, D.; Eccles, S.A. Advances in Establishment and Analysis of Three-Dimensional Tumor Spheroid-Based Functional Assays for Target Validation and Drug Evaluation. *BMC Biol.* **2012**, *10*, 29. [CrossRef] [PubMed]

83. Kim, J.B. Three-Dimensional Tissue Culture Models in Cancer Biology. *Semin. Cancer Biol.* **2005**, *15*, 365–377. [CrossRef]

84. Foty, R. A Simple Hanging Drop Cell Culture Protocol for Generation of 3D Spheroids. *J. Vis. Exp. JoVE* **2011**. [CrossRef]

85. Weiswald, L.-B.; Bellet, D.; Dangles-Marie, V. Spherical Cancer Models in Tumor Biology. *Neoplasia* **2015**, *17*, 1–15. [CrossRef] [PubMed]

86. Kelm, J.M.; Timmins, N.E.; Brown, C.J.; Fussenegger, M.; Nielsen, L.K. Method for Generation of Homogeneous Multicellular Tumor Spheroids Applicable to a Wide Variety of Cell Types. *Biotechnol. Bioeng.* **2003**, *83*, 173–180. [CrossRef] [PubMed]

87. Friedrich, J.; Ebner, R.; Kunz-Schughart, L.A. Experimental Anti-Tumor Therapy in 3-D: Spheroids–Old Hat or New Challenge? *Int. J. Radiat. Biol.* **2007**, *83*, 849–871. [CrossRef]

88. Hamilton, G.; Rath, B. Applicability of Tumor Spheroids for in Vitro Chemosensitivity Assays. *Expert Opin. Drug Metab. Toxicol.* **2019**, *15*, 15–23. [CrossRef]

89. Grimes, D.R.; Kelly, C.; Bloch, K.; Partridge, M. A Method for Estimating the Oxygen Consumption Rate in Multicellular Tumour Spheroids. *J. R. Soc. Interface* **2014**, *11*, 20131124. [CrossRef]

90. Gong, X.; Lin, C.; Cheng, J.; Su, J.; Zhao, H.; Liu, T.; Wen, X.; Zhao, P. Generation of Multicellular Tumor Spheroids with Microwell-Based Agarose Scaffolds for Drug Testing. *PLoS ONE* **2015**, *10*, e0130348. [CrossRef]

91. Dubessy, C.; Merlin, J.M.; Marchal, C.; Guillemin, F. Spheroids in Radiobiology and Photodynamic Therapy. *Crit. Rev. Oncol. Hematol.* **2000**, *36*, 179–192. [CrossRef]

92. Mó, I.; Alves, C.G.; de Melo-Diogo, D.; Lima-Sousa, R.; Correia, I.J. Assessing the Combinatorial Chemo-Photothermal Therapy Mediated by Sulfobetaine Methacrylate-Functionalized Nanoparticles in 2D and 3D In Vitro Cancer Models. *Biotechnol. J.* **2020**, *15*, e2000219. [CrossRef] [PubMed]

93. Imamura, Y.; Mukohara, T.; Shimono, Y.; Funakoshi, Y.; Chayahara, N.; Toyoda, M.; Kiyota, N.; Takao, S.; Kono, S.; Nakatsura, T.; et al. Comparison of 2D- and 3D-Culture Models as Drug-Testing Platforms in Breast Cancer. *Oncol. Rep.* **2015**, *33*, 1837–1843. [CrossRef] [PubMed]

94. Nunes, A.S.; Costa, E.C.; Barros, A.S.; de Melo-Diogo, D.; Correia, I.J. Establishment of 2D Cell Cultures Derived From 3D MCF-7 Spheroids Displaying a Doxorubicin Resistant Profile. *Biotechnol. J.* **2019**, *14*, e1800268. [CrossRef]

95. Azharuddin, M.; Roberg, K.; Dhara, A.K.; Jain, M.V.; Darcy, P.; Hinkula, J.; Slater, N.K.H.; Patra, H.K. Dissecting Multi Drug Resistance in Head and Neck Cancer Cells Using Multicellular Tumor Spheroids. *Sci. Rep.* **2019**, *9*, 20066. [CrossRef] [PubMed]

96. Minchinton, A.I.; Tannock, I.F. Drug Penetration in Solid Tumours. *Nat. Rev. Cancer* **2006**, *6*, 583–592. [CrossRef]

97. Rebelo, S.P.; Pinto, C.; Martins, T.R.; Harrer, N.; Estrada, M.F.; Loza-Alvarez, P.; Cabeçadas, J.; Alves, P.M.; Gualda, E.J.; Sommergruber, W.; et al. 3D-3-Culture: A Tool to Unveil Macrophage Plasticity in the Tumour Microenvironment. *Biomaterials* **2018**, *163*, 185–197. [CrossRef]

98. Herter, S.; Morra, L.; Schlenker, R.; Sulcova, J.; Fahrni, L.; Waldhauer, I.; Lehmann, S.; Reisländer, T.; Agarkova, I.; Kelm, J.M.; et al. A Novel Three-Dimensional Heterotypic Spheroid Model for the Assessment of the Activity of Cancer Immunotherapy Agents. *Cancer Immunol. Immunother.* **2017**, *66*, 129–140. [CrossRef]

99. Alonso-Nocelo, M.; Abuín, C.; López-López, R.; de la Fuente, M. Development and Characterization of a Three-Dimensional Co-Culture Model of Tumor T Cell Infiltration. *Biofabrication* **2016**, *8*, 025002. [CrossRef]

100. Courau, T.; Bonnereau, J.; Chicoteau, J.; Bottois, H.; Remark, R.; Assante Miranda, L.; Toubert, A.; Blery, M.; Aparicio, T.; Allez, M.; et al. Cocultures of Human Colorectal Tumor Spheroids with Immune Cells Reveal the Therapeutic Potential of MICA/B and NKG2A Targeting for Cancer Treatment. *J. Immunother. Cancer* **2019**, *7*, 74. [CrossRef]

101. Hickman, J.A.; Graeser, R.; de Hoogt, R.; Vidic, S.; Brito, C.; Gutekunst, M.; van der Kuip, H.; IMI PREDECT Consortium. Three-Dimensional Models of Cancer for Pharmacology and Cancer Cell Biology: Capturing Tumor Complexity in Vitro/Ex Vivo. *Biotechnol. J.* **2014**, *9*, 1115–1128. [CrossRef]

102. Ji, D.-B.; Wu, A.-W. Organoid in Colorectal Cancer: Progress and Challenges. *Chin. Med. J.* **2020**, *133*, 1971–1977. [CrossRef] [PubMed]

103. Li, X.; Larsson, P.; Ljuslinder, I.; Öhlund, D.; Myte, R.; Löfgren-Burström, A.; Zingmark, C.; Ling, A.; Edin, S.; Palmqvist, R. Ex Vivo Organoid Cultures Reveal the Importance of the Tumor Microenvironment for Maintenance of Colorectal Cancer Stem Cells. *Cancers* **2020**, *12*, 923. [CrossRef] [PubMed]

104. Van de Wetering, M.; Francies, H.E.; Francis, J.M.; Bounova, G.; Iorio, F.; Pronk, A.; van Houdt, W.; van Gorp, J.; Taylor-Weiner, A.; Kester, L.; et al. Prospective Derivation of a Living Organoid Biobank of Colorectal Cancer Patients. *Cell* **2015**, *161*, 933–945. [CrossRef] [PubMed]

105. Sasaki, N.; Clevers, H. Studying Cellular Heterogeneity and Drug Sensitivity in Colorectal Cancer Using Organoid Technology. *Curr. Opin. Genet. Dev.* **2018**, *52*, 117–122. [CrossRef] [PubMed]

106. Roerink, S.F.; Sasaki, N.; Lee-Six, H.; Young, M.D.; Alexandrov, L.B.; Behjati, S.; Mitchell, T.J.; Grossmann, S.; Lightfoot, H.; Egan, D.A.; et al. Intra-Tumour Diversification in Colorectal Cancer at the Single-Cell Level. *Nature* **2018**, *556*, 457–462. [CrossRef]

107. Riedhammer, C.; Halbritter, D.; Weissert, R. Peripheral Blood Mononuclear Cells: Isolation, Freezing, Thawing, and Culture. *Methods Mol. Biol.* **2016**, *1304*, 53–61. [CrossRef]

108. Bar-Ephraim, Y.E.; Kretzschmar, K.; Clevers, H. Organoids in Immunological Research. *Nat. Rev. Immunol.* **2020**, *20*, 279–293. [CrossRef]

109. Gonzalez-Exposito, R.; Semiannikova, M.; Griffiths, B.; Khan, K.; Barber, L.J.; Woolston, A.; Spain, G.; von Loga, K.; Challoner, B.; Patel, R.; et al. CEA Expression Heterogeneity and Plasticity Confer Resistance to the CEA-Targeting Bispecific Immunotherapy Antibody Cibisatamab (CEA-TCB) in Patient-Derived Colorectal Cancer Organoids. *J. Immunother. Cancer* **2019**, *7*, 101. [CrossRef]

110. Dijkstra, K.K.; Cattaneo, C.M.; Weeber, F.; Chalabi, M.; van de Haar, J.; Fanchi, L.F.; Slagter, M.; van der Velden, D.L.; Kaing, S.; Kelderman, S.; et al. Generation of Tumor-Reactive T Cells by Co-Culture of Peripheral Blood Lymphocytes and Tumor Organoids. *Cell* **2018**, *174*, 1586–1598.e12. [CrossRef]

111. Usui, T.; Sakurai, M.; Umata, K.; Yamawaki, H.; Ohama, T.; Sato, K. Preparation of Human Primary Colon Tissue-Derived Organoid Using Air Liquid Interface Culture. *Curr. Protoc. Toxicol.* **2018**, *75*, 22.6.1–22.6.7. [CrossRef]

112. Li, X.; Ootani, A.; Kuo, C. An Air-Liquid Interface Culture System for 3D Organoid Culture of Diverse Primary Gastrointestinal Tissues. *Methods Mol. Biol.* **2016**, *1422*, 33–40. [CrossRef]

113. Finnberg, N.K.; Gokare, P.; Lev, A.; Grivennikov, S.I.; MacFarlane, A.W.; Campbell, K.S.; Winters, R.M.; Kaputa, K.; Farma, J.M.; Abbas, A.E.-S.; et al. Application of 3D Tumoroid Systems to Define Immune and Cytotoxic Therapeutic Responses Based on Tumoroid and Tissue Slice Culture Molecular Signatures. *Oncotarget* **2017**, *8*, 66747–66757. [CrossRef]

114. Neal, J.T.; Li, X.; Zhu, J.; Giangarra, V.; Grzeskowiak, C.L.; Ju, J.; Liu, I.H.; Chiou, S.-H.; Salahudeen, A.A.; Smith, A.R.; et al. Organoid Modeling of the Tumor Immune Microenvironment. *Cell* **2018**, *175*, 1972–1988.e16. [CrossRef]

115. You, D.; Hillerman, S.; Locke, G.; Chaudhry, C.; Stromko, C.; Murtaza, A.; Fan, Y.; Koenitzer, J.; Chen, Y.; Briceno, S.; et al. Enhanced Antitumor Immunity by a Novel Small Molecule HPK1 Inhibitor. *J. Immunother. Cancer* **2021**, *9*. [CrossRef]

116. Selby, M.J.; Engelhardt, J.J.; Johnston, R.J.; Lu, L.-S.; Han, M.; Thudium, K.; Yao, D.; Quigley, M.; Valle, J.; Wang, C.; et al. Preclinical Development of Ipilimumab and Nivolumab Combination Immunotherapy: Mouse Tumor Models, In Vitro Functional Studies, and Cynomolgus Macaque Toxicology. *PLoS ONE* **2016**, *11*, e0161779. [CrossRef]

117. Goggi, J.L.; Tan, Y.X.; Hartimath, S.V.; Jieu, B.; Hwang, Y.Y.; Jiang, L.; Boominathan, R.; Cheng, P.; Yuen, T.Y.; Chin, H.X.; et al. Granzyme B PET Imaging of Immune Checkpoint Inhibitor Combinations in Colon Cancer Phenotypes. *Mol. Imaging Biol.* **2020**, *22*, 1392–1402. [CrossRef]

118. Um, W.; Ko, H.; You, D.G.; Lim, S.; Kwak, G.; Shim, M.K.; Yang, S.; Lee, J.; Song, Y.; Kim, K.; et al. Necroptosis-Inducible Polymeric Nanobubbles for Enhanced Cancer Sonoimmunotherapy. *Adv. Mater.* **2020**, *32*, e1907953. [CrossRef] [PubMed]

119. Kristensen, L.K.; Fröhlich, C.; Christensen, C.; Melander, M.C.; Poulsen, T.T.; Galler, G.R.; Lantto, J.; Horak, I.D.; Kragh, M.; Nielsen, C.H.; et al. CD4+ and CD8a+ PET Imaging Predicts Response to Novel PD-1 Checkpoint Inhibitor: Studies of Sym021 in Syngeneic Mouse Cancer Models. *Theranostics* **2019**, *9*, 8221–8238. [CrossRef] [PubMed]

120. Napolitano, S.; Matrone, N.; Muddassir, A.L.; Martini, G.; Sorokin, A.; De Falco, V.; Giunta, E.F.; Ciardiello, D.; Martinelli, E.; Belli, V.; et al. Triple Blockade of EGFR, MEK and PD-L1 Has Antitumor Activity in Colorectal Cancer Models with Constitutive Activation of MAPK Signaling and PD-L1 Overexpression. *J. Exp. Clin. Cancer Res.* **2019**, *38*, 492. [CrossRef] [PubMed]

121. Schweickert, P.G.; Yang, Y.; White, E.E.; Cresswell, G.M.; Elzey, B.D.; Ratliff, T.L.; Arumugam, P.; Antoniak, S.; Mackman, N.; Flick, M.J.; et al. Thrombin-PAR1 Signaling in Pancreatic Cancer Promotes an Immunosuppressive Microenvironment. *J. Thromb. Haemost.* **2021**, *19*, 161–172. [CrossRef]

122. Kim, J.S.; Kim, E.J.; Lee, S.; Tan, X.; Liu, X.; Park, S.; Kang, K.; Yoon, J.-S.; Ko, Y.H.; Kurie, J.M.; et al. MiR-34a and MiR-34b/c Have Distinct Effects on the Suppression of Lung Adenocarcinomas. *Exp. Mol. Med.* **2019**, *51*, 1–10. [CrossRef]

123. Juneja, V.R.; McGuire, K.A.; Manguso, R.T.; LaFleur, M.W.; Collins, N.; Haining, W.N.; Freeman, G.J.; Sharpe, A.H. PD-L1 on Tumor Cells Is Sufficient for Immune Evasion in Immunogenic Tumors and Inhibits CD8 T Cell Cytotoxicity. *J. Exp. Med.* **2017**, *214*, 895–904. [CrossRef] [PubMed]

124. Vandeveer, A.J.; Fallon, J.K.; Tighe, R.; Sabzevari, H.; Schlom, J.; Greiner, J.W. Systemic Immunotherapy of Non-Muscle Invasive Mouse Bladder Cancer with Avelumab, an Anti-PD-L1 Immune Checkpoint Inhibitor. *Cancer Immunol. Res.* **2016**, *4*, 452–462. [CrossRef] [PubMed]

125. Rigo, V.; Emionite, L.; Daga, A.; Astigiano, S.; Corrias, M.V.; Quintarelli, C.; Locatelli, F.; Ferrini, S.; Croce, M. Combined Immunotherapy with Anti-PDL-1/PD-1 and Anti-CD4 Antibodies Cures Syngeneic Disseminated Neuroblastoma. *Sci. Rep.* **2017**, *7*, 14049. [CrossRef]
126. Yakkundi, P.; Gonsalves, E.; Galou-Lameyer, M.; Selby, M.J.; Chan, W.K. Aryl Hydrocarbon Receptor Acts as a Tumor Suppressor in a Syngeneic MC38 Colon Carcinoma Tumor Model. *Hypoxia* **2019**, *7*, 1–16. [CrossRef]
127. Zhong, W.; Myers, J.S.; Wang, F.; Wang, K.; Lucas, J.; Rosfjord, E.; Lucas, J.; Hooper, A.T.; Yang, S.; Lemon, L.A.; et al. Comparison of the Molecular and Cellular Phenotypes of Common Mouse Syngeneic Models with Human Tumors. *BMC Genom.* **2020**, *21*, 2. [CrossRef] [PubMed]
128. Castle, J.C.; Loewer, M.; Boegel, S.; de Graaf, J.; Bender, C.; Tadmor, A.D.; Boisguerin, V.; Bukur, T.; Sorn, P.; Paret, C.; et al. Immunomic, Genomic and Transcriptomic Characterization of CT26 Colorectal Carcinoma. *BMC Genom.* **2014**, *15*, 190. [CrossRef]
129. Gentles, A.J.; Newman, A.M.; Liu, C.L.; Bratman, S.V.; Feng, W.; Kim, D.; Nair, V.S.; Xu, Y.; Khuong, A.; Hoang, C.D.; et al. The Prognostic Landscape of Genes and Infiltrating Immune Cells across Human Cancers. *Nat. Med.* **2015**, *21*, 938–945. [CrossRef]
130. Mosely, S.I.S.; Prime, J.E.; Sainson, R.C.A.; Koopmann, J.-O.; Wang, D.Y.Q.; Greenawalt, D.M.; Ahdesmaki, M.J.; Leyland, R.; Mullins, S.; Pacelli, L.; et al. Rational Selection of Syngeneic Preclinical Tumor Models for Immunotherapeutic Drug Discovery. *Cancer Immunol. Res.* **2017**, *5*, 29–41. [CrossRef]
131. Liu, T.; Zhang, X.; Du, L.; Wang, Y.; Liu, X.; Tian, H.; Wang, L.; Li, P.; Zhao, Y.; Duan, W.; et al. Exosome-Transmitted MiR-128-3p Increase Chemosensitivity of Oxaliplatin-Resistant Colorectal Cancer. *Mol. Cancer* **2019**, *18*, 43. [CrossRef] [PubMed]
132. Hsu, H.-H.; Chen, M.-C.; Baskaran, R.; Lin, Y.-M.; Day, C.H.; Lin, Y.-J.; Tu, C.-C.; Vijaya Padma, V.; Kuo, W.-W.; Huang, C.-Y. Oxaliplatin Resistance in Colorectal Cancer Cells Is Mediated via Activation of ABCG2 to Alleviate ER Stress Induced Apoptosis. *J. Cell. Physiol.* **2018**, *233*, 5458–5467. [CrossRef] [PubMed]
133. Tesniere, A.; Schlemmer, F.; Boige, V.; Kepp, O.; Martins, I.; Ghiringhelli, F.; Aymeric, L.; Michaud, M.; Apetoh, L.; Barault, L.; et al. Immunogenic Death of Colon Cancer Cells Treated with Oxaliplatin. *Oncogene* **2010**, *29*, 482–491. [CrossRef] [PubMed]
134. Taniura, T.; Iida, Y.; Kotani, H.; Ishitobi, K.; Tajima, Y.; Harada, M. Immunogenic Chemotherapy in Two Mouse Colon Cancer Models. *Cancer Sci.* **2020**, *111*, 3527–3539. [CrossRef] [PubMed]
135. Combès, E.; Andrade, A.F.; Tosi, D.; Michaud, H.-A.; Coquel, F.; Garambois, V.; Desigaud, D.; Jarlier, M.; Coquelle, A.; Pasero, P.; et al. Inhibition of Ataxia-Telangiectasia Mutated and RAD3-Related (ATR) Overcomes Oxaliplatin Resistance and Promotes Antitumor Immunity in Colorectal Cancer. *Cancer Res.* **2019**, *79*, 2933–2946. [CrossRef] [PubMed]
136. Golchin, S.; Alimohammadi, R.; Rostami Nejad, M.; Jalali, S.A. Synergistic Antitumor Effect of Anti-PD-L1 Combined with Oxaliplatin on a Mouse Tumor Model. *J. Cell. Physiol.* **2019**, *234*, 19866–19874. [CrossRef]
137. Eng, C.; Kim, T.W.; Bendell, J.; Argilés, G.; Tebbutt, N.C.; Di Bartolomeo, M.; Falcone, A.; Fakih, M.; Kozloff, M.; Segal, N.H.; et al. Atezolizumab with or without Cobimetinib versus Regorafenib in Previously Treated Metastatic Colorectal Cancer (IMblaze370): A Multicentre, Open-Label, Phase 3, Randomised, Controlled Trial. *Lancet Oncol.* **2019**, *20*, 849–861. [CrossRef]
138. Yu, J.; Green, M.D.; Li, S.; Sun, Y.; Journey, S.N.; Choi, J.E.; Rizvi, S.M.; Qin, A.; Waninger, J.J.; Lang, X.; et al. Liver Metastasis Restrains Immunotherapy Efficacy via Macrophage-Mediated T Cell Elimination. *Nat. Med.* **2021**, *27*, 152–164. [CrossRef]
139. Pietrocola, F.; Pol, J.; Vacchelli, E.; Rao, S.; Enot, D.P.; Baracco, E.E.; Levesque, S.; Castoldi, F.; Jacquelot, N.; Yamazaki, T.; et al. Caloric Restriction Mimetics Enhance Anticancer Immunosurveillance. *Cancer Cell* **2016**, *30*, 147–160. [CrossRef] [PubMed]
140. Liu, X.; Pu, Y.; Cron, K.; Deng, L.; Kline, J.; Frazier, W.A.; Xu, H.; Peng, H.; Fu, Y.-X.; Xu, M.M. CD47 Blockade Triggers T Cell-Mediated Destruction of Immunogenic Tumors. *Nat. Med.* **2015**, *21*, 1209–1215. [CrossRef]
141. Lu, W.; Yu, W.; He, J.; Liu, W.; Yang, J.; Lin, X.; Zhang, Y.; Wang, X.; Jiang, W.; Luo, J.; et al. Reprogramming Immunosuppressive Myeloid Cells Facilitates Immunotherapy for Colorectal Cancer. *EMBO Mol. Med.* **2020**, e12798. [CrossRef]
142. Mestas, J.; Hughes, C.C.W. Of Mice and Not Men: Differences between Mouse and Human Immunology. *J. Immunol.* **2004**, *172*, 2731–2738. [CrossRef]
143. Shultz, L.D.; Brehm, M.A.; Garcia-Martinez, J.V.; Greiner, D.L. Humanized Mice for Immune System Investigation: Progress, Promise and Challenges. *Nat. Rev. Immunol.* **2012**, *12*, 786–798. [CrossRef]
144. Huntington, N.D.; Di Santo, J.P. Humanized Immune System (HIS) Mice as a Tool to Study Human NK Cell Development. *Curr. Top. Microbiol. Immunol.* **2008**, *324*, 109–124. [CrossRef]
145. Wege, A.K. Humanized Mouse Models for the Preclinical Assessment of Cancer Immunotherapy. *BioDrugs Clin. Immunother. Biopharm. Gene Ther.* **2018**, *32*, 245–266. [CrossRef] [PubMed]
146. Okada, S.; Vaeteewoottacharn, K.; Kariya, R. Establishment of a Patient-Derived Tumor Xenograft Model and Application for Precision Cancer Medicine. *Chem. Pharm. Bull.* **2018**, *66*, 225–230. [CrossRef]
147. Walsh, N.C.; Kenney, L.L.; Jangalwe, S.; Aryee, K.-E.; Greiner, D.L.; Brehm, M.A.; Shultz, L.D. Humanized Mouse Models of Clinical Disease. *Annu. Rev. Pathol.* **2017**, *12*, 187–215. [CrossRef] [PubMed]
148. Ali, N.; Flutter, B.; Sanchez Rodriguez, R.; Sharif-Paghaleh, E.; Barber, L.D.; Lombardi, G.; Nestle, F.O. Xenogeneic Graft-versus-Host-Disease in NOD-Scid IL-2Rγnull Mice Display a T-Effector Memory Phenotype. *PLoS ONE* **2012**, *7*, e44219. [CrossRef] [PubMed]
149. Ito, R.; Takahashi, T.; Ito, M. Humanized Mouse Models: Application to Human Diseases. *J. Cell. Physiol.* **2018**, *233*, 3723–3728. [CrossRef] [PubMed]
150. De La Rochere, P.; Guil-Luna, S.; Decaudin, D.; Azar, G.; Sidhu, S.S.; Piaggio, E. Humanized Mice for the Study of Immuno-Oncology. *Trends Immunol.* **2018**, *39*, 748–763. [CrossRef]

151. Allen, T.M.; Brehm, M.A.; Bridges, S.; Ferguson, S.; Kumar, P.; Mirochnitchenko, O.; Palucka, K.; Pelanda, R.; Sanders-Beer, B.; Shultz, L.D.; et al. Humanized Immune System Mouse Models: Progress, Challenges and Opportunities. *Nat. Immunol.* **2019**, *20*, 770–774. [CrossRef] [PubMed]
152. Wang, M.; Yao, L.-C.; Cheng, M.; Cai, D.; Martinek, J.; Pan, C.-X.; Shi, W.; Ma, A.-H.; De Vere White, R.W.; Airhart, S.; et al. Humanized Mice in Studying Efficacy and Mechanisms of PD-1-Targeted Cancer Immunotherapy. *FASEB J.* **2018**, *32*, 1537–1549. [CrossRef] [PubMed]
153. Klichinsky, M.; Ruella, M.; Shestova, O.; Lu, X.M.; Best, A.; Zeeman, M.; Schmierer, M.; Gabrusiewicz, K.; Anderson, N.R.; Petty, N.E.; et al. Human Chimeric Antigen Receptor Macrophages for Cancer Immunotherapy. *Nat. Biotechnol.* **2020**, *38*, 947–953. [CrossRef]
154. Yao, L.-C.; Aryee, K.-E.; Cheng, M.; Kaur, P.; Keck, J.G.; Brehm, M.A. Creation of PDX-Bearing Humanized Mice to Study Immuno-Oncology. *Methods Mol. Biol.* **2019**, *1953*, 241–252. [CrossRef] [PubMed]
155. Lazzari, L.; Corti, G.; Picco, G.; Isella, C.; Montone, M.; Arcella, P.; Durinikova, E.; Zanella, E.R.; Novara, L.; Barbosa, F.; et al. Patient-Derived Xenografts and Matched Cell Lines Identify Pharmacogenomic Vulnerabilities in Colorectal Cancer. *Clin. Cancer Res.* **2019**, *25*, 6243–6259. [CrossRef]
156. Inoue, A.; Deem, A.K.; Kopetz, S.; Heffernan, T.P.; Draetta, G.F.; Carugo, A. Current and Future Horizons of Patient-Derived Xenograft Models in Colorectal Cancer Translational Research. *Cancers* **2019**, *11*, 1321. [CrossRef]
157. Zhang, Y.; Lee, S.H.; Wang, C.; Gao, Y.; Li, J.; Xu, W. Establishing Metastatic Patient-Derived Xenograft Model for Colorectal Cancer. *Jpn. J. Clin. Oncol.* **2020**, *50*, 1108–1116. [CrossRef]
158. Hidalgo, M.; Amant, F.; Biankin, A.V.; Budinská, E.; Byrne, A.T.; Caldas, C.; Clarke, R.B.; de Jong, S.; Jonkers, J.; Mælandsmo, G.M.; et al. Patient-Derived Xenograft Models: An Emerging Platform for Translational Cancer Research. *Cancer Discov.* **2014**, *4*, 998–1013. [CrossRef]
159. Gitto, S.B.; Kim, H.; Rafail, S.; Omran, D.K.; Medvedev, S.; Kinose, Y.; Rodriguez-Garcia, A.; Flowers, A.J.; Xu, H.; Schwartz, L.E.; et al. An Autologous Humanized Patient-Derived-Xenograft Platform to Evaluate Immunotherapy in Ovarian Cancer. *Gynecol. Oncol.* **2020**, *156*, 222–232. [CrossRef]
160. Capasso, A.; Lang, J.; Pitts, T.M.; Jordan, K.R.; Lieu, C.H.; Davis, S.L.; Diamond, J.R.; Kopetz, S.; Barbee, J.; Peterson, J.; et al. Characterization of Immune Responses to Anti-PD-1 Mono and Combination Immunotherapy in Hematopoietic Humanized Mice Implanted with Tumor Xenografts. *J. Immunother. Cancer* **2019**, *7*, 37. [CrossRef]
161. Chan, T.; Wiltrout, R.H.; Weiss, J.M. Immunotherapeutic Modulation of the Suppressive Liver and Tumor Microenvironments. *Int. Immunopharmacol.* **2011**, *11*, 879–889. [CrossRef] [PubMed]
162. Halama, N.; Spille, A.; Lerchl, T.; Brand, K.; Herpel, E.; Welte, S.; Keim, S.; Lahrmann, B.; Klupp, F.; Kahlert, C.; et al. Hepatic Metastases of Colorectal Cancer Are Rather Homogeneous but Differ from Primary Lesions in Terms of Immune Cell Infiltration. *Oncoimmunology* **2013**, *2*, e24116. [CrossRef] [PubMed]
163. Skardal, A.; Shupe, T.; Atala, A. Organoid-on-a-Chip and Body-on-a-Chip Systems for Drug Screening and Disease Modeling. *Drug Discov. Today* **2016**, *21*, 1399–1411. [CrossRef]
164. Duzagac, F.; Saorin, G.; Memeo, L.; Canzonieri, V.; Rizzolio, F. Microfluidic Organoids-on-a-Chip: Quantum Leap in Cancer Research. *Cancers* **2021**, *13*, 737. [CrossRef]
165. Skardal, A.; Devarasetty, M.; Forsythe, S.; Atala, A.; Soker, S. A Reductionist Metastasis-on-a-Chip Platform for in Vitro Tumor Progression Modeling and Drug Screening. *Biotechnol. Bioeng.* **2016**, *113*, 2020–2032. [CrossRef] [PubMed]
166. Parlato, S.; De Ninno, A.; Molfetta, R.; Toschi, E.; Salerno, D.; Mencattini, A.; Romagnoli, G.; Fragale, A.; Roccazzello, L.; Buoncervello, M.; et al. 3D Microfluidic Model for Evaluating Immunotherapy Efficacy by Tracking Dendritic Cell Behaviour toward Tumor Cells. *Sci. Rep.* **2017**, *7*, 1093. [CrossRef] [PubMed]
167. Miller, C.P.; Shin, W.; Ahn, E.H.; Kim, H.J.; Kim, D.-H. Engineering Microphysiological Immune System Responses on Chips. *Trends Biotechnol.* **2020**, *38*, 857–872. [CrossRef] [PubMed]

MDPI
St. Alban-Anlage 66
4052 Basel
Switzerland
Tel. +41 61 683 77 34
Fax +41 61 302 89 18
www.mdpi.com

Cancers Editorial Office
E-mail: cancers@mdpi.com
www.mdpi.com/journal/cancers

reported literature findings. The authors further investigated the kinase signatures and identified additional targets such as an increased activation of *AKT1*, *PDPK1* and *PRKCA* kinases. Synergy was observed when inhibitors of *AKT1*, *PDPk1* and *PRKCA* were paired with *BRAF*^{V600E} targeting agents. This example of the screening platform introduces a new versatile approach of target-based drug discovery, eventually to be implemented alongside other strategies to improve precision medicine.

3.6. 3D Bioprinting

In the current unmet need of closer cellular and spatial complexity of a tumor in in vitro conditions, Boland et al. first deposed a patent for ink-jet printing of viable cells in 2006 [131]. During the last decade, tissue engineering has known significant advances with the emergence of 3D bioprinting. The latter showed potential to recreate tailored in vitro 3D heterocellular complex structures to replicate the heterogeneity of the native in vivo tissue [132]. High anatomic precise positioning of living cells embedded in a scaffold or scaffold-free based support, make it possible to obtain 3D structures [133]. A primordial component of the bioprinting procedure is the bioink. It consists of decellularized matrix, microcarriers, hydrogels and cells. Scaffold based approach consists in using bioink where cells are loaded in hydrogels (i.e., agarose, alginate, matrigel, etc.) that differ by their crosslinking properties and construct size they can create. Whereas, in scaffold-free models cell density is higher, cells self-organize and deposit an extracellular-matrix, which allows superior cell-cell interactions [134,135].

To date, there are no reports available on the use of 3D bioprinting to mimic intestine models. This could be probably explained by the complex intestine functions, containing absorption and secretion functions. Currently, only pharmacokinetics and toxicity studies have been reported using such technology with the use of CRC cell lines (i.e., Caco-2) [136]. Madden et al. established a 3D in vitro model based on 3D bilayered bioprinitng of human primary intestinal epithelium for the evaluation of pharmacokinetic parameters, i.e., absorption, distribution, metabolism and elimination). In this study, human intestinal epithelial cells (hIEC) were cultured for 21 days with human intestinal myofibroblast and printed on cell culture inserts allowing easy passage of compounds between apical and basolateral surfaces. Tissue architecture obtained with the 3D bioprinted model was compared to monolayer culture of Caco-2 cells, the gold standard cell line model of the intestinal barrier [137]. Cell-specific markers were identified such as CK19 (epithelial) or vimentin (myofibroblasts) and allowed to distinct both compartments; protein involved in tight junction (E-cadherin) and brush border formation (villus). Immunohistochemical staining for mucin-2 confirmed mucous secretion, which indicates normal intestinal function. Furthermore, genomic analysis of this 3D intestinal tissue showed that main intestinal phase I Cytochrome p450, which is the main family enzyme implicated in the metabolism of drugs and xenobiotics [138], especially CYP3A4, and phase II metabolic enzymes, as well as efflux transporters, were expressed at similar levels compared to the native intestine. The same enzymes in Caco-2 monolayers were upregulated, downregulated or undetectable. To confirm these results, CYP450 metabolism in 3D conditions was further evaluated using an inductor of CYP3A4 rifampicin. Its activity significantly induced the metabolization of its substrate Midazolam in the 3D intestine [136]. Those findings create a new venue for 3D bioprinting as preclinical model for evaluation and prediction of drug efficacy in drug development.

Langer et al. reported a study on patient-derived material integrated in the 3D bioprinting platform. Their approach was based on incorporating multiple cell types (fibroblasts, cancer cells including patient-derived cells or endothelial cells) into scaffold-free in vitro tissue of breast or pancreatic cancers. Various parameters including cell signaling, proliferation, and response to therapies were assessed. The 3D bioprinted model was used to create breast tumors using MCF-7 cell line. Ten-day-old tissues containing breast cancer cells were implanted into immunodeficient mice, and xenografts were treated using doxorubicin and targeted therapies (i.e., BEZ235, an mTOR inhibitor and sunitinib, a multi-target inhibitor). Sunitinib reduced significantly the vasculature density in the stromal compartment and increased collagen deposition.